Volum   ers 1-13)

Student Solutions

for

CALCULUS AND ANALYTIC GEOMETRY
Second Edition

by Mizrahi and Syllivan

R. A. Fritz
Moraine Valley Community College

Richard Tucker
Mary Baldwin College

Wadsworth Publishing Company
Belmont, California
A Division of Wadsworth, Inc.

Mathematics Editor: Jim Harrison
Editorial Coordinator: Anne Scanlan-Rohrer
Production: Matrix
Illustration: George Omura
Typist: Goolcher Wadia

Printed in the United States of America

1 2 3 4 5 6 7 8 9 10---90 89 88 87 86

ISBN 0-534-05456-0

CONTENTS

This solutions manual contains solutions to every other odd-numbered problem.

Chapter One

Exercise 1, pp. 13-14

3. $3 = \sqrt{9}$ by definition of the $\sqrt{}$ symbol

7. $3x + 5 \geqslant 2$

$3x \geqslant -3$

$x \geqslant -1$

11. $14x - 21x + 16 \leqslant 3x - 2$

$-7x + 16 \leqslant 3x - 2$

$-10x \leqslant -18$

$x \geqslant \dfrac{9}{5}$

15. $x^2 + 7x + 12 < 0$

$(x+4)(x+3) < 0$

	-4		-3		
$---$	$+$		$+ + + +$		$x+4$
$---$		$-$	$+ + +$		$x+3$

$-4 < x < -3,$ or $(-4,-3)$

19. $|x| \leqslant 3$

$\{x: -3 \leqslant x \leqslant 3\},$ or $[-3,3]$

23. $|2x-4| + 5 \leqslant 9$

$|2x-4| \leqslant 4$

$-4 \leqslant 2x-4 \leqslant 4$

$0 \leqslant 2x \leqslant 8$

$0 \leqslant x \leqslant 4,$ or $[0,4]$

27. $|2x-4| \geqslant -8$

$(-\infty, +\infty)$ since $|u|$

is always $\geqslant 0$

31. $\left|\dfrac{1}{2} - 2x\right| \leqslant 4$

$-4 \leqslant \dfrac{1}{2} - 2x \leqslant 4$

$-\dfrac{9}{2} \leqslant -2x \leqslant \dfrac{7}{2}$

$\dfrac{9}{4} \geqslant x \geqslant -\dfrac{7}{4}$

$-\dfrac{7}{4} \leqslant x \leqslant \dfrac{9}{4},$ or $[-\dfrac{7}{4}, \dfrac{9}{4}]$

35. $\dfrac{2}{x-2} \leqslant -5,$ $x \neq 2,$ or the left side is undefined

$\dfrac{2}{x-2} + 5 \leqslant 0$

$\dfrac{2 + 5(x-2)}{x-2} \leqslant 0$

$\dfrac{5x - 8}{x-2} \leqslant 0$

	$\dfrac{8}{5}$		2		
$-----$	$+ +$		$+ +$		$5x-8$
$---$		$-$	$+ +$		$x-2$

$\{x: \dfrac{8}{5} \leqslant x < 2\},$ or $[\dfrac{8}{5}, 2)$

39. $\dfrac{2}{3-x} < 1$

$\dfrac{2 - (3-x)}{3-x} < 0$

| | --- | 1 | + + | 3 | + + + | x-1 |
| | + + | | + + | | - - - | 3-x |

$\dfrac{x-1}{3-x} < 0$

$\{x:\ x < 1\ \text{ or }\ x > 3\}$,

or $(-\infty, 1) \cup (3, +\infty)$

43. $\left|\dfrac{1}{x}\right| < 2$ implies $\dfrac{1}{|x|} < 2$, or $|x| > \dfrac{1}{2}$

$x < -\dfrac{1}{2}$ or $x > \dfrac{1}{2}$

$\{x:\ x < -\dfrac{1}{2}\ \text{ or }\ x > \dfrac{1}{2}\}$, or $(-\infty, -\dfrac{1}{2}) \cup (\dfrac{1}{2}, +\infty)$

47. $\sqrt[4]{x^2 - 3x + 2}$

| | --- | 1 | - - | 2 | + + + | x-2 |
| | --- | | + + | | + + | x-1 |

$x^2 - 3x + 2 \geqslant 0$

$(x-2)(x-1) \geqslant 0$

$(-\infty, 1] \cup [2, +\infty)$

51. If $a < b$, prove $a < \dfrac{a + b}{2} < b$.

 Proof: If $a < b$, then $\dfrac{a}{2} < \dfrac{b}{2}$.

 Thus, $\dfrac{a}{2} + \dfrac{a}{2} < \dfrac{a}{2} + \dfrac{b}{2} < \dfrac{b}{2} + \dfrac{b}{2}$, so that

 $a < \dfrac{a + b}{2} < b$. Since, $\left|\dfrac{a + b}{2} - a\right| = \left|\dfrac{b}{2} - \dfrac{a}{2}\right|$

 $= \left|\dfrac{a}{2} - \dfrac{b}{2}\right| = \left|\dfrac{a + b}{2} - b\right|$, the harmonic mean is

 equidistant from a and b.

55. If the perimeter is fixed, say P, then for a circle $2\pi R = P$ and for a square $4x = P$, or $R = \dfrac{P}{2\pi}$ and

 $x = \dfrac{P}{4}$. The area of the circle is $A_c = \pi R^2 = \pi(\dfrac{P}{2\pi})^2$,

 or $A_c = \dfrac{P^2}{4\pi}$. For the square, $A_s = x^2 = \dfrac{P^2}{16}$. Since

 $4\pi < 16$, then $\dfrac{P^2}{4\pi} > \dfrac{P^2}{16}$, so that $A_c > A_s$.

59. Prove that if $0 < a \leqslant b$, then $a^2 \leqslant b^2$.

   <u>Proof</u>: $0 < a \leqslant b$ implies $a \cdot a \leqslant a \cdot b$ and $ab \leqslant b \cdot b$. Thus, $a^2 \leqslant ab$ and $ab \leqslant b^2$. This implies $a^2 \leqslant b^2$.

<u>Exercise 2</u>, pp. 22-24

3. $|P_1P_2| = \sqrt{(-0.6 + 0.4)^2 + (2 + 0.2)^2}$

   $= \sqrt{(0.2)^2 + (2.2)^2} = \sqrt{0.04 + 4.84}$

   $= \sqrt{4.88} \approx 2.209$

7. $y = 2x - 3$

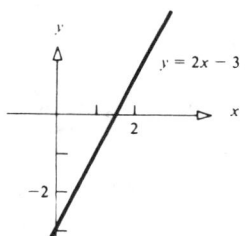

11. $y = x^2 + 4$

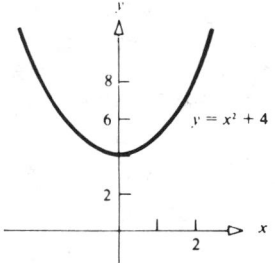

15. If $(h,k) = (1,-2)$ and $R = 1$, the equation is $(x-1)^2 + (y+2)^2 = 1$, or $x^2 + y^2 - 2x + 4y + 4 = 0$.

19. If $x^2 + 4x + y^2 - 6y = 3$, then $(x + 2)^2 + (y - 3)^2 = 16$. Thus, $(h,k) = (-2,3)$ and $R = 4$.

23. $x = \dfrac{0 + 1}{2} = \dfrac{1}{2}$ and $y = \dfrac{1 + 0}{2} = \dfrac{1}{2}$

   $M = (\dfrac{1}{2}, \dfrac{1}{2})$

27. For symmetry with respect to the origin:

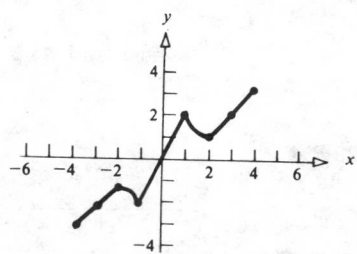

31. $x^2 - 2xy = y^4$ has symmetry with respect to the origin, since replacing x by $-x$, and y by $-y$ gives $(-x)^2 - 2(-x)(-y) = (-y)^4$, or $x^2 - 2xy = y^4$. It has no symmetry with respect to the x-axis or y-axis, since replacing:

  a. x by $-x$ gives $x^2 + 2xy = y^4$.

  b. y by $-y$ gives $x^2 + 2xy = y^4$.

35. $y = -x^5 + 3x$ is symmetric with respect to the origin since replacing x by $-x$ and y by $-y$ gives $-y = -(-x)^5 + 3(-x)$, or $-y = +x^5 - 3x$, which is equivalent to $y = -x^5 + 3x$. The graph is not symmetric to the x-axis or y-axis.

39.

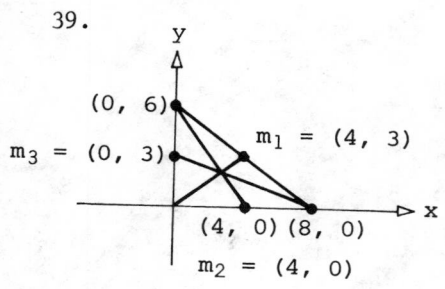

$m_3 = (0, 3)$

$(0, 6)$

$m_1 = (4, 3)$

$(4, 0)$ $(8, 0)$

$m_2 = (4, 0)$

Lengths of medians:

$M_1 = \sqrt{16 + 9} = 5$

$M_2 = \sqrt{64 + 9} = \sqrt{73}$

$M_3 = \sqrt{36 + 16} = \sqrt{52}$

43. $P_1(-2,-1)$; $P_2(0,7)$; $P_3(3,2)$

$|P_1P_2| = d_1 = \sqrt{4 + 64} = \sqrt{68}$

$|P_1P_2| = d_2 = \sqrt{25 + 9} = \sqrt{34}$

$|P_2P_3| = d_3 = \sqrt{9 + 25} = \sqrt{34}$

$d_2^2 + d_3^2 = d_1^2 \implies$ right triangle

$d_2 = d_3 \implies$ isosceles triangle

47. If P divides the line segment P_1P_2 in the ratio $|P_1P|/|P_1P_2| = r$, then from the figure we use the fact that $\triangle P_1RP_2$ is similar to $\triangle P_1QP$ to conclude that

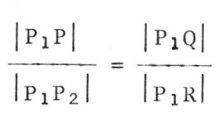

$$\frac{|P_1P|}{|P_1P_2|} = \frac{|P_1Q|}{|P_1R|}$$

and

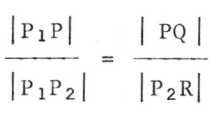

$$\frac{|P_1P|}{|P_1P_2|} = \frac{|PQ|}{|P_2R|}$$

From the figure:

$|P_1Q| = x - x_1$; $|P_1R| = x_2 - x_1$; $|PQ| = y - y_1$;

$|P_2R| = y_2 - v_1$.

Combining these results and $r = \dfrac{|P_1P|}{|P_1P_2|}$ yields

$r = \dfrac{x - x_1}{x_2 - x_1}$ and $r = \dfrac{y - y_1}{y_2 - y_1}$

Solving for x and y, we have

$x - x_1 = rx_2 - rx_1$ $y - y_1 = ry_2 - ry_1$

$x = x_1 - rx_1 + rx_2$ $y = y_1 - ry_1 + ry_2$

$= x_1(1-r) + rx_2$ $= y_1(1-r) + ry_2$

51. The point $P(x,y)$ has the property that P_2 is the midpoint of P_1 and P. Hence, $(\dfrac{x+1}{2}, \dfrac{y+4}{2})$ = $(5,6)$, so that $x = 9$ and $y = 8$. Thus, $P = (9,8)$.

Exercise 3, pp. 31-33

3. $y + 1 = -\dfrac{2}{3}(x - 1)$

    $3y + 3 = -2x + 2$

    $2x + 3y = -1$, or $2x + 3y + 1 = 0$

7. If $m = -3$ and the y-intercept is 3, then $y = -3x + 3$, or $3x + y - 3 = 0$.

11. If m is undefined and the line passes through $(1,4)$, the line is vertical and its equation is $x = 1$, or $x - 1 = 0$.

15. $2y = -x + 4$

    $y = -\dfrac{1}{2}x + 2$

    $m = -\dfrac{1}{2}$; y-intercept = 2

19. If the line is parallel to $2x - y = 6$, or $y = 2x - 6$, then $m = 2$. Given the point $(1,2)$, we have $y - 2 = 2(x-1)$, or $2x - y = 0$.

23. If $x + 2y = -4$, then $y = -\dfrac{1}{2}x - 2$, so that

    $m = -1/2$ and $m_1 = 2$, where m_1 is the slope of the perpendicular line.

27. $\begin{aligned}2x - 3y + 6 &= 0\\4x - 6y + 7 &= 0\end{aligned}$ $\Longrightarrow$ $\begin{aligned}4x - 6y + 12 &= 0\\4x - 6y + 7 &= 0\end{aligned}$ $\Longrightarrow$ Lines parallel; no intersection

31. $\begin{aligned}3x - 3y + 10 &= 0\\x + y - 2 &= 0\end{aligned}$ $\Longrightarrow$ $\begin{aligned}3x - 3y + 10 &= 0\\3x + 3y - 6 &= 0\end{aligned}$

    $$\begin{aligned}6x \qquad\quad + 4 &= 0\\x \qquad\qquad &= -2/3\end{aligned}$$

Solving for y in L_2 and substituting gives
$y = 2 - x = 2 + \frac{2}{3} = \frac{8}{3}$. The point of intersection is
$(-\frac{2}{3}, \frac{8}{3})$.

35. Let x be the amount invested at 14%. Then we have
$0.14x + 0.10(100,000-x) = 12,000$, or
$0.04x = 12,000-10,000 = 2,000$. Thus, $x = \frac{2000}{0.04}$
$= 50,000$. He invests $50,000 at 14% and $50,000 at 10%.

39. Let m_1 and m_2 be the slopes of two lines and suppose $m_1 m_2 = -1$. Then $m_1 = -1/m_2$ so one slope, say m_1, is positive and the other is negative. Let $m_1 = \tan \theta_1$ and $m_2 = \tan \theta_2$ so that $m_1 > 0$ implies $0 < \theta_1 < \frac{\pi}{2}$, and $m_2 < 0$ implies $\frac{\pi}{2} < \theta_2 < \pi$. Since $m_1 = -1/m_2$, we have $\tan \theta_1 = -1/\tan \theta_2$. Considering the identity $\cot(\alpha + \beta) = \frac{\cot \alpha \cot \beta - 1}{\cot \alpha + \cot \beta}$, we have $\cot(\theta_1 + \frac{\pi}{2})$ $= -1/\cot \theta_1$, since $\cot \frac{\pi}{2} = 0$. Hence, $1/\tan(\theta_1 + \frac{\pi}{2})$ $= -\tan \theta_1$, or $\tan \theta_1 = -1/\tan(\theta_1 + \pi_2)$. Hence, we can conclude that $\tan \theta_2 = \tan(\theta_1 + \frac{\pi}{2})$ so that $\theta_1 + \frac{\pi}{2}$ equals either θ_2 or $\theta_2 + \pi$. From our definition of measuring the angle between lines, we conclude $\theta_1 + \frac{\pi}{2} = \theta_2$. Hence, the lines are perpendicular since $\theta_2 - \theta_1 = \frac{\pi}{2}$.

43. We have $m_1 = -\frac{2}{3}$ and $m_2 = 1$ so that
$$\tan \theta = \frac{1 + \frac{2}{3}}{1 + (-\frac{2}{3})(1)} = \frac{\frac{5}{3}}{\frac{1}{3}} = 5.$$

Exercise 4, pp. 42-44

3. $y = x^2 + 2x + 1$ is a function.

7. $y^2 = 1 - x^2$ is not a function. If $x = 0$, then $y = \pm 1$.

11. $x^2 y^2 = 5 \Rightarrow y = \pm\sqrt{\dfrac{5}{x^2}}$, so this is not a function.

15. $\{(1,5), (2,5), (5,1)\}$ is a function.

19. dom f = $[1, +\infty)$, since we must have $x - 1 \geqslant 0$.

23. dom f $= \{x : x \neq 2\}$.

27. $f(x) = \begin{cases} 2x - 3 & \text{if } x < 0 \\ x - 3 & \text{if } 0 \leqslant x < 5 \end{cases}$

dom f = $(-\infty, 5)$

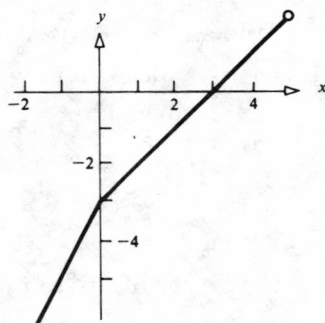

31. $f(x) = \begin{cases} x^2 & \text{if } x \leqslant 0 \\ \sqrt{x + 1} & \text{if } x > 0 \end{cases}$

dom f = $(-\infty, +\infty) = \mathbb{R}$

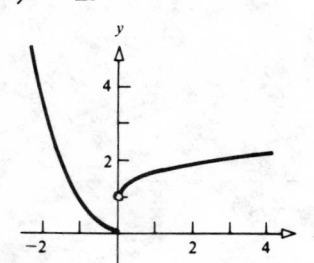

35. $f(x) = |x + 4|$

dom $f = (-\infty, +\infty) = \mathbb{R}$

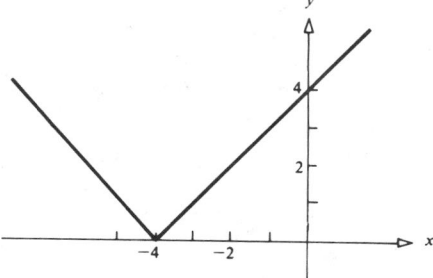

39. $f(x) = x^2 + 3x + 4$

(a) $f(x + \Delta x) = x^2 + 2x\Delta x + (\Delta x)^2 + 3x + 3\Delta x + 4$

(b) $f(x + \Delta x) - f(x) = 2x\Delta x + (\Delta x)^2 + 3\Delta x$

(c) $\dfrac{f(x + \Delta x) - f(x)}{\Delta x} = 2x + 3 + \Delta x$

43. If $f(x) = 2x^3 + Ax^2 + Bx - 5$, and if $f(2) = 3$ and $f(-2) = -37$, find $A + B$. We have

$f(2) = 3 = 16 + 4A + 2B - 5 \implies 4A + 2B = -8$

$f(-2) = -37 = -16 + 4A - 2B - 5 \implies \underline{4A - 2B = -16}$

$ 8A = -24$

So $A = -3$ and substituting gives $4(-3) + 2B = -8$, or $B = 2$. Thus, $A + B = (-3) + 2 = -1$.

47.

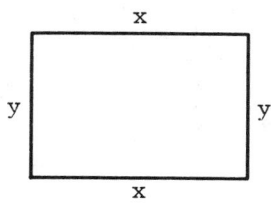

$P = 2x + 2y$

$3000 = 2x + 2y$

$1500 - x = y$

$A = xy = x(1500 - x)$

$ = 1500x - x^2$

dom $A = 0 < x < 1500$

3. If $f(x) = \sqrt{x+1}$ and $g(x) = x+1$, then
    domain $f = [-1, +\infty)$, domain $g = (-\infty, +\infty)$.

    (a) $(f+g)(x) = \sqrt{x+1} + x+1$ domain $(f+g) = [-1, +\infty)$

    (b) $(f-g)(x) = \sqrt{x+1} - x-1$ domain $(f-g) = [-1, +\infty)$

    (c) $(f \cdot g)(x) = (x+1)^{\frac{3}{2}}$ domain $(f \cdot g) = [-1, +\infty)$

    (d) $\left(\dfrac{f}{g}\right)(x) = \dfrac{\sqrt{x+1}}{x+1} = \dfrac{1}{\sqrt{x+1}}$ domain $\left(\dfrac{f}{g}\right) = (-1, +\infty)$

7. $f(x) = 2x^5 - 3x + 4$ is a polynomial.

11. $f(x) = \sqrt{x} - 2$ is not a polynomial.

15. If $f(x) = \dfrac{2}{x^2 - 4} = \dfrac{2}{(x+2)(x-2)}$, then

    dom $f = \{x: x \neq -2, x \neq 2\}$.

19. If $f(x) = x^2 + x$, then
    we have $f(x) = (x + \frac{1}{2})^2 - \frac{1}{4}$
    so that the graph of f is
    the graph of $y = x^2$ shifted
    horizontally to the left
    $\frac{1}{2}$ unit and downward $\frac{1}{4}$ unit.

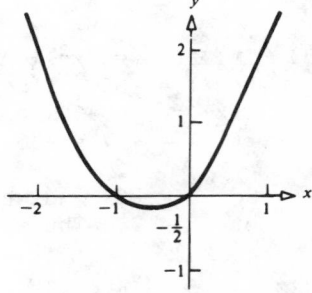

23. $f(-x) = 4(-x)^3 = -4x^3 = -f(x)$, so that f is odd.

27. $f(-x) = \dfrac{-x - 1}{-x + 1} = -\dfrac{x + 1}{1 - x}$, so that f is neither
    even nor odd.

31. Let $f(x) = x^2$ and $g(x) = x^4$, then $f(x) \cdot g(x)$
    $= x^6$, which is an even function.

35. If $f(x) = ax^2 + bx + c$, then $f(x+y) = a(x+y)^2$
    $+ b(x+y) + c = ax^2 + 2axy + ay^2 + bx + by + c$.
    Thus, $f(x+y) = f(x) + f(y)$ implies
    $ax^2 + 2axy + ay^2 + bx + by + c = ax^2 + bx + c + ay^2$
    $+ by + c$, or $2axy - c = 0$ (for all choices of
    x and y). Thus, we must have $a = 0$ and $c = 0$.
    Then b can have any value, so that $f(x) = bx$.

3. If $f(x) = \sqrt{x}$ and $g(x) = x^2 - 1$, then

    (a) $(f \circ g)(x) = f[g(x)] = \sqrt{x^2 - 1}$

    (b) $(g \circ f)(x) = g[f(x)] = \overset{2}{f}(x) - 1 = (\sqrt{x})^2 - 1 = x - 1$

    (c) $(g \circ g)(x) = g[g(x)] = (g(x))^2 - 1 = (x^2 - 1)^2 - 1$
    $$= x^4 - 2x^2$$

    (d) $(f \circ f)(x) = f[f(x)] = \sqrt{f(x)} = \sqrt{\sqrt{x}} = \sqrt[4]{x} = x^{\frac{1}{4}}$

7. If $f(x) = 3x^4 - 2x^2$ and $g(x) = \dfrac{2}{\sqrt{x}}$, then

    (a) $(f \circ g)(x) = f[g(x)] = 3g^4(x) - 2g^2(x) = \dfrac{48}{x^2} - \dfrac{8}{x}$
    $$= \dfrac{8(6-x)}{x^2}$$

    (b) $(g \circ f)(x) = g[f(x)] = \dfrac{2}{\sqrt{3x^4 - 2x^2}}$

    (c) $(g \circ g)(x) = g[g(x)] = \dfrac{2}{\sqrt{g(x)}} = \dfrac{2}{\sqrt{\dfrac{2}{\sqrt{x}}}} = \sqrt{2} \; x^{\frac{1}{4}}, \; x \neq 0$

    (d) $(f \circ f)(x) = 3f^4(x) - 2f^2(x) = 3(3x^4 - 2x^2)^4$
    $$- 2(3x^4 - 2x^2)^2$$

11. If $f(x) = x^7$ and $g(x) = x^2 - 1$, then $f \circ g = H$.

15. If $g(x) = \sqrt[3]{x}$ and $f(x) = x^3$, then $f(g(x)) = x$.

19. If $f(x) = 2x^3 + 3x^2 + 4x + 5$ and $g(x) = 2$, then
    (a) $g(f(x)) = 2$, since for all x, $g(x) = 2$
    (b) $f(g(x)) = f(2) = 2(2)^3 + 3(2)^2 + 4(2) + 5$
    $$= 16 + 12 + 8 + 5 = 41$$

3. If $y = x^2 + 3$, the horizontal line $y = 6$ intersects the graph more than once. The function is not one-one.

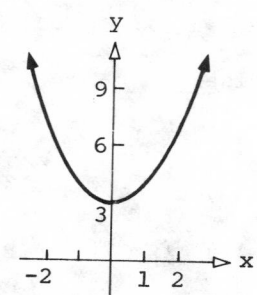

7. If $y = x^n$, n even, the graph is shaped as shown in the figure. Any horizontal line $y = a$ with $a > 0$ intersects the graph in more than one point. The function is not one-one.

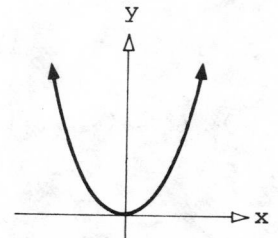

11. Let $y = x^3$ and interchange x and y to arrive at $x = y^3$.

    Hence, $y = \sqrt[3]{x}$, so that $f^{-1}(x) = \sqrt[3]{x}$.

    The graph of f^{-1} is shown in the figure.

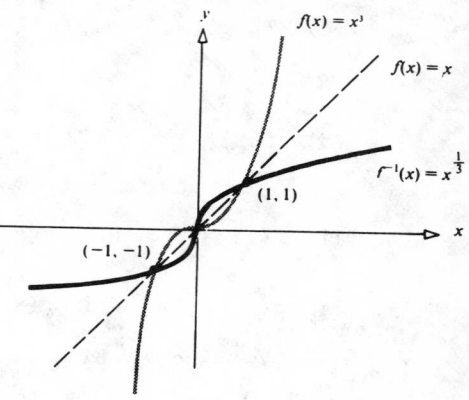

15. Let $y = x/(x+1)$ and interchange x and y so that $x = y/(y+1)$. This gives $xy + x = y$, or $y(x-1) = -x$, so that $y = x/(1-x)$. Hence, $f^{-1}(x) = x/(1-x)$.

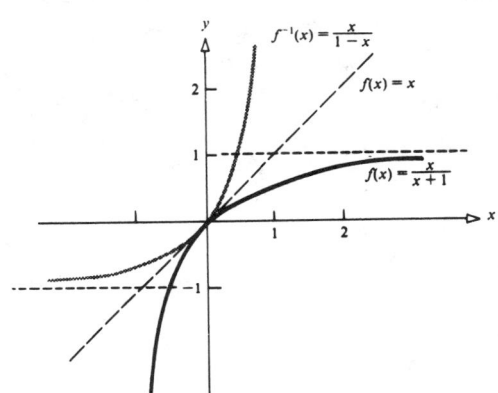

19. If $y = f(x) = \frac{9}{5}x + 32$, interchange x and y to

arrive at $x = \frac{9}{5}y + 32$. Hence, $\frac{9}{5}y = x - 32$, or

$y = \frac{5}{9}(x-32)$. Let $f(x) = \frac{9}{5}x + 32$, and

$g(x) = \frac{5}{9}(x-32)$. Then

(a) $f[g(x)] = \frac{9}{5}g(x) + 32 = \frac{9}{5} \cdot \frac{5}{9}(x-32) + 32$

$\qquad = x - 32 + 32 = x$

(b) $g[f(x)] = \frac{5}{9}(f(x)-32) = \frac{5}{9}(\frac{9}{5}x+32-32) = \frac{5}{9}(\frac{9}{5}x) = x$

Thus, $g = f^{-1}$ and $f = g^{-1}$.

Miscellaneous Exercises, pp. 54-57

3. $\dfrac{1}{x^2 - 4} < 0$

$\dfrac{1}{(x-2)(x+2)} < 0$

	-2		2		
--		- - -		+ +	x-2
--		+ + +		+ + +	x+2

$\{x: -2 < x < 2\}$ or $(-2,2)$.

7. $A(1,-6)$, $B(8,8)$, and $C(-7,-2)$ are the given points. Then

    (a) $m_{AB} = 14/7 = 2$, $m_{AC} = 4/-8 = -\frac{1}{2}$, and $m_{BC} = 10/15$. Since $m_{AB} = -1/m_{AC}$, the line segments $\overline{AB}$ and $\overline{AC}$ are perpendicular and the triangle is a right triangle.

    (b) $\begin{aligned} |AB|^2 &= (8-1)^2 + (8+6)^2 = 7^2 + 14^2 \\ &= 49 + 196 = 245 \end{aligned}$

        $\begin{aligned} |BC|^2 &= (8+7)^2 + (8+2)^2 = 15^2 + 10^2 \\ &= 225 + 100 = 325 \end{aligned}$

        $\begin{aligned} |AC|^2 &= (1+7)^2 + (-6+2)^2 = 8^2 + 4^2 \\ &= 64 + 16 = 80 \end{aligned}$

    Since $|AB|^2 + |BC|^2 = |AC|^2$, the triangle is a right triangle.

11. If (a,b) is 4 units from $15x + 8y - 34 = 0$. $\Rightarrow$ $8y = -15x + 34 \Rightarrow y = -\frac{15}{8}x + \frac{17}{4}$. Also, (a,b) is on a parallel line 4 units from the line L. Let this line be L_1 and thus L_1 has an equation $y = -\frac{15}{8}x + b_1$, where b_1 is suitably chosen so that L_1 is 4 units from L. Using the results of Problem 10, b_1 satisfies

$$d = \frac{|b_1 - b_2|}{\sqrt{m^2 + 1}} \Rightarrow 4 = \frac{|b_1 - \frac{17}{4}|}{\sqrt{\frac{225}{64} + 1}} \Rightarrow |b_1 - \frac{17}{4}| = 4\sqrt{\frac{289}{64}}$$

$$= 4(\frac{17}{8}) = \frac{17}{2}, \quad \text{and} \quad |b_1 - \frac{17}{4}| = \frac{17}{2} \Rightarrow b_1 = \frac{17}{4} \pm \frac{17}{2}$$

$$= \frac{17 \pm 34}{4} \Rightarrow b_1 = -\frac{17}{4} \quad \text{or} \quad \frac{51}{4}$$

There are two lines L_1 parallel to L:

$$y = -\frac{15}{8}x - \frac{17}{4} \qquad 8y + 15x = -34$$
$$\Rightarrow$$
$$y = -\frac{15}{8}x + \frac{51}{4} \qquad 8y + 15x = 102$$

Thus, the equations involving a and b are $15a + 8b = -34$ and $15a + 8b = 102$.

15. Letting L_1 be the line through $(x,4)$ and $(-2,1)$, and L_2 be the line through $(2,3)$ and $(-1,y)$, we have L_1 is perpendicular to L_2, and $m_1 = \dfrac{4-1}{x+2} = \dfrac{3}{x+2}$ and $m_2 = \dfrac{y-3}{-1-2} = \dfrac{y-3}{-3}$. Then L_1 perpendicular to L_2 implies $m_2 = -1/m_1$, so that $\dfrac{y-3}{-3} = -\dfrac{x+2}{3}$. Thus, $y - 3 = x + 2$, or $y = x + 5$.

19. The circle with center at $(-3,-2)$ and tangent to the line $y = 5$ has radius 7. Thus, the equation is $(x+3)^2 + (y+2)^2 = 49$, or $x^2 + y^2 + 6x + 4y - 36 = 0$.

23. We will use the point-slope equation for a line and we have a point (x_0,y_0), so we find the slope m_T.

    Note that center of circle is at $(0,0)$, so the slope of the line from $(0,0)$ to (x_0,y_0) is $m = y_0/x_0$. Our tangent line is perpendicular to this line, so $m_T = -1/m$, or $m_T = -x_0/y_0$. Hence, the equation is $y - y_0 = \dfrac{-x_0}{y_0}(x-x_0)$, $yy_0 - y_0^2 = -x_0x + x_0^2$
    $\Rightarrow xx_0 + yy_0 = x_0^2 + y_0^2$, or $xx_0 + yy_0 - R^2 = 0$.

27. If f is even and g is odd, then

    (a) $(f+g)(-x) = f(-x) + g(-x)$
    $$= f(x) - g(x)$$
    Thus, $f + g$ is neither even nor odd.

    (b) $(f \cdot g)(-x) = f(-x)g(-x)$
    $$= -f(x)g(x) = -(f \cdot g)(x)$$
    Thus, $f \cdot g$ is odd.

    (c) $(f \circ f)(-x) = f[f(-x)] = f[f(x)] = (f \circ f)(x)$
    Thus, $f \circ f$ is even.

    (d) $(g \circ g)(-x) = g[g(-x)] = g[-g(x)] = -g[g(x)]$
    $$= -(g \circ g)(x)$$
    Thus, $g \circ g$ is odd.

    (e) Let $h(x) = (f \circ g)(x) = f[g(x)]$; then $h(-x)$
    $= f[g(-x)] = f[-g(x)] = f[g(x)] = h(x)$
    Thus, $f \circ g$ is even.

    (f) Let $h(x) = (g \circ f)(x) = g[f(x)]$. Then $h(-x)$
    $= g[f(-x)] = g[f(x)]$. Hence, $g \circ f$ is even.

31. In order that $(f \circ g)(x) = (g \circ f)(x)$, where
$f(x) = 3x + 2$ and $g(x) = 2x - p$, we have the
following:

(a) $(f \circ g)(x) = f[g(x)] = 3g(x) + 2 = 3(2x-p) + 2$
$= 6x - 3p + 2$

(b) $(g \circ f)(x) = g[f(x)] = 2f(x) - p = 2(3x+2) - p$
$= 6x + 4 - p$

Hence, $-3p + 2 = 4 - p$, or $2p = -2$, so that
$p = -1$.

35. $f(x) \begin{cases} -x & \text{if } x < 0 \\ -2 & \text{if } x = 0 \\ \dfrac{-1}{x+1} & \text{if } x > 0 \end{cases}$

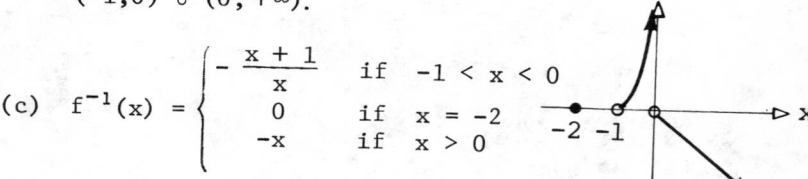

(a) f is one-one by the
horizontal line test,
so that f^{-1} exists.

(b) dom $f = (-\infty, +\infty)$ and
dom $f^{-1} = \{-2\} \cup$
$(-1,0) \cup (0, +\infty)$.

(c) $f^{-1}(x) = \begin{cases} -\dfrac{x+1}{x} & \text{if } -1 < x < 0 \\ 0 & \text{if } x = -2 \\ -x & \text{if } x > 0 \end{cases}$

If $x < 0$, $y = -x \Rightarrow x = -y$ and $y > 0$. Thus,
$f^{-1}(x) = -x$ if $x > 0$. If $x = 0$, $y = -2$
$\Rightarrow f^{-1}(-2) = 0$. If $x > 0$, $y = \dfrac{-1}{x+1}$ and
$-1 < y < 0$. Thus, $xy + y = -1 \Rightarrow xy = -(y+1)$
$\Rightarrow x = -(y+1)/y$. Thus, $f^{-1}(x) = -(x+1)/x$ if
$-1 < x < 0$. This verifies the f^{-1} formula given.

39. If $|x-2| < \frac{1}{5}$ and $|y-3| < \frac{1}{10}$, then we have

$-\frac{1}{5} < x - 2 < \frac{1}{5}$ and $-\frac{1}{10} < y - 3 < \frac{1}{10}$, or

$\frac{9}{5} < x < \frac{11}{5}$ and $\frac{29}{10} < y < \frac{31}{10}$. Hence, $\frac{261}{50} < xy < \frac{341}{50}$,

so that $\frac{261-300}{50} < xy - 6 < \frac{341-300}{50}$, which implies

$-\frac{41}{50} < -\frac{39}{50} < xy - 6 < \frac{41}{50}$. Hence, we conclude that

$|xy-6| < 41/50$.

43. We have V = (height)(width)(length)
$= x(16-2x)(200-2x) = 4x(8-x)(100-x)$

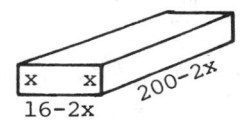

47. Using the points $(70, 29 \times 10^6)$, $(900, 25 \times 10^6)$, we find
the slope is $m = (4 \times 10^6)/-830$. The equation of the
line through
the points is $e - (29 \times 10^6) = -\frac{4 \times 10^6}{830}(T-70)$. Hence,

the formula is $e = (29 \times 10^6) - (\frac{4 \times 10^6}{830})(T-70)$, where
e is elasticity and T is temperature.

51. Prove that if a^2 is even, then a is even.

Proof: Suppose a is an odd positive integer. Then
there is some natural number k such that $a = 2k + 1$.
Then we have $a^2 = (2k+1)^2 = 4k^2 + 4k + 1$
$= 2(2k^2+2k) + 1 = 2M + 1$. That is, dividing a^2 by 2
leaves a remainder of 1. Thus, a odd implies that
a^2 is odd. This implies that whenever a^2 is even,
we must have a even, otherwise a^2 would not be
even.

Chapter Two

Exercise 1, pp. 66-69

3.

x	0.1	0.01	0.001	−0.001	−0.01	−0.1
$f(x) = x^2 + 2$	2.01	2.0001	2.000001	2.000001	2.0001	2.01

$$\lim_{x \to 0} (x^2 + 2) = 2$$

7.

x (radian)	−0.2	−0.1	−0.01	0.01	0.1	0.2
$f(x) = \dfrac{\tan x}{x}$	1.01355	1.00335	1.00003	1.00003	1.00335	1.01355

$$\lim_{x \to 0} \frac{\tan x}{x} = 1$$

11. $\lim_{x \to c} f(x)$ exists   15. $\lim_{x \to c} f(x)$ does not exist

19.
$$f(x) = \begin{cases} 3x - 1 & \text{if } x < 1 \\ 4 & \text{if } x = 1 \quad c = 1 \\ 2x & \text{if } x > 1 \end{cases}$$

$$\lim_{x \to 1} f(x) = 2$$

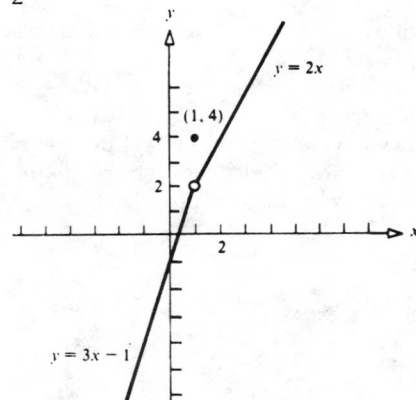

23. $f(x) = \begin{cases} x^2 & \text{if } x \leqslant 0 \\ 2x+1 & \text{if } x > 0 \end{cases}$

   $\lim\limits_{x \to 0} f(x)$ does not exist

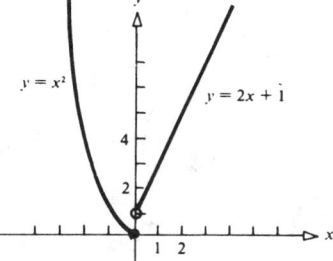

27. $\lim\limits_{x \to -1} (x) = -1$

31. $f(x) = \frac{1}{2}x^2 - 1$

   (a) $P = (2, f(2))$ $Q = (2+h, f(2+h))$

   $P = (2, 1)$ $Q = (2+h, \frac{h^2+4h+2}{2})$

   $= (2+h, \frac{1}{2}h^2 + 2h + 1)$

   $m_{sec} = \dfrac{(\frac{1}{2}h^2 + 2h + 1) - 1}{(2+h) - 2} = \dfrac{\frac{1}{2}h^2 + 2h}{h}$

   $= \frac{1}{2}h + 2 = 2 + \frac{h}{2}$

   (b) m_{sec}(for $h = -0.5$) $= 2 - 0.25 = 1.75$
   m_{sec}(for $h = +0.5$) $= 2 + 0.25 = 2.25$

   (c)

h	-0.5	-0.1	-0.001	0.001	0.1	0.5
m_{sec}	1.75	1.95	1.9995	2.0005	2.05	2.25

   (d) $\lim\limits_{h \to 0} m_{sec} = 2.0$

   (e) Tangent line at $P = (2, f(2))$: $m = 2$ and
   $f(2) = 1$. The equation of the tangent line
   is $y - 1 = 2(x-2)$, or $y = 2x - 3$.

(f)

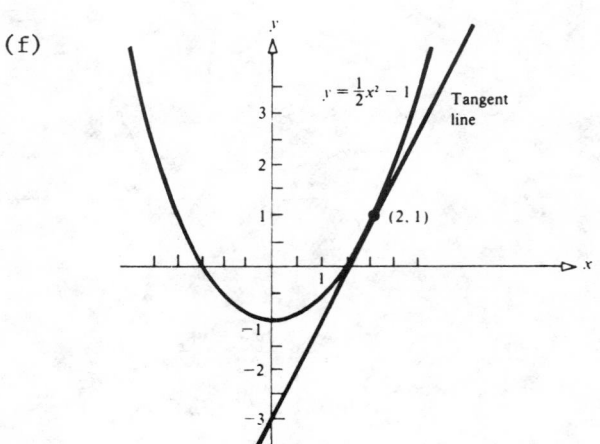

$y = \frac{1}{2}x^2 - 1$ Tangent line

(2.1)

<u>Exercise 2</u>, pp. 75-76

3. $\lim_{x\to 1} (3x^2-2x+4) = 3(1)^2 - 2(1) + 4 = 5$

7. $\lim_{x\to 1/2} (2x^4-8x^3+4x-5) = 2(\frac{1}{2})^4 - 8(\frac{1}{2})^3 + 4(\frac{1}{2}) - 5$

$$= \frac{1}{8} - 1 + 2 - 5 = -\frac{31}{8}$$

11. $\lim_{x\to 3} \sqrt{x^2+x+4} = \sqrt{(3)^2+3+4} = \sqrt{16} = 4$

15. $\lim_{x\to 2} (\sqrt{x^3+1} - \sqrt{x^2+5}) = \sqrt{(2)^3+1} - \sqrt{(2)^2+5} = \sqrt{9} - \sqrt{9} = 0$

19. $\lim_{x\to -2} \frac{2x^3+5x}{3x-2} = \frac{-16-10}{-8} = \frac{-26}{-8} = \frac{13}{4}$

23. $\lim_{x\to 2} \frac{x^3-8}{x^2+x-6} = \lim_{x\to 2} \frac{(x-2)(x^2+2x+4)}{(x-2)(x+3)} = \frac{12}{5}$

27. $\lim_{x\to -3} \frac{x+3}{x^2-x-12} = \lim_{x\to -3} \frac{x+3}{(x+3)(x-4)} = \lim_{x\to -3} \frac{1}{x-4} = -\frac{1}{7}$

31. $\lim_{x\to -8} \frac{2x+16}{x+8} = \lim_{x\to -8} (2) = 2$

20   CHAPTER TWO

35. $\lim\limits_{x \to 2} \dfrac{(\sqrt{x} - \sqrt{2})}{(\sqrt{x} - \sqrt{2})(\sqrt{x} + \sqrt{2})} = \lim\limits_{x \to 2} \left(\dfrac{1}{\sqrt{x} + \sqrt{2}} \right) = \dfrac{1}{2\sqrt{2}}$ or $\dfrac{\sqrt{2}}{4}$

39. $\lim\limits_{h \to 0} \dfrac{\sqrt{x+h} - \sqrt{x}}{h} \cdot \dfrac{\sqrt{x+h} + \sqrt{x}}{\sqrt{x+h} + \sqrt{x}} = \lim\limits_{h \to 0} \dfrac{x + h - x}{h(\sqrt{x+h} + \sqrt{x})}$

$= \lim\limits_{h \to 0} \dfrac{h}{h(\sqrt{x+h} + \sqrt{x})} = \lim\limits_{h \to 0} \dfrac{1}{\sqrt{x+h} + \sqrt{x}} = \dfrac{1}{2\sqrt{x}}$

43. $\lim\limits_{x \to c} g^3 = 2^3 = 8$

47. If $f(x) = 3x^2 + 4x + 1$, then $\lim\limits_{h \to 0} \dfrac{f(x+h) - f(x)}{h}$

$= \lim\limits_{h \to 0} \dfrac{[3(x+h)^2 + 4(x+h) + 1] - (3x^2 + 4x + 1)}{h}$

$= \lim\limits_{h \to 0} \dfrac{3x^2 + 6xh + 3h^2 + 4x + 4h + 1 - 3x^2 - 4x - 1}{h} = \lim\limits_{h \to 0} \dfrac{h(6x + 3h + 4)}{h}$

$= \lim\limits_{h \to 0} (6x + 3h + 4) = 6x + 4$

51. $\lim\limits_{x \to 2} \dfrac{x^2 - 4}{3 - \sqrt{x^2 + 5}} \cdot \dfrac{3 + \sqrt{x^2 + 5}}{3 + \sqrt{x^2 + 5}} = \lim\limits_{x \to 2} \dfrac{(x^2 - 4)(3 + \sqrt{x^2 + 5})}{9 - (x^2 + 5)}$

$= \lim\limits_{x \to 2} \dfrac{(x^2 - 4)(3 + \sqrt{x^2 + 5})}{4 - x^2} = \lim\limits_{x \to 2} \dfrac{(x^2 - 4)(3 + \sqrt{x^2 + 5})}{-(x^2 - 4)}$

$= \lim\limits_{x \to 2} \dfrac{3 + \sqrt{x^2 + 5}}{-1} = \dfrac{3 + \sqrt{9}}{-1} = -6$

55. $\lim\limits_{x \to 1} \dfrac{x^m - 1}{x^n - 1} = \lim\limits_{x \to 1} \dfrac{\left(\dfrac{x^m - 1}{x - 1} \right)}{\left(\dfrac{x^n - 1}{x - 1} \right)} = \lim\limits_{x \to 1} \dfrac{m(1)^{m-1}}{n(1)^{n-1}} = \lim\limits_{x \to 1} \dfrac{m}{n} = \dfrac{m}{n}$

[Using the results of problem 53]

Clearly, we must have both m, n positive integers.

59. For $-\dfrac{\pi}{2} < x < \dfrac{\pi}{2}$, $1 - x^2 \le f(x) \le \cos x$. Also,

$\lim\limits_{x \to 0} \cos x = 1$ and $\lim\limits_{x \to 0} (1 - x^2) = 1$. Then, using the

Squeezing Theorem, we have $\lim\limits_{x \to 0} f(x) = 1$.

63. Show that $\lim\limits_{x\to 0} x^n \sin\frac{1}{x} = 0$, n a positive integer, we have

$-1 \le \sin\frac{1}{x} \le 1$ for all x with $x \ne 0$. Hence,

$-|x|^n \le |x|^n \sin\frac{1}{x} \le |x|^n$. Also, we have that

$\lim\limits_{x\to 0} (-|x|^n)$ and $\lim\limits_{x\to 0} (|x|^n) = 0$, so $\lim\limits_{x\to 0} [|x|^n \sin\frac{1}{x}]$

$= 0$. Since $x^n \to 0$ whenever $|x|^n \to 0$, we conclude

that $\lim\limits_{x\to 0} [x^n \sin\frac{1}{x}] = 0$.

Exercise 3, pp. 78-80

3. $\lim\limits_{x\to -2^+} (2x^3 + x - 1) = 2(-8) + (-2) - 1 = -19$

7. $\lim\limits_{x\to 3^-} \dfrac{x^2 - 9}{x - 3} = \lim\limits_{x\to 3^-} \dfrac{(x-3)(x+3)}{(x-3)} = 6$

11. $\lim\limits_{x\to 5^+} \dfrac{|x-5|}{x-5} = \lim\limits_{x\to 5^+} \dfrac{x-5}{x-5} = 1$

15. $\lim\limits_{x\to 2/3^-} [[2x]] = 1$, since $\frac{1}{2} \le x < 1 \Rightarrow [[x]] = 1$

19. $\lim\limits_{x\to 2^+} \sqrt[3]{[[x]] - x} = 0$, since $2 < x < 3 \Rightarrow [[x]] = 2$.

23. If $f(x) = \begin{cases} \dfrac{x^2-9}{x-3} & \text{if } x \ne 3 \\[2mm] 6 & \text{if } x = 3 \end{cases}$ and $c = 3$, then

   (a) $\lim\limits_{x\to 3^-} f(x) = \lim\limits_{x\to 3^-} \dfrac{x^2-9}{x-3} = \lim\limits_{x\to 3^-} \dfrac{(x-3)(x+3)}{x-3}$

   $= \lim\limits_{x\to 3^-} (x+3) = 6$

   (b) $\lim\limits_{x\to 3^+} f(x) = \lim\limits_{x\to 3^+} \dfrac{x^2-9}{x-3} = \lim\limits_{x\to 3^+} (x+3) = 6$

   (c) yes, $\lim\limits_{x\to 3} f(x) = 6$

27. If $f(x) = \begin{cases} 3x - 1 & \text{if} \quad x < 1 \\ 4 & \text{if} \quad x = 1 \\ 2x & \text{if} \quad x > 1 \end{cases}$ and $c = 1$, then

(a) $\lim\limits_{x \to 1^-} f(x) = \lim\limits_{x \to 1^-} (3x-1) = 2$

(b) $\lim\limits_{x \to 1^+} f(x) = \lim\limits_{x \to 1^+} (2x) = 2$

(c) Yes, $\lim\limits_{x \to 1} f(x) = 2$.

31. If $f(x) = \begin{cases} \sqrt{x^2-9} & \text{if} \quad x \geq 3 \\ \sqrt{9-x^2} & \text{if} \quad x < 3 \end{cases}$ and $c = 3$, then

(a) $\lim\limits_{x \to 3^-} f(x) = \lim\limits_{x \to 3^-} \sqrt{9-x^2} = 0$

(b) $\lim\limits_{x \to 3^+} f(x) = \lim\limits_{x \to 3^+} \sqrt{x^2-9} = 0$

(c) Yes, $\lim\limits_{x \to 3} f(x) = 0$.

35. $\lim\limits_{x \to 3^-} f(x) = \lim\limits_{x \to 3^-} \sqrt{9-x^2} = \sqrt{9-9} = 0$

39. $\lim\limits_{h \to 0^-} \dfrac{f(2+h) - f(2)}{h} = \lim\limits_{h \to 0^-} \dfrac{[3(2+h)+5] - 11}{h}$

$= \lim\limits_{h \to 0^-} \dfrac{3h}{h} = 3$

43. Yes, $\lim\limits_{x \to 3} f(x) = 0$.

47. As an example that $\lim\limits_{x \to c} |f(x)|$ may exist even though

$\lim\limits_{x \to c} f(x)$ does not exist, let $f(x) = \begin{cases} x - 2 & \text{if} \quad x \geq 0 \\ -x+2 & \text{if} \quad x < 0 \end{cases}$.

Since $\lim\limits_{x \to 0^-} f(x) = \lim\limits_{x \to 0^-} (-x+2) = 2$, and $\lim\limits_{x \to 0^+} f(x)$

$= \lim\limits_{x \to 0^+} (x-2) = -2$, we have $\lim\limits_{x \to 0} f(x)$ does not exist.

However, $\lim_{x \to 0^-} |f(x)| = \lim_{x \to 0^-} |-x+2| = 2$ and $\lim_{x \to 0^+} |f(x)|$

$= \lim_{x \to 0^+} |x-2| = |-2| = 2$. Thus, $\lim_{x \to 0} |f(x)|$ exists.

Exercise 4, pp. 88-89

3. $f(x) = \dfrac{x^2-9}{x-3}$, $c = -3$

$\lim_{x \to -3} f(x) = \lim_{x \to -3} \dfrac{x^2-9}{x-3} = \lim_{x \to -3} (x+3) = 0 = f(-3)$

$\Rightarrow$ f continuous at -3.

7. $f(x) = \begin{cases} 2x+5 & \text{if } x \le 2 \\ 4x+1 & \text{if } x > 2 \end{cases}$ $\quad c = 2$

$\lim_{x \to 2^-} f(x) = 9$, $\lim_{x \to 2^+} f(x) = 9$, and $f(2) = 9$, so f

is continuous at 2.

11. f is not defined at $x = 1$ so f is not continuous at $x = 1$.

15. $f(x) = \begin{cases} 4 - 3x^2 & \text{if } x < 0 \\ 4 & \text{if } x = 0 \\ \sqrt{16-x^2} & \text{if } 0 < x < 4 \end{cases}$ $\quad c = 0$

$\lim_{x \to 0^-} f(x) = \lim_{x \to 0^+} f(x) = f(0) = 4 \Rightarrow$ f is continuous

at $x = 0$.

19. $\lim_{x \to 2} \dfrac{x^2-4}{x-2} = \lim_{x \to 2} \dfrac{(x-2)(x+2)}{x-2} = \lim_{x \to 2} (x+2) = 4$

Therefore, define $f(2) = 4$ for continuity.

23. $f(x) = 3x^5 - 2x^3 + x - 2$ as a polynomial is continuous for all x.

27. $f(x) = \dfrac{x}{x-2}$ is continuous for all x except $x = 2$ (the denominator is zero).

31. $f(x) = |x|$ is continuous for all x.

35. $f(x) = \sqrt{\dfrac{x^2+1}{2-x}}$ is continuous only for $x < 2$. (Note: $x > 2$ is excluded since the fraction is negative, and $x = 2$ is excluded since the denominator is zero.)

39. f is not continuous at $x = 2$ since $\lim\limits_{x \to 2^-} f(x) = \sqrt{9} = 3$ and $\lim\limits_{x \to 2^+} f(x) = \sqrt{5}$, which implies $\lim\limits_{x \to 2} f(x)$ does not exist.

43. $f(x) = \dfrac{x}{(x+1)^2} - 1$ on $[10, 20]$. We have $f(10) = \dfrac{10}{121} - 1 = \dfrac{-111}{121}$ and $f(20) = \dfrac{20}{441} - 1 = \dfrac{-421}{441}$. No conclusion can be drawn, since both f values have the same sign.

47. $f(x) = \begin{cases} \dfrac{\sqrt{2x+5} - \sqrt{x+7}}{x - 2} & \text{if } x \neq 2 \\ k & \text{if } x = 2 \end{cases}$

We have $\dfrac{\sqrt{2x+5} - \sqrt{x+7}}{x - 2} \cdot \dfrac{\sqrt{2x+5} + \sqrt{x+7}}{\sqrt{2x+5} + \sqrt{x+7}}$

$= \dfrac{(2x+5) - (x+7)}{(x-2)[\sqrt{2x+5} + \sqrt{x+7}]} = \dfrac{x - 2}{(x-2)[\sqrt{2x+5} + \sqrt{x+7}]}$

$= \dfrac{1}{\sqrt{2x+5} + \sqrt{x+7}}$ if $x \neq 2$. Then, $\lim\limits_{x \to 2} f(x)$

$= \lim\limits_{x \to 2} \dfrac{1}{\sqrt{2x+5} + \sqrt{x+7}} = \dfrac{1}{\sqrt{9} + \sqrt{9}} = \dfrac{1}{6}$. We have that f is continuous at $x = 2$ if $k = 1/6$.

51. $f(x) = x$ and $g(x) = x^2$ are clearly continuous for all x. However, $\left(\dfrac{f}{g}\right)(x) = \dfrac{x}{x^2} = \dfrac{1}{x}$ is not defined at $x = 0$. Therefore, $\dfrac{f}{g}$ is not continuous at $x = 0$.

Exercise 5, pp. 98-99

3. If $f(x) = \dfrac{x^2-9}{x+3}$, then $\lim\limits_{x \to -3} f(x) = -6$. We can

observe that if $x \neq 3$, then $\left|\dfrac{x^2-9}{x+3} - (-6)\right|$

$= \left|(x-3) + 6\right| = \left|x+3\right|$. Thus, for any given $\varepsilon > 0$,
we can choose $\delta = \varepsilon$.

(a) If $\varepsilon = 0.1$, choose $\delta = 0.1$.

(b) If $\varepsilon = 0.01$, choose $\delta = 0.01$.

(c) For any given $\varepsilon > 0$, choose $\delta = \varepsilon$. Thus, we
have that if $0 < \left|x+3\right| < \delta$, then
$\left|f(x) - (-6)\right| < \varepsilon$.

7. If $\lim\limits_{x \to 2} (6x-1) = 11$ and $\varepsilon = 1/2$, we see that
$\left|(6x-1) - 11\right| = \left|6x-12\right| = \left|6(x-2)\right| = 6\left|x-2\right| < 1/2$
whenever $\left|x-2\right| < 1/12$. The largest δ for $\varepsilon = 1/2$
is $\delta = 1/12$.

11. $\lim\limits_{x \to 0} (2x+5) = 5$

Proof: Let $\varepsilon > 0$ be given. We have $\left|(2x+5) - 5\right|$
$= \left|2x\right| = 2\left|x\right|$. Thus, we can choose $\delta = \varepsilon/2$ and we
have the conclusion that if $0 < \left|x\right| < \delta = \varepsilon/2$, then
we have $\left|(2x+5) - 5\right| = 2\left|x\right| < 2\delta = \varepsilon$.

15. $\lim\limits_{x \to 2} (x^2-2x) = 0$

Proof: Let $\varepsilon > 0$ be given and consider $\left|(x^2-2x) - 0\right|$.
We have $\left|(x^2-2x) - 0\right| = \left|x(x-2)\right| = \left|x\right|\left|x-2\right|$. Since
we are considering x near 2, we can assume $\left|x-2\right| < 1$,
which gives $-1 < x-2 < 1$ or $1 < x < 3$. Thus, $\left|x-2\right| < 1$
$\Rightarrow \left|x\right| < 3$. Hence, if $\left|x-2\right| < 1$, we have $\left|x^2-2x\right|$
$= \left|x\right|\left|x-2\right| < 3\left|x-2\right|$. Choosing $\delta = \min(1, \varepsilon/3)$, we
have that if $\left|x-2\right| < \delta$ then $\left|x^2-2x\right| < 3\delta < \varepsilon$.

19. $\lim\limits_{x\to 0} \sqrt[3]{x} = 0$

Proof: Let $\epsilon > 0$ be given and consider $\left|\sqrt[3]{x} - 0\right|$.
We have $\left|\sqrt[3]{x} - 0\right| = \left|\sqrt[3]{x}\right| = \sqrt[3]{|x|}$. We see that if
$|x-0| = |x| < \delta$, then $\sqrt[3]{|x|} < \sqrt[3]{\delta}$. Hence, choose
$\delta = \epsilon^3$. Thus, if $|x-0| < \delta$, we have $\left|\sqrt[3]{x} - 0\right|$
$= \sqrt[3]{|x|} < \sqrt[3]{\delta} = \sqrt[3]{\epsilon^3} = \epsilon$.

23. $\left|\dfrac{1}{x^2+9} - \dfrac{1}{18}\right| = \left|\dfrac{18 - (x^2+9)}{18(x^2+9)}\right| = \left|\dfrac{-x^2+9}{18(x^2+9)}\right| = \left|\dfrac{x^2-9}{18(x^2+9)}\right|$

$= \left|\dfrac{x+3}{18(x^2+9)}\right| \cdot |x-3|$

Now, restricting our attention to $2 < x < 4$ we have

$2 < x < 4 \Rightarrow 5 < x+3 < 7 \Rightarrow |x+3| < 7$ and

$2 < x < 4 \Rightarrow 4 < x^2 < 16 \Rightarrow 13 < x^2+9 < 25$

$\Rightarrow 234 < 18(x^2+9) < 450 \Rightarrow 234 < \left|18(x^2+9)\right|$

$\Rightarrow \dfrac{1}{234} > \dfrac{1}{\left\lceil 18(x^2+9)\right\rceil}.$

Putting these two results together gives

$\dfrac{|x+3|}{\left\lceil 18(x^2+9)\right\rceil} < \dfrac{7}{234}.$ Hence, $\left|\dfrac{1}{x^2+9} - \dfrac{1}{18}\right|$

$= \left|\dfrac{x+3}{18(x^2+9)}\right| \, |x-3| < \dfrac{7}{234}|x-3|$. To show that

$\lim\limits_{x\to 3} \dfrac{1}{x^2+9} = \dfrac{1}{18}$ we must show that for any $\epsilon > 0$,

there exists a $\delta > 0$ such that $\left|\dfrac{1}{x^2+9} - \dfrac{1}{18}\right| < \epsilon$

whenever $0 < |x-3| < \delta$. Since we are only
interested in x near 3, we will restrict our
attention to $|x-3| < 1$; that is, $2 < x < 4$. Using
our earlier result, we see that if we choose
$\delta \leq \min\{1, \dfrac{234}{7}\epsilon\}$. We will have

$|x-3| < \delta \Rightarrow |x-3| \leq \dfrac{234}{7}\epsilon \Rightarrow \dfrac{7}{234}|x-3| \leq \epsilon$; thus,

$\left|\dfrac{1}{x^2+9} - \dfrac{1}{18}\right| < \dfrac{7}{234}|x-3| \leq \epsilon$. That is, $\left|\dfrac{1}{x^2+9} - \dfrac{1}{18}\right| < \epsilon$

whenever $|x-3| < \delta$ for $\delta \leq \min\{1, \dfrac{234}{7}\epsilon\}$.

3. Given $y = x^2$ and
    P(2,4). We have that
    A = 4 and V = 0.
    Thus, B = -4. We find
    the equation of the
    line through P(2,4)
    and B(0,-4). We have
    $m = \dfrac{4 + 4}{2 - 0} = \dfrac{8}{2} = 4,$ so
    the equation is
    y - 4 = 4(x-2), or
    y = 4x - 4.

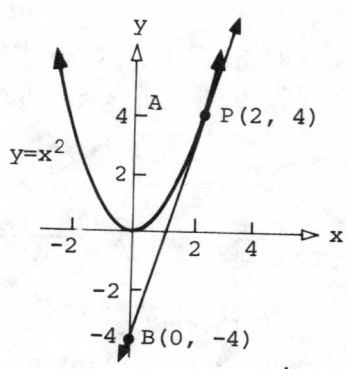

7. Given $x = y^2 + 1$ and P(2,1). The tangent line is
    y = mx + b and P(2,1) on the line gives
    1 = 2m + b, or b = 1 - 2m. The line then has the

    equation y = mx + (1-2m), or $x = \dfrac{1}{m}y + \dfrac{2m - 1}{m}$.

    We wish (2,1) to be the unique solution to the

    system $\begin{cases} x = y^2 + 1 \\ x = \dfrac{1}{m}y + \dfrac{2m - 1}{m} \end{cases}$. This gives

    $y^2 + 1 = \dfrac{1}{m}y + \dfrac{2m - 1}{m}$, $my^2 - y + m - (2m-1) = 0$,

    or $my^2 - y - (m-1) = 0$. Solving for y gives

    $y = \dfrac{1 \pm \sqrt{1 + 4m(m-1)}}{2m} = \dfrac{1 \pm \sqrt{4m^2 - 4m + 1}}{2m}$. This has

    a unique solution if $4m^2 - 4m + 1 = 0$, or

    $(2m-1)^2 = 0$. This implies m = 1/2. The equation

    of the tangent line [using m = 1/2, P(2,1)] is

    given by $y - 1 = \dfrac{1}{2}(x-2)$, or $y = \dfrac{1}{2}x$.

11. Let $y = x^2$ and $P(3,9)$. Choose $Q(x_0, y_0)$ on the tangent line below P. Then $y_0 < x_0^2$ and

$$\frac{9}{y_0} > \frac{9}{x_0^2}. \quad \text{But}$$

$y_0 = 9 - e$, so that

$$\frac{9}{9 - e} > \frac{9}{x_0^2} = \left(\frac{3}{x_0}\right)^2.$$

However, $\dfrac{3}{x_0} = \dfrac{9 + a}{9 + a - e}$.

This gives

$$\frac{9}{9 - e} > \left(\frac{9 + a}{9 + a - e}\right)^2$$

$$= \frac{81 + 18a + a^2}{81 + 18a + a^2 - 2e(9+a) + e^2}.$$

When Q is near P, then we have

$$\frac{9}{9 - e} \approx \frac{a^2 + 18a + 81}{a^2 + 18a + 81 - 18e - 2ae + e^2},$$

$$9a^2 + 162a + 729 - 162e - 18ae + 9e^2$$

$$\approx 9a^2 + 162a + 729 - a^2e - 18ae - 81e \Longrightarrow$$

$$-162e + 9e^2 \approx -a^2e - 81e,$$

$$\Longrightarrow a^2e - 81e + 9e^2 \approx 0 \Longrightarrow$$

$$e(a^2 - 81 + 9e) \approx 0.$$

Dividing by e gives $a^2 - 81 + 9e \approx 0$, and letting $e = 0$ gives $a^2 \approx 81$, or $a \approx 9$. Thus, the y-intercept is $(0,-9)$. The slope of the line between $(3,9)$ and $(0,-9)$ is $m = 18/3 = 6$, so the equation is $y + 9 = 6(x-0)$, or $y = 6x - 9$.

Miscellaneous Exercises, pp. 103-105

3. $\lim\limits_{x \to 1} \dfrac{x^8 - 1}{x - 1} = \lim\limits_{x \to 1} \dfrac{(x^4+1)(x^2+1)(x+1)(x-1)}{(x-1)}$

   $= (2)(2)(2) = 8$

7. $f(x) = \dfrac{\sqrt{x} - 2}{x - 4}$

x	3.9	3.99	3.999	4.001	4.01	4.1
f(x)	0.2516	0.2502	0.2500	0.2499	0.2498	0.2485

$$\lim_{x \to 4} f(x) = \lim_{x \to 4} \frac{\sqrt{x} - 2}{(\sqrt{x}-2)(\sqrt{x}+2)} = \lim_{x \to 4} \frac{1}{\sqrt{x} + 2} = \frac{1}{4}$$

11. $\displaystyle\lim_{x \to 0^+} \frac{|x|}{x}(1-x) = \lim_{x \to 0^+} (1-x) = 1$ and

$$\lim_{x \to 0^-} \frac{|x|}{x}(1-x) = \lim_{x \to 0^-} -(1-x) = -1, \quad \text{so}$$

$\displaystyle\lim_{x \to 0} \frac{|x|}{x}(1-x)$ is not defined.

15. If $y = x^3$ and $y = 1 - x^2$, prove they intersect
for some $x \in (0,1)$. Set $x^3 = 1 - x^2$, so that
$x^3 + x^2 - 1 = 0$. Let $g(x) = x^3 + x^2 - 1$; $g(0) = -1$,
$g(1) = 1$ and g is continuous on $[0,1]$. Applying
the intermediate value theorem, we conclude that
$g(x) = 0$ for some $x \in (0,1)$.

19. To find K (Lipschitz constant) for $f(x) = x^3$
on $(0,2)$, we have

$$\left| f(x) - f(c) \right| = \left| x^3 - c^3 \right| \quad \text{and}$$

$$\frac{\left| f(x) - f(c) \right|}{\left| x - c \right|} = \left| \frac{x^3 - c^3}{x - c} \right| = \left| \frac{(x-c)(x^2+cx+c^2)}{x - c} \right|$$

$$= \left| x^2 + cx + c^2 \right| \quad \text{if} \quad x \neq c.$$

If $x, c \in (0,2)$, then $\left| x^2 + cx + c^2 \right| \leq 4 + 4 + 4 = 12$;
thus, $K = 12$.

23. Prove that $\displaystyle\lim_{x \to 2} \frac{1}{x^2 + 9} = \frac{1}{13}$.

<u>Proof</u>: Let $\varepsilon > 0$ be given and consider

$$\left| \frac{1}{x^2 + 9} - \frac{1}{13} \right| = \left| \frac{13 - (x^2+9)}{13(x^2+9)} \right| = \left| \frac{4 - x^2}{13(x^2+9)} \right|$$

$$= \left| \frac{x + 2}{13(x^2+9)} \right| \, |x - 2|$$

$|x - 2| < 1 \Rightarrow -1 < x - 2 < 1 \Rightarrow 1 < x < 3$

$\qquad\qquad \Rightarrow 3 < x + 2 < 5 \Rightarrow |x + 2| < 5$ and

$|13(x^2+9)| = 13(x^2+9) > 13(10) = 130.$

Thus, $\dfrac{|x + 2|}{|13(x^2+9)|} < \dfrac{5}{13(10)} = \dfrac{1}{26}$.

Then, letting $\delta = \min(1, 26\varepsilon)$, we have that if $|x - 2| < \delta$, then

$$\left|\frac{1}{x^2 + 9} - \frac{1}{13}\right| = \left|\frac{x + 2}{13(x^2+9)}\right| \, |x - 2| < \frac{1}{26} \, |x - 2|$$

$$< \frac{1}{26} \cdot \delta \leq \frac{1}{26} \cdot 26\varepsilon = \varepsilon.$$

Thus, $\displaystyle\lim_{x \to 2} \frac{1}{x^2 + 9} = \frac{1}{13}$.

27. Prove that $\displaystyle\lim_{x \to 0} (4-x^2) = 4.$

Proof: Let $\varepsilon > 0$ be given and consider

$|(4-x^2) - 4| = |-x^2| = |x|^2.$ Letting $\delta = \sqrt{\varepsilon}$, then

if $|x| < \delta = \sqrt{\varepsilon}$, we have

$|(4-x^2) - 4| = |x|^2 < \delta^2 = \varepsilon.$

31. Prove that $\cos x$ is continuous at each real number c.

Proof: Let c be any real number. To show continuity at $x = c$, we need to show $\displaystyle\lim_{x \to c} \cos x$ exists, and that the value of the limit is $\cos c$. We have

$\cos x = \cos(x-c+c) = \cos(x-c) \cos c - \sin(x-c) \sin c.$

$\displaystyle\lim_{x \to c} \cos(x-c) = \lim_{u \to 0} \cos u = 1$ and

$\displaystyle\lim_{x \to c} \sin(x-c) = \lim_{u \to 0} \sin u = 0,$ so that

$$\lim_{x \to c} \cos x = \lim_{x \to c} [\cos(x-c) \cos c - \sin(x-c) \sin c]$$

$$= \cos c \cdot \lim_{x \to c} \cos(x-c) - \sin c \cdot \lim_{x \to c} \sin(x-c)$$

$$= (\cos c)(1) - (\sin c)(0)$$

$$= \cos c.$$

Thus, $\cos x$ is continuous at each real number c.

Chapter Three

Exercise 1, pp. 110-112

3. If $f(x) = \dfrac{x^2}{x + 3}$, we have:

(a) Change from $x = -1$ to $x = 1$ is

$\Delta y = f(1) - f(-1) = \dfrac{1}{4} - \dfrac{1}{2} = -1/4.$

(b) Change from $x = 0$ to $x = 4$ is

$\Delta y = f(4) - f(0) = \dfrac{16}{7} - 0 = 16/7.$

(c) Average change in (a) is $\dfrac{\Delta y}{\Delta x} = \dfrac{-1/4}{2} = -\dfrac{1}{8}$.

(d) Average change in (b) is $\dfrac{\Delta y}{\Delta x} = \dfrac{16/7}{4} = \dfrac{4}{7}$.

(e) Since $f(-2) = \dfrac{4}{1} = 4,$ and $f(2) = \dfrac{4}{5}$, then

$m_{sec} = \dfrac{\dfrac{4}{1} - \dfrac{4}{5}}{-2 - 2} = \dfrac{\dfrac{20 - 4}{5}}{-4} = \dfrac{16}{-20} = -\dfrac{4}{5}.$

7. If $f(x) = x^2 + 4$, then $f(0) = 4$, and $f(2) = 8$,

so that $m_{sec} = \dfrac{8 - 4}{2 - 0} = \dfrac{4}{2} = 2.$

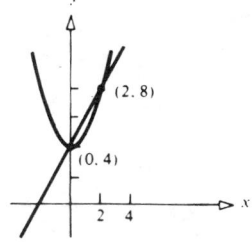

11. If $f(t) = 16t^2$, then

$\dfrac{\Delta s}{\Delta t} = \dfrac{f(t_0 + \Delta t) - f(t_0)}{\Delta t} = \dfrac{16(t_0 + \Delta t)^2 - 16t_0^2}{\Delta t}$

$= \dfrac{32t_0 \Delta t + 16(\Delta t)^2}{\Delta t} = 32t_0 + 16\Delta t$

(a) $t_0 = 3$, $\Delta t = 0.5 \Rightarrow \dfrac{\Delta s}{\Delta t} = 32(3) + 16(0.5)$

$$= 96 + 8 = 104 \text{ ft/sec}$$

(b) $t_0 = 3$, $\Delta t = 0.1 \Rightarrow \dfrac{\Delta s}{\Delta t} = 96 + 16(0.1)$

$$= 96 + 1.6 = 97.6 \text{ ft/sec}$$

15. If $f(x) = 3x^2 - 2$ and $c = -1$, then

$$\frac{f(-1+\Delta x) - f(-1)}{\Delta x} = \frac{3(1-2\Delta x+(\Delta x)^2) - 2 - 1}{\Delta x} = -6 + 3\Delta x.$$

19. If $f(x) = x + \dfrac{1}{x}$ and $c = 1$, then

$$\frac{f(1+\Delta x) - f(1)}{\Delta x} = \frac{1 + \Delta x + \dfrac{1}{1 + \Delta x} - 2}{\Delta x} = \frac{\dfrac{\Delta x - 1}{1} + \dfrac{1}{\Delta x + 1}}{\Delta x}$$

$$= \frac{\dfrac{(\Delta x)^2 - 1 + 1}{\Delta x + 1}}{\dfrac{\Delta x}{1}} = \frac{\Delta x}{\Delta x + 1}.$$

23. The car travels half the distance at 45 mph and half
at 55 mph. Let d be the distance from New York to
Miami. The average velocity is distance/time. Using
the formula distance = (rate) × (time), we have

$$t_1 = \frac{\frac{1}{2}d}{45} = \frac{d}{90}, \quad \text{and} \quad t_2 = \frac{\frac{1}{2}d}{55} = \frac{d}{110}. \quad \text{Then the}$$

average velocity is $\dfrac{d}{t_1 + t_2} = \dfrac{d}{\dfrac{d}{90} + \dfrac{d}{110}} = \dfrac{d}{\dfrac{110d + 90d}{9900}}$

$$= \frac{d}{\dfrac{200d}{9900}} = \frac{9900}{200} = 49.5 \text{ mph}.$$

27. The average rate of
increase in yield
from:

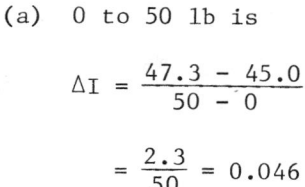

(a) 0 to 50 lb is

$$\Delta I = \frac{47.3 - 45.0}{50 - 0}$$

$$= \frac{2.3}{50} = 0.046$$

bushel/lb.

(b) 10 to 40 lb is $\Delta I = \dfrac{47.6 - 46.2}{30} = \dfrac{1.4}{30}$

$$\approx 0.0467 \text{ bushel/lb.}$$

(c) 10 to 30 lb is $\Delta I = \dfrac{48.2 - 46.2}{20} = \dfrac{2}{20}$

$$= 0.1 \text{ bushel/lb.}$$

Exercise 2, pp. 118-119

3. If $f(x) = x^2 - 2$, then $f'(0) = \lim\limits_{x \to 0} \dfrac{(x^2-2) - (-2)}{x - 0}$

$$= \lim\limits_{x \to 0} \frac{x^2}{x} = \lim\limits_{x \to 0} x = 0.$$

7. If $f(x) = \sqrt{x}$, then $f'(4) = \lim\limits_{x \to 4} \dfrac{\sqrt{x} - \sqrt{4}}{x - 4}$

$$= \lim\limits_{x \to 4} \frac{(\sqrt{x}-2)}{(\sqrt{x}-2)(\sqrt{x}+2)} = \lim\limits_{x \to 4} \frac{1}{\sqrt{x} + 2} = \frac{1}{\sqrt{4} + 2}$$

$$= \frac{1}{2 + 2} = \frac{1}{4}.$$

11. If $f(x) = x^2 - 2$, then

$$f'(x) = \lim\limits_{h \to 0} \frac{[(x+h)^2-2] - (x^2-2)}{h}$$

$$= \lim\limits_{h \to 0} \frac{x^2 + 2xh + h^2 - 2 - x^2 + 2}{h}$$

$$= \lim_{h \to 0} \frac{h(2x+h)}{h} = \lim_{h \to 0} (2x+h) = 2x.$$

15. If $f(x) = 5$, then $f'(x) = \lim_{h \to 0} \frac{f(x+h) - f(x)}{h}$

$$= \lim_{h \to 0} \frac{5 - 5}{h} = \lim_{h \to 0} (0) = 0.$$

19. If $f(x) = mx + b$, then

$$f'(x) = \lim_{h \to 0} \frac{[m(x+h)+b] - (mx+b)}{h}$$

$$= \lim_{h \to 0} \frac{mh}{h} = \lim_{h \to 0} (m) = m.$$

23. If $s(t) = 10t^2$, then $\frac{\Delta s}{\Delta t} = \frac{s(t+\Delta t) - s(t)}{\Delta t}$

$$= \frac{10(t+\Delta t)^2 - 10t^2}{\Delta t}$$

$$= \frac{10t^2 + 20t(\Delta t) + 10(\Delta t)^2 - 10t^2}{\Delta t}$$

$$= \frac{20t(\Delta t) + 10(\Delta t)^2}{\Delta t} = 20t + 10(\Delta t).$$

If $t = 3$, then $\frac{s(3+\Delta t) - s(3)}{\Delta t}$

$$= 20(3) + 10(\Delta t) = 60 + 10(\Delta t).$$

(a) Letting $\Delta t = 0.1$ implies $\Delta s/\Delta t = 60 + 1$
= 61 cm/sec.

(b) Letting $\Delta t = 0.01$ implies $\Delta s/\Delta t = 60 + 0.1$
= 60.1 cm/sec.

(c) Letting $\Delta t = 0.001$ implies $\Delta s/\Delta t = 60 + 0.01$
= 60.01 cm/sec.

(d) The limiting value seems to be 60 cm/sec.

27. f even function $\rightarrow$ f(-c) = f(c)
     f differentiable at c $\rightarrow$

$$f'(c) = \lim_{h \to 0} \frac{f(c+h) - f(c)}{h}$$

$$= \lim_{h \to 0} \frac{f[-(c+h)] - f(-c)}{h} \quad \text{since f is even}$$

$$= \lim_{h \to 0} \frac{f\ -c-h\ -\ f(-c)}{-(-h)}$$

Now substitute k for -h in this right side:

$$f'(c) = \lim_{-k \to 0} \frac{f(-c+k) - f(-c)}{-k}$$

Now the fact that -k$\rightarrow$0 implies that k$\rightarrow$0 so factor out the minus sign in the denominator:

$$f'(c) = -\lim_{k \to 0} \frac{f(-c+k) - f(-c)}{k}$$

$$= f\ (-c), \text{ by definition.}$$
     Thus f' is odd.

Exercise 3, pp. 121-123

3. $f(x) = x^2 + 2x + 1$ at   (1,4)

$$m = f'(1) = \lim_{x \to 1} \frac{(x^2+2x+1) - 4}{x - 1} = \lim_{x \to 1} \frac{(x-1)(x+3)}{x - 1}$$

$$= \lim_{x \to 1} (x+3) = 4.$$
     The equation is   y - 4 = 4(x-1)   or   y = 4x.

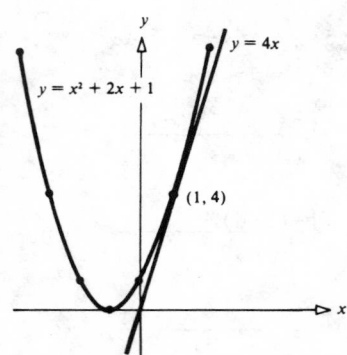

$y = 4x$

$y = x^2 + 2x + 1$

$(1, 4)$

7. Let $y = f(x) = x^2$. $m = f'(1) = \lim\limits_{x \to 1} \dfrac{x^2 - 1}{x - 1}$

$= \lim\limits_{x \to 1} (x+1) = 2.$

The equation of the tangent line at $(1,1)$ is
$y - 1 = 2(x-1)$, or $y = 2x-1$. The point $(2,5)$ is
not on the line $y = 2x - 1.$

11. $f(x) = x^2 + 1$ at $(1,2)$

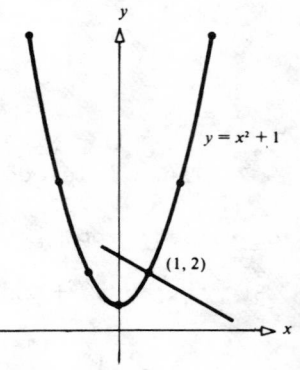

$y = x^2 + 1$

$(1, 2)$

$m_T = f'(1) = \lim\limits_{x \to 1} \dfrac{(x^2+1) - 2}{x - 1}$

$= \lim\limits_{x \to 1} \dfrac{(x-1)(x+1)}{x - 1}$

$= \lim\limits_{x \to 1} (x+1) = 2$

Hence, $m_N = -1/2$ and the
equation of the normal line

is $y - 2 = -\dfrac{1}{2}(x-1) \Rightarrow y = -\dfrac{1}{2}x + \dfrac{5}{2}$, or $x + 2y = 5.$

15. $f(x) = \frac{1}{x}$ at $(1,1)$.

$$m_T = f'(1) = \lim_{x \to 1} \frac{\frac{1}{x} - 1}{x - 1}$$

$$= \lim_{x \to 1} \left(\frac{-1}{x}\right) = -1.$$

Hence, $m_N = -\left(\frac{1}{-1}\right) = +1$

and the equation of the normal line is
$y - 1 = 1(x-1)$ or $y = x$.

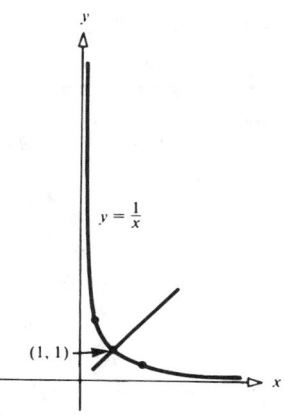

$y = \frac{1}{x}$

$(1, 1)$

19. If $A = \pi R^2$ and $C = 2\pi R$, and if the radius changes from R to $R+\Delta R$, then we have the following:

(a) Change in area is $\Delta A = \pi(R+\Delta R)^2 - \pi R^2$
$= 2\pi R \Delta R + \pi(\Delta R)^2.$

(b) Change in circumference is $\Delta C = 2\pi(R+\Delta R) - 2\pi R$
$= 2\pi \Delta R.$

(c) Average rate of change of area with respect to radius is

$$\frac{A(R+\Delta R) - A(R)}{\Delta R} = 2\pi R + \pi \Delta R.$$

(d) Average rate of change of circumference with respect to radius is

$$\frac{C(R+\Delta R) - C(R)}{\Delta R} = 2\pi.$$

(e) $\displaystyle \lim_{\Delta R \to 0} \frac{\Delta C}{\Delta R} = \lim_{\Delta R \to 0} \frac{2\pi \Delta R}{\Delta R} = 2\pi$

23. From (3.12), $y - f(c) = f'(c)(x-c)$
That is, $y = f'(c)x - f'(c)c + f(c)$.
We have $y = -3x+12$ as the equation of the tangent line at $(2,6) = (c,f(c))$. Thus $f'(2) = f'(c) = -3$, by comparing terms in the two tangent line equations.

3. $f(x) = 3x + 2 \Rightarrow f' = 3$

7. $f(x) = 8x^5 - 5x + 1 \Rightarrow f' = 40x^4 - 5$

11. $f(x) = \pi x^3 + \frac{3}{2}x^2 \Rightarrow f' = 3\pi x^2 + 3x$

15. $f(x) = ax^2 + bx + c \Rightarrow f' = 2ax + b$

19. $A = \pi R^2 \Rightarrow \frac{dA}{dR} = 2\pi R$

23. If $f(x) = x^3 + 3x - 1$ at $(0,-1)$, then $f'(x) = 3x^2 + 3$, so $m_T = f'(0) = 3$. The equation of the tangent line is $y + 1 = 3(x-0)$, or $y = 3x - 1$.

27. $f(x) = x^3 - 3x + 2$
    $f'(x) = 3x^2 - 3 = 3(x^2-1)$
    $f' = 0$ when $x = -1$, $x = 1$

31. $f(x) = \frac{1}{3}x^3 - x^2 \to f'(x) = x^2 - 2x$

    To be parallel to $y - 3x + 2 = 0$ (i.e., $y = 3x - 2$), we need $f'(x) = x^2 - 2x = 3$: that is, $x^2 - 2x - 3 = 0$ $\to$ $x = 3$ or $x = -1$. Two possible tangent lines exist: (i) At the point $[3,f(3)] = (3,0)$, we have $(y-0) = 3(x-3)$, or $y = 3x - 9$.

    (ii) At the point $[-1,f(-1)] = (-1,-\frac{4}{3})$, we have

    $$[y-(-\frac{4}{3})] = 3[x-(-1)], \text{ or } y = 3x + \frac{5}{3}.$$

35. Let $f(x) = 5x^8$. Then $f'(a) = \lim_{h\to 0} \frac{f(a+h) - f(a)}{h}$

    $$= \lim_{h\to 0} \frac{5(a+h)^8 - 5a^8}{h} .$$

    Hence, by comparison, $\lim_{h\to 0} \frac{5(\frac{1}{2}+h)^8 - 5(\frac{1}{2})^8}{h} = f'(\frac{1}{2})$.

    Since $f'(x) = 40x^7$, the value of the limit is $40(\frac{1}{2})^7 = \frac{40}{2^7} = \frac{5}{16}$.

39. Let $f(x) = x^n$, where n is a positive integer, and use (3.10):

$$f'(c) = \lim_{x \to c} \frac{f(x) - f(c)}{x - c} = \lim_{x \to c} \frac{x^n - c^n}{x - c}$$

$x^n - c^n$ can be written as

$$x^n - c^n = (x-c)\left[x^{n-1} + cx^{n-2} + c^2 x^{n-3} + \ldots + c^{n-1}\right].$$

Thus,

$$f'(c) = \lim_{x \to c}\left[x^{n-1} + cx^{n-2} + c^2 x^{n-3} + \ldots + c^{n-1}\right]$$

$$= c^{n-1} + c^{n-1} + c^{n-1} + \ldots + c^{n-1}$$

$$= nc^{n-1} \quad \text{since there were n terms.}$$

Exercise 5, pp. 134-135

3. $f(x) = (3x^2-5)(2x+1)$
   $f'(x) = (3x^2-5)(2) + (2x+1)(6x) = 6x^2 - 10 + 12x^2 + 6x$
   $= 18x^2 + 6x - 10$

7. $f(t) = t^{-3}$
   $f'(t) = -3t^{-4} = -3/t^4$

11. $f(s) = \dfrac{2s}{s + 1}$

   $f'(s) = \dfrac{(s+1)(2) - (2s)(1)}{(s+1)^2} = \dfrac{2}{(s+1)^2}$

15. $f(t) = 3t + \dfrac{1}{3} t^{-1}$

   $f'(t) = 3 - \dfrac{1}{3} t^{-2} = 3 - \dfrac{1}{3t^2} = \dfrac{9t^2 - 1}{3t^2}$

19. $f(x) = 3x^3 - \dfrac{1}{3} x^{-2}$

   $f'(x) = 9x^2 + \dfrac{2}{3} x^{-3} = 9x^2 + \dfrac{2}{3x^3} = \dfrac{27x^5 + 2}{3x^3}$

23. $f(w) = \dfrac{1}{w^3 - 1}$

$f'(w) = \dfrac{(w^3-1)(0) - 1(3w^2)}{(w^3-1)^2} = \dfrac{-3w^2}{(w^3-1)^2}$

27. $f(x) = \dfrac{x^2}{x + 1}$

$f'(x) = \dfrac{(x+1)(2x) - x^2(1)}{(x+1)^2} = \dfrac{2x^2 + 2x - x^2}{(x+1)^2}$

$= \dfrac{x^2 + 2x}{(x+1)^2} = \dfrac{x(x+2)}{(x+1)^2}$

$f'(x) = 0$ when $x = 0$ and $x = -2$.

31. We have $I = k/r^2$, and if $I = 1000$ when $r = 1$, then $1000 = k/1$, or $k = 1000$. Thus, $I = 1000\, r^{-2}$. Then $dI/dr = -2000\, r^{-3} = -2000/r^3$. At $r = 10$ meters, $dI/dr = -2000/1000 = -2$ units/m.

35. $y = (x-1)(x^2+5)(x^3-1)$
$y' = (x-1)(x^2+5)(3x^2) + (x-1)(2x)(x^3-1) + 1(x^2+5)(x^3-1)$
$= (3x^5-3x^4+15x^3-15x^2) + (2x^5-2x^4-2x^2+2x)$
$\quad + (x^5+5x^3-x^2-5)$
$= 6x^5 - 5x^4 + 20x^3 - 18x^2 + 2x - 5$

39. $y = (1-\dfrac{1}{x})(1-\dfrac{1}{x^2})(1-\dfrac{1}{x^3})$

$y' = (1-\dfrac{1}{x})(1-\dfrac{1}{x^2})(\dfrac{3}{x^4}) + (1-\dfrac{1}{x})(1-\dfrac{1}{x^3})(\dfrac{2}{x^3})$

$\quad + (1-\dfrac{1}{x^2})(1-\dfrac{1}{x^3})(\dfrac{1}{x^2})$

Exercise 6, pp. 139-140

3. $f(x) = 3x^2 + x - 2$

$f'(x) = 6x + 1$

$f''(x) = 6$

7. $f(t) = \dfrac{t}{t + 1}$

42   CHAPTER THREE

$$f'(t) = \frac{(t+1) - t}{(t+1)^2} = \frac{1}{(t+1)^2} = \frac{1}{t^2 + 2t + 1}$$

$$f''(t) = \frac{(t^2+2t+1)(0) - 1(2t+2)}{(t+1)^4} = \frac{-2}{(t+1)^3}$$

11. $f(x) = \dfrac{4}{3x + 5}$

$$f'(x) = \frac{(3x+5)(0) - 4(3)}{(3x+5)^2} = \frac{-12}{9x^2 + 30x + 25}$$

$$f''(x) = \frac{(9x^2+30x+25)(0) - (-12)(18x+30)}{(3x+5)^4} = \frac{72}{(3x+5)^3}$$

15. (a) $y = 4x^3 - 3x^2 + x$
$y' = 12x^2 - 6x + 1$
$y'' = 24x - 6$
$y''' = 24$

(b) $y = ax^3 + bx^2 + cx + d$
$y' = 3ax^2 + 2bx + c$
$y'' = 6ax + 2b$
$y''' = 6a$

19. $\dfrac{d^{20}}{dx^{20}}(8x^{19}-2x^{14}+2x^5) = 0$

23. $s = 16t^2 + 20t$

$v(t) = 32t + 20$

$a(t) = 32$

27. $f = x^2 g$
$f' = x^2 g' + 2xg$
$f'' = x^2 g'' + 2xg' + 2xg' + 2g$
$\quad = x^2 g'' + 4xg' + 2g$

31. For simplicity, let's use the function notation:
$F = fg$. We will assume that f', g', f'', g'' exist.
Since $F' = f'g + fg'$, we have
$F'' = (f''g + f'g') + (f'g' + fg'')$ or
$F'' = f''g + 2f'g' + fg''$.

Exercise 7, pp. 146-147

3. $y = \sin x \cos x$
    $y' = \cos x \cos x + \sin x(-\sin x) = \cos^2 x - \sin^2 x$

7. $y = \dfrac{\cot x}{x}$

    $y' = \dfrac{x(-\csc^2 x) - \cot x(1)}{x^2} = \dfrac{-x \csc^2 x - \cot x}{x^2}$

11. $y = x \tan x - 3 \sec x$
    $y' = 1 \cdot \tan x + x \cdot \sec^2 x - 3 \sec x \tan x$
    $y' = \tan x + x \sec^2 x - 3 \sec x \tan x$

15. $y = \dfrac{\sin x}{1 + x}$

    $y' = \dfrac{(1+x)\cos x - \sin x(1)}{(1+x)^2} = \dfrac{\cos x + x \cos x - \sin x}{(1+x)^2}$

19. $y = \dfrac{\sec x}{1 + x \sin x}$

    $y' = \dfrac{(1+x \sin x)\sec x \tan x - \sec x (1 \cdot \sin x + x \cos x)}{(1+x \sin x)^2}$

    $y' = \dfrac{\sec x \tan x + x \tan^2 x - \tan x - x}{(1+x \sin x)^2}$

    $\left[\begin{array}{l}\text{using } \sin x \sec x = \sin x/\cos x = \tan x \\ \text{and } \sec x \cos x = \cos x/\cos x = 1\end{array}\right]$

23. $y = \dfrac{1 + \tan x}{1 - \tan x}$

    $y' = \dfrac{(1 - \tan x)(\sec^2 x) - (1+\tan x)(-\sec^2 x)}{(1-\tan x)^2}$

    $y' = \dfrac{\sec^2 x - \tan x \sec^2 x + \sec^2 x + \tan x \sec^2 x}{(1-\tan x)^2}$

    $y' = \dfrac{2 \sec^2 x}{(1-\tan x)^2}$

27. $y = \sec x$
    $y' = \sec x \tan x$

44   CHAPTER THREE

$$y'' = \sec x \tan x \cdot \tan x + \sec x \cdot \sec^2 x$$
$$= \sec x \tan^2 x + \sec^3 x$$

31. $y = 2 \sin x - 3 \cos x$
$y' = 2 \cos x + 3 \sin x$
$y'' = -2 \sin x + 3 \cos x$

35. $y = a \sin x + b \cos x$
$y' = a \cos x - b \sin x$
$y'' = -a \sin x - b \cos x$

39. $f = \dfrac{\cos x}{1 + \sin x}$; $c = \pi/3$

$$f'(x) = \frac{(1 + \sin x)(-\sin x) - \cos x (\cos x)}{(1 + \sin x)^2}$$

$$= \frac{-\sin x - 1}{(1 + \sin x)^2} = \frac{-1}{1 + \sin x}, \text{ using } \sin^2 x + \cos^2 x = 1$$

$$f'(\tfrac{\pi}{3}) = \frac{-1}{1 + (\sqrt{3}/2)} = \frac{-2}{2 + \sqrt{3}} = \frac{-2(2 - \sqrt{3})}{1} = 2\sqrt{3} - 4$$

43. $\displaystyle\lim_{x \to 0} \frac{\cos x}{1 + \sin x} = \frac{1}{1} = 1$

47. $\displaystyle\lim_{x \to \pi} \frac{\sin x}{\pi - x} = \lim_{t \to 0} \frac{\sin(\pi - t)}{t}$

$$= \lim_{t \to 0} \frac{\sin \pi \cos t - \cos \pi \sin t}{t}$$

$$= \lim_{t \to 0} \frac{\sin t}{t} = 1, \quad \text{letting } t = \pi - x$$

51. $f(x) = \cos x \rightarrow f'(x) = -\sin x$

$$\rightarrow f'(\tfrac{\pi}{3}) = -\frac{\sqrt{3}}{2}$$

Tangent at $(\tfrac{\pi}{3}, \tfrac{1}{2})$ is given by $(y - \tfrac{1}{2}) = -\dfrac{\sqrt{3}}{2}(x - \tfrac{\pi}{3})$, or

$$y = -\frac{\sqrt{3}}{2}x + \frac{3 + \pi\sqrt{3}}{6}.$$

55. $f(x) = \sin x + \cos x \rightarrow f'(x) = \cos x - \sin x$

$$f'\left(\frac{\pi}{4}\right) = \frac{\sqrt{2}}{2} - \frac{\sqrt{2}}{2} = 0$$

Tangent at $\left(\frac{\pi}{4}, \sqrt{2}\right)$ is given by $y - \sqrt{2}) = 0\left(x - \frac{\pi}{4}\right)$, or

$y = \sqrt{2}$.

59. $\lim\limits_{h \to 0} \dfrac{\cos\left(\frac{\pi}{2} + h\right) - \cos\frac{\pi}{2}}{h} = f'\left(\frac{\pi}{2}\right)$

for $f(x) = \cos x$. Since $f'(x) = -\sin x$, the limit is

$f'\left(\frac{\pi}{2}\right) = -\sin\left(\frac{\pi}{2}\right) = -1$.

63. If $y = A\sin t + B\cos t$
$y' = A\cos t - B\sin t$
and $y'' = -A\sin t - B\cos t$
(for A,B constants).
Hence $y'' + y = 0$.

67. $\lim\limits_{x \to 0} \dfrac{\sin ax}{\sin bx} = \lim\limits_{x \to 0} \dfrac{\dfrac{\sin ax}{x}}{\dfrac{\sin bx}{x}}$

$= \dfrac{\lim\limits_{x \to 0}\left(\dfrac{\sin ax}{x}\right)}{\lim\limits_{x \to 0}\left(\dfrac{\sin bx}{x}\right)} = \dfrac{\lim\limits_{\frac{m}{a} \to 0}\left(\dfrac{\sin m}{\frac{m}{a}}\right)}{\lim\limits_{\frac{n}{b} \to 0}\left(\dfrac{\sin n}{\frac{n}{b}}\right)}$

$= \dfrac{a\,\lim\limits_{\frac{m}{a} \to 0}\left(\dfrac{\sin m}{m}\right)}{b\,\lim\limits_{\frac{n}{b} \to 0}\left(\dfrac{\sin n}{n}\right)} = \dfrac{a\,\lim\limits_{m \to 0}\left(\dfrac{\sin m}{m}\right)}{b\,\lim\limits_{n \to 0}\left(\dfrac{\sin n}{n}\right)}$

$= \dfrac{a \cdot 1}{b \cdot 1} = \dfrac{a}{b}$

Within this string of equalities, we

(1) substituted $m = ax$ and $n = bx$;

(2) factored constants a and b out of limits; and

(3) used the facts that $\frac{m}{a} \to 0$ implies $m \to 0$ and $\frac{n}{b} \to 0$ implies $n \to 0$.

<u>Exercise 8</u>, pp. 154-156

3. $f(x) = (6x-5)^{-3}$

$f'(x) = -3(6x-5)^{-4}(6) = -18(6x-5)^{-4} = \dfrac{-18}{(6x-5)^4}$

7. $f(t) = (t^5-t^2+t)^7$
$f'(t) = 7(5t^4-2t+1)(t^5-t^2+t)^6$

11. $f(z) = (\dfrac{z}{z+1})^3$

$f'(z) = 3(\dfrac{z}{z+1})^2 [\dfrac{(z+1)1 - z(1)}{(z+1)^2}]$

$= \dfrac{3z^2}{(z+1)^2} \cdot \dfrac{1}{(z+1)^2} = \dfrac{3z^2}{(z+1)^4}$

15. $f(t) = \sin^2 t - \cos^2 t$
$f'(t) = 2 \sin t \cos t - 2 \cos t(-\sin t)$
$= 4 \sin t \cos t$

19. $f(x) = (x^2+4)^2(2x^3-1)^3$
$f'(x) = (x^2+4)^2[3(2x^3-1)^2 \cdot 6x^2] + (2x^3-1)^3[2(x^2+4)(2x)]$
$= 2x(x^2+4)(2x^3-1)^2[9x(x^2+4) + 2(2x^3-1)]$
$= 2x(x^2+4)(2x^3-1)^2(13x^3+36x-2)$

23. $y = f(u) = u^5$, $u = g(x) = x^3 + 1$

$\dfrac{dy}{dx} = \dfrac{dy}{du} \cdot \dfrac{du}{dx} = 5u^4(3x^2) = 15x^2(x^3+1)^4$,

since $u = x^3 + 1$

27. $y = (u+1)^2$, $u = \dfrac{1}{x}$

$\dfrac{dy}{dx} = \dfrac{dy}{du} \cdot \dfrac{du}{dx} = 2(u+1)(\dfrac{-1}{x^2}) = \dfrac{-2}{x^2}(\dfrac{1}{x}+1) = \dfrac{-2(x+1)}{x^3}$

31. $y = \sin 4x$
$y' = 4 \cos 4x$

35. $y = \sin(3x^2+4)$
$y' = 6x \cos(3x^2+4)$

39. $y = 2 \sin(x^2+2x-1)$
$y' = 4(x+1)\cos(x^2+2x-1)$

43. $y = x^2\sin 4x$
$y' = x^2(4 \cos 4x) + 2x \sin 4x$
$ = 4x^2\cos 4x + 2x \sin 4x$

47. $y = x^2 \sin^2 x$
$y' = x^2[2 \sin x \cos x] + 2x \sin^2 x$
$ = 2x \sin x[x \cos x + \sin x]$

51. If $y = u^3$, $u = 3v^2 + 1$, and $v = 4/x^2 = 4x^{-2}$,

then $\dfrac{dy}{dx} = \dfrac{dy}{du} \cdot \dfrac{du}{dv} \cdot \dfrac{dv}{dx} = (3u^2)(6v)(\dfrac{-8}{x^3})$

$= \dfrac{-8}{x^3}(6 \cdot \dfrac{4}{x^2})[3(3v^2+1)^2] = \dfrac{-192}{x^5}[3(\dfrac{48}{x^4}+1)^2]$

$= \dfrac{-576}{x^5}(\dfrac{48}{x^4}+1)^2 = \dfrac{-576(48+x^4)^2}{x^{13}}.$

55. $y = \tan^3 2x$
Let $y = u^3$, $u = \tan v$, $v = 2x$.

$\dfrac{dy}{dx} = \dfrac{dy}{du} \cdot \dfrac{du}{dv} \cdot \dfrac{dv}{dx} = 3u^2 \cdot \sec^2 v \cdot 2$

$\phantom{\dfrac{dy}{dx}} = 3 \tan^2 v \cdot \sec^2 v \cdot 2$

$\phantom{\dfrac{dy}{dx}} = 6 \tan^2(2x)\sec^2(2x)$

59. $f(x) = (2x+3)^n$

$f'(x) = 2n(2x+3)^{n-1}$

$$f''(x) = 2^2 n(n-1)(2x+3)^{n-2}$$

$$f^{(n)}(x) = 2^n n!$$

63. $\dfrac{d}{dx} f(x^2+1) = 2xf'(x^2+1)$

67. $\dfrac{d}{dx} f(\sin x) = f'(\sin x)\cos x$

71. If $h = f \circ g$, $f'(2) = 6$, $f(1) = 4$, $g(1) = 2$, and
    $g'(1) = -2$, then
    $h(x) = f[g(x)] \Rightarrow h'(x) = f'[g(x)] \cdot g'(x)$. Thus,
    $h'(1) = f'[g(1)] \cdot g'(1) = f'(2)(-2)$
    $\qquad = -2(6) = -12$.

75. $s(t) = 8 - (2-t)^3$; $0 \le t \le 2$
    $v(t) = -3(2-t)^2 (-1) = 3(2-t)^2$
    $v(1) = 3(2-1)^2 = 3 \ \dfrac{m}{sec}$
    $a(t) = 3(2)(2-t)^1 (-1) = -6(2-t)$
    $\qquad = -12 + 6t \ \dfrac{m}{sec^2}$

79. Since $\cos x = \sin(\frac{\pi}{2}-x)$, then using the chain rule,
    we have

    $$\dfrac{d}{dx} \cos x = \dfrac{d}{dx}[\sin(\tfrac{\pi}{2}-x)] = \cos(\tfrac{\pi}{2}-x) \cdot (-1)$$

    Again using the identity $\cos x = \sin(\frac{\pi}{2}-x)$, we have

    $$\cos(\tfrac{\pi}{2}-x) = \sin[\tfrac{\pi}{2}-(\tfrac{\pi}{2}-x)] = \sin(x).$$

    Therefore, $\dfrac{d}{dx} \cos x = \sin x(-1) = -\sin x$.

Exercise 9, pp. 160-161

3. $x^2 y = 5$

    $x^2 \dfrac{dy}{dx} + 2xy = 0$

    $\dfrac{dy}{dx} = \dfrac{-2y}{x}$

7. $x^2 - 4xy + y^2 = y$

    $2x - 4xy' - 4y + 2yy' = y'$

    $(2y-4x-1)y' = 4y - 2x$

    $y' = \dfrac{4y - 2x}{2y - 4x - 1}$

11. $4x^3 + 2y^3 = x$

$12x^2 + 6y^2y' = 1$

$y' = \dfrac{1 - 12x^2}{6y^2}$

15. $x^{-1} + y^{-1} = 1$

$-x^{-2} - y^{-2}y' = 0$

$y' = -\dfrac{x^{-2}}{y^{-2}} = -\dfrac{y^2}{x^2}$

19. $\dfrac{x}{y} + \dfrac{y}{x} = 4$

$\dfrac{y - xy'}{y^2} + \dfrac{xy' - y}{x^2} = 0$

$x^2y - x^3y' + xy^2y' - y^3 = 0$

$y'(xy^2 - x^3) = y^3 - x^2y$

$y' = \dfrac{y(y^2 - x^2)}{x(y^2 - x^2)} = \dfrac{y}{x}$

23. $y = x \sin y$

$y' = x(y' \cos y) + \sin y$

$y'[1 - x \cos y] = \sin y$

$y' = \dfrac{\sin y}{1 - x \cos y}$

27. $y = \tan(x-y)$

$y' = \sec^2(x-y)(1-y')$

$y'[1 + \sec^2(x-y)] = \sec^2(x-y)$

$y' = \dfrac{\sec^2(x-y)}{1 + \sec^2(x-y)}$

31. $x^2 + y^2 = 4$; $2x + 2yy' = 0$; $y' = -\dfrac{x}{y}$

$y'' = \dfrac{y(-1) - (-x)y'}{y^2} = \dfrac{-y + x\left(\dfrac{-x}{y}\right)}{y^2} = \dfrac{-y^2 - x^2}{y^3} = \dfrac{-4}{y^3}$

35. $x^2 + y^2 = 5$ at $(1,2)$

$2x + 2yy' = 0$

$y' = \dfrac{-x}{y} \Rightarrow m_T = y' = -\dfrac{1}{2}$ at $(1,2)$

The equation of the tangent line is

$y - 2 = \dfrac{-1}{2}(x-1) \Rightarrow 2y - 4 = -x + 1$, or $y = -\dfrac{1}{2}x + \dfrac{5}{2}$.

39. $x + xy + 2y^2 = 6$

(a) $1 + xy' + y + 4yy' = 0$

$y'(x+4y) = -1 - y$

$m_T = y' = -\dfrac{1 + y}{x + 4y}$

(b) At $(2,1)$, $m_T = -\dfrac{1 + 1}{2 + 4} = -\dfrac{1}{3}$ so the equation is

$y - 1 = -\dfrac{1}{3}(x-2) \Rightarrow 3y - 3 = -x + 2 \Rightarrow y = -\dfrac{1}{3}x + \dfrac{5}{3}.$

(c) $y' = -\dfrac{1}{3} \Rightarrow \dfrac{-(1+y)}{x + 4y} = \dfrac{-1}{3} \Rightarrow 3 + 3y = x + 4y$

$\Rightarrow y = 3 - x$

The line $y = 3 - x$ intersects $x + xy + 2y^2 = 6$

when $x + x(3-x) + 2(3-x)^2 = 6$, or

$x^2 - 8x + 12 = (x-2)(x-6) = 0$. Thus, $x = 2$ or

$x = 6$. When $x = 6$, $y = -3$; therefore, $P(6,-3)$.

Exercise 10, pp. 166-167

3. $y = x^{2/3}$

$y' = \dfrac{2}{3}x^{-1/3} = \dfrac{2}{3x^{1/3}}$

7. $y = x^{1/3} - x^{-1/3}$

$y' = \dfrac{1}{3}x^{-2/3} + \dfrac{1}{3}x^{-4/3}$

$= \dfrac{1}{3x^{2/3}} + \dfrac{1}{3x^{4/3}}$

11. $y = (x^3-1)^{1/2}$

$y' = \dfrac{1}{2}(x^3-1)^{-1/2}(3x^2)$

$= \dfrac{3x^2}{2(x^3-1)^{1/2}}$

15. $y = \sec\sqrt{x} = \sec(x)^{1/2}$

$\dfrac{dy}{dx} = \sec\sqrt{x}\ \tan\sqrt{x}\ \cdot\ \dfrac{1}{2}x^{-1/2} = \dfrac{\sec\sqrt{x}\ \tan\sqrt{x}}{2\sqrt{x}}$

19. $y = x(x^3+1)^{1/2}$

$y' = x[\dfrac{1}{2}(x^3+1)^{-1/2}(3x^2)] + (x^3+1)^{1/2}$

$\quad = \dfrac{3x^3}{2(x^3+1)^{1/2}} + (x^3+1)^{1/2} = \dfrac{5x^3 + 2}{2(x^3+1)^{1/2}}$

23. $y = (x^2+1)^{1/2}$

$y' = \dfrac{1}{2}(x^2+1)^{-1/2}(2x) = \dfrac{x}{(x^2+1)^{1/2}}$

27. $y = \sqrt{x^3(8x+1)} = (8x^4+x^3)^{1/2}$

$y' = \dfrac{1}{2}(8x^4+x^3)^{-1/2}(32x^3+3x^2) = \dfrac{x^2(32x+3)}{2(8x^4+x^3)^{1/2}}$

$\quad = \dfrac{x^2(32x+3)}{2\sqrt{x^2}\sqrt{x(8x+1)}} = \dfrac{|x|(32x+3)}{2\sqrt{x(8x+1)}}$

31. $y = (x^2\cos x)^{3/2}$

$y' = (\dfrac{3}{2})(x^2\cos x)^{1/2}[-x^2\sin x + 2x\cos x]$

$\quad = (\dfrac{3}{2})(2x\cos x - x^2\sin x(x^2\cos x)^{1/2}$

$\quad = \dfrac{3x}{2}(2\cos x - x\sin x)\sqrt{x^2\cos x}$

35. $y = (x^2-3)^{1/2}(6x+1)^{1/3}$

$\dfrac{dy}{dx} = \dfrac{1}{2}(x^2-3)^{-1/2}(2x)(6x+1)^{1/3}$

$\quad + (x^2-3)^{1/2}\quad(1/3)(6x+1)^{-2/3}\cdot 6$

$\quad = x(x^2-3)^{-1/2}(6x+1)^{1/3} + 2(x^2-3)^{1/2}(6x+1)^{-2/3}$

$\quad = (x^2-3)^{-1/2}(6x+1)^{-2/3}[x(6x+1) + 2(x^2-3)]$

$$= \frac{8x^2 + x - 6}{\sqrt{x^2-3} \; \sqrt[3]{(6x+1)^2}}$$

39. $y = (2x^3+x)^{1/3} (x^2+1)^{1/4}$

$$\frac{dy}{dx} = \frac{1}{3}(2x^3+x)^{-2/3}(6x^2+1)(x^2+1)^{1/4}$$

$$+ (2x^3+x)^{1/3}(1/4)(x^2+1)^{-3/4}(2x)$$

$$= \frac{1}{3}(6x^2+1)(2x^3+x)^{-2/3}(x^2+1)^{1/4}$$

$$+ \frac{1}{2}x(2x^3+x)^{1/3}(x^2+1)^{-3/4}$$

$$\frac{dy}{dx} = \frac{1}{6}(2x^3+x)^{-2/3}(x^2+1)^{-3/4}[2(6x^2+1)(x^2+1) + 3x(2x^3+x)]$$

$$= \frac{18x^4 + 17x^2 + 2}{6\sqrt[3]{(2x^3+x)^2} \; \sqrt[4]{(x^2+1)^3}}$$

43. $x^{1/3} + y^{1/3} = 1$ at $(8, -1)$

$$\frac{1}{3}x^{-2/3} + \frac{1}{3}y^{-2/3}y' = 0$$

$$y' = -\frac{x^{-2/3}}{y^{-2/3}} = -(\frac{y}{x})^{2/3}$$

$m_T = -(\frac{-1}{8})^{2/3} = -\frac{1}{4}$. The equation of the tangent

line is $y + 1 = -\frac{1}{4}(x-8) = -\frac{1}{4}x + 2$, or $y = -\frac{1}{4}x + 1$.

47. If $f(0) = 4$, $f'(0) = -2$, and $g = f^{-1}$, find $g'(4)$. We

have $g'(4) = g'[f(0)] = \frac{1}{f'(0)} = \frac{1}{-2} = -\frac{1}{2}$.

51. If $f(x) = x^3 + 2x$, then $f'(x) = 3x^2 + 2$. Hence,
$g'(3) = g'[f(1)] = 1/f'(1)$ and $g'(0) = g'[f(0)]$
$= 1/f'(0)$.

$$f(x) = x^3 + 2x = 3 \Rightarrow x = 1 \qquad f(x) = x^3 + 2x = 0 \Rightarrow x = 0$$
$$f'(1) = 3 + 2 = 5 \qquad\qquad f'(0) = 2$$
$$g'(3) = 1/5 \qquad\qquad\qquad g'(0) = 1/2$$

Exercise 11. p. 170

3. $f(x) = 2x^3 + 3x^2 + 4x - 1$, $f(0) = -1$,
   and $f(1) = 8$, so let $c_1 = 0$. $f'(x) = 6x^2 + 6x + 4$

   $$c_2 = 0 - \frac{f(0)}{f'(0)} = 0 - \frac{-1}{4} = 0.25$$

   $$c_3 = 0.25 - \frac{f(0.25)}{f'(0.25)} = 0.25 - \frac{0.21875}{5.875} = 0.2128$$

7. $f(x) = x^4 - 2x^3 + 21x - 23$, $f(1) = -3$, and
   $f(2) = 19$, so let $c_1 = 1$. $f'(x) = 4x^3 - 6x^2 + 21$

   $$c_2 = 1 - \frac{f(1)}{f'(1)} = 1 - \frac{-3}{19} \approx 1.15789$$

   $$c_3 = c_2 - \frac{f(c_2)}{f'(c_2)} \approx 1.15789 - \frac{0.008494}{19.16533} \approx 1.15745$$

11. $12 = \frac{1}{3} \pi h^2 [3(4) - h] \Rightarrow \pi h^3 - 12\pi h^2 + 36 = 0$

    $f(h) = \pi h^3 - 12\pi h^2 + 36$, $f(1) > 0$, and $f(2) < 0$

    We will try $c_1 = 1$.

    $$f'(h) = 3\pi h^2 - 24\pi h, \quad c_2 = 1 - \frac{f(1)}{f'(1)} \approx 1 - \frac{1.4425}{-65.9734}$$

    $$\approx 1.022 \text{ ft}$$

3. $f(x) = \begin{cases} 2x - 4 & \text{if } x \leqslant 1/2 \\ 4x^2 - 4 & \text{if } x > 1/2 \end{cases}$, $c = 1/2$

$\lim\limits_{x \to \frac{1}{2}^-} \left[\dfrac{f(x) - f(\frac{1}{2})}{x - \frac{1}{2}} \right]$

$= \lim\limits_{x \to \frac{1}{2}^-} \left[\dfrac{2x - 4 + 3}{x - \frac{1}{2}} \right]$

$= \lim\limits_{x \to \frac{1}{2}^-} \left[\dfrac{2(2x-1)}{2x - 1} \right] = 2$

$\lim\limits_{x \to \frac{1}{2}^+} \left[\dfrac{f(x) - f(\frac{1}{2})}{x - \frac{1}{2}} \right] = \lim\limits_{x \to \frac{1}{2}^+} \left[\dfrac{4x^2 - 1}{x - \frac{1}{2}} \right]$

$= 2 \lim\limits_{x \to \frac{1}{2}^+} \left[\dfrac{(2x-1)(2x+1)}{(2x-1)} \right]$

$= 2(2) = 4$

Thus, $f'(1/2)$ does not exist.

7. $f(x) = \begin{cases} 2x^2 + 1 & \text{if } x < -1 \\ 2 + 2x & \text{if } x \geqslant -1 \end{cases}$, $c = -1$

$\lim\limits_{x \to -1^-} f(x) = 3$ and

$\lim\limits_{x \to -1^+} f(x) = 0$

Since f is not continuous at $x = -1$, $f'(-1)$ cannot exist.

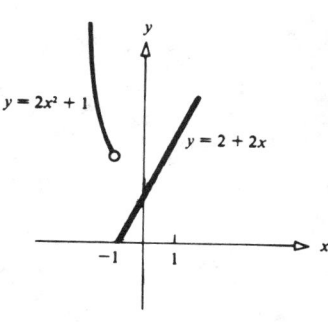

11.   If $s(t) = \begin{cases} t^3 & \text{if } 0 \leqslant t < 5 \\ 125 & \text{if } t \geqslant 5 \end{cases}$,   then we can write

$$v(t) = \begin{cases} 3t^2 & \text{if } 0 < t < 5 \\ 0 & \text{if } t > 5. \end{cases}$$

Velocity just before impact $= \displaystyle\lim_{t \to 5^-} \frac{s(t) - s(5)}{t - 5}$

$$= \lim_{t \to 5^-} \frac{t^3 - 125}{t - 5}$$

$$= \lim_{t \to 5^-} (t^2 + 5t + 25)$$

$$= 25 + 25 + 25$$

$$= 75 \text{ ft/sec}$$

Velocity after impact $= \displaystyle\lim_{t \to 5^+} \frac{s(t) - s(5)}{t - 5}$

$$= \lim_{t \to 5^+} \frac{125 - 125}{t - 5}$$

$$= \lim_{t \to 5^+} (0) = 0$$

$v(5)$ is not defined.

Miscellaneous Exercises, pp. 177-181

3. $y = x\sqrt{1 - x} = x(1-x)^{1/2}$

$y' = x[\frac{1}{2}(1-x)^{-1/2}(-1)] + \sqrt{1 - x} = \dfrac{-x}{2\sqrt{1 - x}} + \sqrt{1 - x}$

$= \dfrac{2 - 3x}{2\sqrt{1 - x}}$

7. $u = \dfrac{r}{a^2\sqrt{a^2 - r^2}}$

$u = \dfrac{a^2\sqrt{a^2 - r^2} - a^2 r \,\frac{1}{2}(a^2-r^2)^{-1/2}(-2r)}{a^4(a^2-r^2)}$

$$= \frac{a^2\sqrt{a^2 - r^2} + \dfrac{a^2 r^2}{\sqrt{a^2 - r^2}}}{a^4(a^2-r^2)} = \frac{a^2 - r^2 + r^2}{a^2(a^2-r^2)^{3/2}} = \frac{1}{(a^2-r^2)^{3/2}}$$

11. $y = 3x^{-2} + 2x^{-1} + 1$

$\quad y' = -6x^{-3} - 2x^{-2}$

$$\quad\quad = \frac{-6}{x^3} - \frac{2}{x^2} = \frac{-6 - 2x}{x^3}$$

15. $f = \dfrac{x^2}{\sqrt{x^2 - 1}}$

$$f' = \frac{\sqrt{x^2 - 1}\,(2x) - x^2\,\dfrac{x}{\sqrt{x^2 - 1}}}{(x^2-1)} = \frac{2x(x^2-1) - x^3}{(x^2-1)^{3/2}}$$

$$\quad\quad = \frac{x^3 - 2x}{(x^2-1)^{3/2}} = \frac{x(x^2-2)}{(x^2-1)^{3/2}}$$

19. $y = x^2(a^2+x^2)^{3/2}$

$$y' = x^2\left[\frac{3}{2}(a^2+x^2)^{1/2}2x\right] + 2x(a^2+x^2)^{3/2}$$

$$\quad = x(a^2+x^2)^{1/2}[3x^2 + 2(a^2+x^2)]$$

$$\quad = x(5x^2+2a^2)(a^2+x^2)^{1/2}$$

23. $u = \dfrac{y}{\sqrt{y^2 + 9}}$

$$u' = \frac{\sqrt{y^2 + 9} - \dfrac{y^2}{\sqrt{y^2 + 9}}}{(y^2+9)} = \frac{9}{(y^2+9)^{3/2}}$$

27. $w = \dfrac{1}{1 - z + z^2}$

$$w = \frac{-(-1+2z)}{(1-z+z^2)^2}$$

$$\quad = \frac{1 - 2z}{(1-z+z^2)^2}$$

31. $v = (1+u)^{3/2}$

$$v' = \frac{3}{2}(1+u)^{1/2}$$

35. $u = (a^{1/2} - x^{1/2})^2$

$u' = 2(a^{1/2} - x^{1/2})(-\frac{1}{2}x^{-1/2})$

$= \dfrac{x^{1/2} - a^{1/2}}{x^{1/2}} = 1 - \sqrt{\dfrac{a}{x}}$

39. $v = \tan u + \sec u$

$v' = \sec^2 u + \sec u \tan u$

$= \sec u (\sec u + \tan u)$

43. $f(x) = \dfrac{x - 1}{x + 1}$

$f'(x) = \dfrac{(x+1) - (x-1)}{(x+1)^2} = \dfrac{2}{(x+1)^2}$

$f'(1) = \dfrac{2}{2^2} = \dfrac{1}{2}$

47. $y = x + \sin xy$

$y' = 1 + \cos xy \, (xy' + y)$

$y'(1 - x \cos xy) = 1 + y \cos xy$

$y' = \dfrac{1 + y \cos xy}{1 - x \cos xy}$

51. If the line $y = -\dfrac{1}{3}x + b$ is normal to $y = x^3$, find
b ($b > 0$). $y' = f'(x) = 3x^2$; m_T at $(x, y) = 3x^2$;
$m_N = -1/3x^2 = -1/3 \Rightarrow x^2 = 1 \Rightarrow x = \pm 1$. Thus, the
possible points are $(1,1)$ and $(-1,-1)$. The equation
of the normal line at $(1,1)$ is $y - 1 = -\dfrac{1}{3}(x-1)$
$\Rightarrow y = -\dfrac{1}{3}x + \dfrac{4}{3}$. The equation of the normal line
at $(-1,-1)$ is $y + 1 = -\dfrac{1}{3}(x+1) \Rightarrow y = -\dfrac{1}{3}x - \dfrac{4}{3}$.
Since $b > 0$, then $b = 4/3$.

55. $\dfrac{d}{dx}(f_1 f_2 f_3 f_4) = (f_1 f_2) \cdot \dfrac{d}{dx}(f_3 f_4) + \dfrac{d}{dx}(f_1 f_2) \cdot (f_3 f_4)$

$$= f_1 f_2 (f_3 f_4' + f_3' f_4) + (f_1 f_2' + f_1' f_2) f_3 f_4$$

$$= f_1 f_2 f_3 f_4' + f_1 f_2 f_3' f_4 + f_1 f_2' f_3 f_4$$

$$+ f_1' f_2 f_3 f_4$$

59. $s = \dfrac{1}{8} \cos 4\pi t$

$v(t) = \dfrac{1}{8}(-\sin 4\pi t)(4\pi) = \dfrac{-\pi}{2} \sin 4\pi t$

$a(t) = \dfrac{-\pi}{2} (\cos 4\pi t)(4\pi) = -2\pi^2 \cos 4\pi t$

63. $h = \dfrac{fg}{f + g}$

$h' = \dfrac{(f+g)(fg'+gf') - fg(f'+g')}{(f+g)^2}$

$= \dfrac{f^2 g' + fgf' + fgg' + g^2 f' - fgf' - fgg'}{(f+g)^2}$

$= \dfrac{f^2 g' + g^2 f'}{(f+g)^2}$

67. n is even $\Rightarrow n = 2k$, $k \geqslant 1$; $y_1 = \sqrt[n]{x} = x^{\frac{1}{n}} = x^{\frac{1}{2k}}$;

$y_2 = x^n = x^{2k}$

$y_1' = \dfrac{1}{2k} x^{\frac{1}{2k} - 1} = \dfrac{1}{2k} x^{\frac{1-2k}{2k}}$ at $(1,1) \Rightarrow m_1 = \dfrac{1}{2k}$

$y_2' = 2k\, x^{2k-1}$ at $(-1,1) \Rightarrow m_2 = 2k(-1)^{2k-1} = -2k$

Since $m_1 = -1/m_2$, the lines are perpendicular.

71. Let $y = ax^3 + bx^2 + cx + d$. If the tangent line at $(1,0)$ is $y = 3x - 3$ and at $(2,9)$ is $y = 18x - 27$, then $y'(1) = 3$ and $y'(2) = 18$. Also, $y' = 3ax^2 + 2bx + c$ and

(a) $(1,0)$ on curve $\Rightarrow 0 = a + b + c + d$

(b) $(2,9)$ on curve $\Rightarrow 9 = 8a + 4b + 2c + d$
Subtracting to eliminate d gives
$9 = 7a + 3b + c$ (1).

(c) $y'(1) = 3 \Rightarrow 3 = 3a + 2b + c$ (2)

(d) $y'(2) = 18 \Rightarrow 18 = 12a + 4b + c$ (3)
Subtracting (3) − (2) gives $15 = 9a + 2b$ and
subtracting (3) − (1) gives $9 = 5a + b$.
Solving these two equations, we find $a = 3$
and $b = -6$. Substituting $a = 3$ and $b = -6$
in (1) gives $c = 9 - 7a - 3b$
$\Rightarrow c = 9 - 21 + 18 = 6$. Using the equation in
(a), we find $d = -a - b - c$
$\Rightarrow d = -3 + 6 - 6 = -3$. Thus,
$y = 3x^3 - 6x^2 + 6x - 3$.

75. If $f(1) = 5$, $f(3) = 21$, $f(a+b) - f(a) = kab + 2b^2$:

(a) Let $a = 1$, $b = 2$ to find k.
$f(3) - f(1) = k(1)(2) + 2(2)^2$
$21 - 5 = 2k + 8 \Rightarrow 2k = 8 \Rightarrow k = 4$.
Thus, $f(a+b) - f(a) = 4ab + 2b^2$.

(b) $f'(3) = \lim_{h \to 0} \dfrac{f(3+h) - f(3)}{h} = \lim_{h \to 0} \dfrac{4(3)h + 2h^2}{h}$

$= \lim_{h \to 0} (12 + 2h) = 12$

(c) $f'(x) = \lim_{h \to 0} \dfrac{f(x+h) - f(x)}{h} = \lim_{h \to 0} \dfrac{4xh + 2h^2}{h}$

$= \lim_{h \to 0} (4x + 2h) = 4x$

Thus, $f'(x) = 4x \Rightarrow f(x) = 2x^2 + c$ where c
is a constant. We have $f(1) = 5 \Rightarrow 5 = 2 + c$
$\Rightarrow c = 3 \Rightarrow f(x) = 2x^2 + 3$.

79. If $s = \dfrac{t}{t^2 - 1}$, then

$$v(t) = s'(t) = \frac{(t^2-1) - t(2t)}{(t^2-1)^2} = \frac{-t^2 - 1}{(t^2-1)^2}$$

and

$$a(t) = v'(t) = \frac{(t^2-1)^2(-2t) + (t^2+1)2(t^2-1)(2t)}{(t^2-1)^4}$$

$$= \frac{-2t^3 + 2t + 4t^3 + 4t}{(t^2-1)^3} = \frac{2t^3 + 6t}{(t^2-1)^3}$$

83. $\displaystyle\lim_{\Delta x \to 0} \dfrac{[4-2(x+\Delta x)]^2 - (4-2x)^2}{\Delta x} = f'(x)$ where
$f(x) = (4-2x)^2$. If $f(x) = (4-2x)^2$, then
$f'(x) = 2(4-2x)(-2) = 8x - 16$.

87. Given the line $y = x - 1$ and the function
$f(x) = y = \sqrt{25 - x^2}$, we have

$$f'(x) = \frac{1}{2}(25-x^2)^{-1/2}(-2x) = \frac{-x}{\sqrt{25 - x^2}}.$$

If $y = x - 1$ is tangent to the graph, then $m = 1$.

Thus, $f'(x) = 1 \Rightarrow \dfrac{-x}{\sqrt{25 - x^2}} = 1 \Rightarrow -x = \sqrt{25 - x^2}$

$\Rightarrow x^2 = 25 - x^2 \Rightarrow x^2 = \dfrac{25}{2} \Rightarrow x = \pm \dfrac{5}{\sqrt{2}}$.

$f(\dfrac{5}{\sqrt{2}}) = \sqrt{25 - \dfrac{25}{2}} = \sqrt{\dfrac{25}{2}} = \dfrac{5}{\sqrt{2}} \Rightarrow$ the point $(\dfrac{5}{\sqrt{2}}, \dfrac{5}{\sqrt{2}})$

and $m = 1$. The equation of the tangent line is
$y - 5/\sqrt{2} = x - 5/\sqrt{2}$, or $y = x$.

$f(-\dfrac{5}{\sqrt{2}}) = \sqrt{25 - \dfrac{25}{2}} = \sqrt{\dfrac{25}{2}} = \dfrac{5}{\sqrt{2}} \Rightarrow$ the point $(\dfrac{-5}{\sqrt{2}}, \dfrac{5}{\sqrt{2}})$

The tangent line is $y - \frac{5}{\sqrt{2}} = (x + \frac{5}{\sqrt{2}}) \Rightarrow y = x + \frac{10}{\sqrt{2}}$;

therefore, at no point is the line $y = x - 1$ tangent

to $y = \sqrt{25 - x^2}$.

Chapter Four

Exercise 1, pp. 189-192

3. $x^3y^2 = 432$

$$2x^3y \frac{dy}{dt} + 3x^2y^2 \frac{dx}{dt} = 0$$

$$2(27)(4)(2) + 3(9)(16) \frac{dx}{dt} = 0 \implies \frac{dx}{dt} = -1$$

7. If $V = (\pi/12)h^3$, then $\frac{dV}{dt} = \frac{\pi}{4} h^2 \frac{dh}{dt}$.

When $h = 8$, $\frac{dh}{dt} = \frac{5\pi}{16}$ and $\frac{dV}{dt} = \frac{\pi}{4}(64)\frac{5\pi}{16} = 5\pi^2$.

11. If $x^2 + y^2 = 45^2 = 2025$ and $dx/dt = 2$, then

$$2x \frac{dx}{dt} + 2y \frac{dy}{dt} = 0 \quad \text{and} \quad \frac{dy}{dt} = -\frac{x}{y}\frac{dx}{dt}$$

When $x = 4$, then $y = \sqrt{2009}$ and $\frac{dy}{dt} = \frac{-4(2)}{\sqrt{2009}}$

$$= \frac{-8}{\sqrt{2009}} \text{ cm/min.}$$

15. Given: $L = 30$ $W = 15$
         $D = 3$ $d = 1$
         $dV/dt = 15m^3/min$

Using similar triangles, we see from the figure that for $y <= 2$, $x/y = 30/2$, and thus $x = 15y$. If $0 \leqslant D \leqslant 2$, $V = (1/2)xy(15) = (1/2)(15y)(15y) = (225/2)y^2$.

$$\frac{dV}{dt} = 225y \frac{dy}{dt}$$

When $y = 2$, $\frac{dy}{dt} = \frac{1}{225(2)} \frac{dV}{dt} = \frac{15}{(15)(15)(2)} = \frac{1}{30}$ m/min.

19. Given that $V = (\pi/3)R^2h$ and $dV/dt = 16m^3/min$, we want to find dh/dt when $h = 8$. We have $4/16 = R/h \implies R = h/4$.

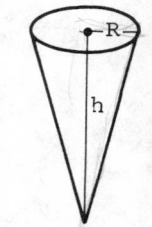

$$V = \frac{\pi}{3}\left(\frac{h}{4}\right)^2 h = \frac{\pi}{48} h^3$$

$$\frac{dV}{dt} = \frac{\pi}{16} h^2 \frac{dh}{dt}$$

$$16 = \frac{\pi}{16}(64)\frac{dh}{dt} \Rightarrow \frac{dh}{dt} = \frac{4}{\pi} \text{ m/min} \approx 1.273 \text{ m/min}$$

23. Letting x = distance from wall to base of ladder and y = height of the wall from the ground to the ladder top, we have $dx/dt = 1/2$ m/sec and $x^2 + y^2 = 64$

$$\Rightarrow 2x \frac{dx}{dt} + 2y \frac{dy}{dt} = 0 \Rightarrow \frac{dy}{dt} = -\frac{x}{y}\frac{dx}{dt}$$

(a) When $x = 3 \Rightarrow y = \sqrt{64 - 9} = \sqrt{55} \Rightarrow \frac{dy}{dt}$

$$= \frac{-3}{\sqrt{55}}\left(\frac{1}{2}\right) = \frac{-3}{2\sqrt{55}} \text{ m/sec.}$$

(b) When $x = 4 \Rightarrow y = \sqrt{64 - 16} = \sqrt{48} \Rightarrow \frac{dy}{dt}$

$$= \frac{-4}{\sqrt{48}}\left(\frac{1}{2}\right) = \frac{-4}{8\sqrt{3}} = \frac{-1}{2\sqrt{3}} \text{ m/sec.}$$

(c) When $x = 6 \Rightarrow y = \sqrt{64 - 36} = \sqrt{28} \Rightarrow \frac{dy}{dt}$

$$= \frac{-6}{\sqrt{28}}\left(\frac{1}{2}\right) = \frac{-3}{2\sqrt{7}} \text{ m/sec.}$$

27. Given that $d\theta/dt = 2°/\text{min} = \pi/90$ radian/min, we want to find dA/dt when $\theta = 30° = \pi/6$ radians.

$\sin\frac{\theta}{2} = \frac{(x/2)}{4}$

$16 = h^2 + \frac{x^2}{4}$

$= \frac{x}{8}$

$h = \sqrt{16 - x^2/4}$

$x = 8\sin\frac{\theta}{2}$

$h = \frac{\sqrt{64 - x^2}}{2}$

Thus, $A = \left(\frac{1}{2}\right)(x)(h) = \frac{x}{2}\frac{\sqrt{64 - x^2}}{2} = \frac{x}{4}\sqrt{64 - x^2}.$

Using the figure, we have

$$A = \frac{x}{4}\sqrt{64 - x^2} = 2 \sin \frac{\theta}{2} \sqrt{64 - 64 \sin^2 \frac{\theta}{2}}$$

$$= 16 \sin \frac{\theta}{2}\sqrt{1 - \sin^2 \frac{\theta}{2}} = 16 \sin \frac{\theta}{2} \cos \frac{\theta}{2}$$

$$= 8 \sin \theta, \quad \text{since} \quad 2 \sin \alpha \cos \alpha = \sin 2\alpha. \quad \text{Thus,}$$

$$\frac{dA}{dt} = 8 \cos \theta \frac{d\theta}{dt}, \quad \text{and if} \quad \theta = \frac{\pi}{6}, \frac{d\theta}{dt} = \frac{\pi}{90}, \quad \text{then}$$

$$\frac{dA}{dt} = \frac{8\pi}{90} \cos \frac{\pi}{6} = \frac{2\sqrt{3}\pi}{45} \; cm^2/min.$$

31. Find $\frac{dy}{dt}$ when $x = 5$

given that $\frac{dx}{dt} = 2$ m/sec.
Since the rope is always
20 m long, $10 - y + r = 20$
or $r = 10 + y$. Applying
the Pythagorean Theorem to
the figure yields
$10^2 + x^2 = r^2$. Substituting
$r = 10 + y$ we have $100 + x^2 = (10+y)^2$. Thus,

$$2x \frac{dx}{dt} = 2(10+y) \frac{dy}{dt} \quad \text{or} \quad \frac{dy}{dt} = \frac{x}{10 + y} \frac{dx}{dt}.$$

Since $x = 5$, we have $100 + (5)^2 = (10+y)^2$ or

$10 + y = \sqrt{125}$. Thus with $\frac{dx}{dt} = 2$ m/sec,

$$\frac{dy}{dt} = \frac{5}{\sqrt{125}}(2) = \frac{10}{5\sqrt{5}} = \frac{2}{\sqrt{5}} = \frac{2\sqrt{5}}{5} \; m/sec.$$

35. The plane in the figure is
flying away from the observer
so θ should be decreasing.
From the figure, $\tan \theta = \frac{2000}{x}$

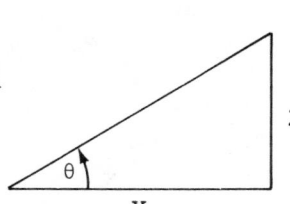

2000 m

$$\Rightarrow \sec^2\theta \frac{d\theta}{dt} = \frac{-2000}{x^2} \frac{dx}{dt}.$$

$$\Rightarrow \frac{d\theta}{dt} = \frac{-2000}{x^2\sec^2\theta} \frac{dx}{dt}$$

We are given that $\frac{dx}{dt} = 300$ m/sec and $\theta = 45° = \pi/4$ rad

$\Rightarrow \sec^2 \theta = 2$ and $\tan \theta = 1$

$\Rightarrow \dfrac{2000}{x} = \tan \theta = 1 \Rightarrow x = 2000.$

Hence $\dfrac{d\theta}{dt} = \dfrac{-2000}{(2000)^2(2)}(300) = \dfrac{-3}{40}$ rad/sec or $3/40$

rad/sec if the plane is approaching the observer.

Exercise 2, pp. 198-200

3. $y = \dfrac{x - 1}{x^2 + 2x - 8}$

$dy = \dfrac{-x^2 + 2x - 6}{(x^2+2x-8)^2}\, dx$

7. $xy = 7$

$x\, dy + y\, dx = 0$

$\dfrac{dy}{dx} = -\dfrac{y}{x} \qquad \dfrac{dx}{dy} = -\dfrac{x}{y}$

11. $x^3 + y^3 = 3x^2 y$

$3x^2\, dx + 3y^2\, dy = 3x^2\, dy + 6xy\, dx$

$dy(3y^2 - 3x^2) = dx(6xy - 3x^2)$

$\dfrac{dy}{dx} = \dfrac{2xy - x^2}{y^2 - x^2} \qquad \dfrac{dx}{dy} = \dfrac{y^2 - x^2}{2xy - x^2}$

15. $d(\sqrt{x}+2) = \dfrac{1}{2\sqrt{x}}\, dx$

19. $f(x) = x^2 - 2x + 1 \Rightarrow f'(x) = 2x - 2$

$\quad x_o = 2,\; f(x_o) = 1,\; f'(x_o) = 2$

$\quad f(x) \approx f(x_o) + f'(x_o)(x-x_o)$

$\quad \Rightarrow f(x) \approx 1 + 2(x-2)$, or, $f(x) \approx 2x - 3$

23. $f(x) = \sin x \Rightarrow f'(x) = \cos x$

$\quad x_o = \dfrac{\pi}{6},\; f(x_o) = \dfrac{1}{2},\; f'(x) = \dfrac{\sqrt{3}}{2}$

$\quad f(x) \approx f(x_o) + f'(x_o)(x-x_o) = \dfrac{1}{2} + \dfrac{\sqrt{3}}{2}\left(x - \dfrac{\pi}{6}\right)$, or

$\quad f(x) \approx \dfrac{\sqrt{3}}{2}x + \dfrac{6 - \pi\sqrt{3}}{12}$

27. (a) $f(x) = x^2 \qquad x = 3;\; dx = 0.001$
$\qquad dy = 2xdx = 6(0.001) = 0.006$

(b) $f(x) = \dfrac{1}{x + 2} \qquad x = 2;\; dx = -0.02$

$$dy = \frac{-1}{(x+2)^2}\, dx = \frac{-1}{4^2}(-0.02) = \frac{0.02}{16} = \frac{0.01}{8}$$

$$= 0.00125 \ (dy > 0 \text{ since } y \text{ is higher at } x = 1.98 \text{ than at } x = 2)$$

31. $V = \frac{4}{3}\pi R^3$ $R = 3$ meters; $dR = 0.1$ meter

    $dV = 4\pi R^2 dR = 4\pi(9)(0.1) = 3.6\pi \text{ m}^3 \approx 11.31 \text{ m}^3$

35. If the diameter is 4 cm, then $R = 2$ cm, but for an actual diameter of 3 cm, the change in radius is $dR = -0.5$ cm. We are given that $h = 4R$; therefore,
    $V = \frac{1}{3}\pi R^2 h = \frac{1}{3}\pi R^2(4R) = \frac{4}{3}\pi R^3$.

    $dV = 4\pi R^2 dR = 4\pi(2^2)(-\frac{1}{2}) = -8\pi$

    Thus, the approximate loss in volume is $8\pi \text{ cm}^3$.

39. If the length ℓ is usually 1 m and the length is increased 10 centimeters, then $\Delta\ell = 0.1$ m, so that

    $\frac{\Delta\ell}{\ell} = \frac{0.1}{1} = 0.1$. $T = 2\pi\sqrt{\ell/g} \implies dT = \pi/(g\sqrt{\ell/g})\, d\ell$

    $\implies \frac{\Delta T}{T} \approx \frac{dT}{T} = \frac{\pi/(g\sqrt{\ell/g})}{2\pi\sqrt{\ell/g}} = \frac{1}{2}\frac{d\ell}{\ell} = \frac{1}{2}(0.1) = 0.05$,

    which is a 5% increase in the period of the pendulum. There are $24(60)(60) = 86,400$ seconds in a day, and the clock will lose $5\% \ (86,400) = 0.05(86,400) = 4320$ sec $= 72$ min each day.

Exercise 3, pp. 208-211

3. $f(x) = 1 - 6x + x^2$ 7. $f(x) = x^4 - 2x^2 + 1$

   $f' = -6 + 2x$ $f' = 4x^3 - 4x = 4x(x^2-1)$

   CN is 3.   CN are $0, -1, 1$.

11. $f(x) = 2x^{1/2}$ dom $f = [0, +\infty)$

    $f' = \frac{1}{x^{1/2}}$

    CN is 0.

15. $f(x) = \dfrac{x^2}{x - 1}$ dom $f = \{x : x \neq 1\}$

$f' = \dfrac{(x-1)(2x) - x^2}{(x-1)^2} = \dfrac{x^2 - 2x}{(x-1)^2} = \dfrac{x(x-2)}{(x-1)^2}$

CN are $0, 2$.

19. $f(x) = \dfrac{(x-3)^{1/3}}{x - 1}$ dom $f = \{x : x \neq 1\}$

$f' = \dfrac{(x-1)\dfrac{1}{3(x-3)^{2/3}} - (x-3)^{1/3}}{(x-1)^2} = \dfrac{(x-1) - 3(x-3)}{3(x-3)^{2/3}(x-1)^2}$

$= \dfrac{8 - 2x}{3(x-3)^{2/3}(x-1)^2}$

CN are $4, 3$.

23. $f(x) = x^2 + 2x$ on $[-3,3]$

$f' = 2x + 2$; CN $= -1$.

$f(-3) = 9 - 6 = 3$

$f(-1) = 1 - 2 = -1$ min

$f(3) = 9 + 6 = 15$ max

27. $f(x) = x^3 - 3x^2$ on $[1,4]$

$f' = 3x^2 - 6x = 3x(x-2)$

CN in $[1,4]$ is 2.

$f(1) = 1 - 3 = -2$

$f(2) = 8 - 12 = -4$ min

$f(4) = 64 - 48 = 16$ max

31. $f(x) = x^{2/3}$ on $[-1,1]$

$f'(x) = \dfrac{2}{3x^{1/3}}$; CN $= 0$.

$f(-1) = 1$ max

$f(0) = 0$ min

$f(1) = 1$ max

35. $f(x) = x(1-x^2)^{1/2}$ on $[-1,1]$

$f' = \dfrac{1 - 2x^2}{\sqrt{1 - x^2}}$

CN $= \pm 1,\ \pm \sqrt{2}/2$

$f(-1) = 0$ $\qquad\qquad$ $f(1) = 0$

$f(-\sqrt{2}/2) = -1/2$ min $\quad$ $f(\sqrt{2}/2) = 1/2$ max

39. $f(x) = (x+3)^2(x-1)^{2/3}$ on $[-4,5]$

$f' = \dfrac{2(x+3)(4x)}{3(x-1)^{1/3}}$

CN $= 1, -3, 0$

$f(-4) = \sqrt[3]{25}$ $\qquad\qquad$ $f(1) = 0$ min

$f(-3) = 0$ min $\qquad\qquad$ $f(5) = 64(4)^{2/3} = 128\sqrt[3]{2}$ max

$f(0) = 9$

43. $f(x) = \dfrac{(x^2-9)^{1/3}}{x}$ on $[3,6]$

$f' = \dfrac{27 - x^2}{3x^2(x^2-9)^{2/3}}$

CN in $[3,6]$ are $3\sqrt{3},\ 3$

$f(3) = 0$ min

$f(6) = \dfrac{\sqrt[3]{27}}{6} = \dfrac{1}{2}$

$f(3\sqrt{3}) = \dfrac{\sqrt[3]{18}}{3\sqrt{3}} \approx 0.504$ max

47. $f(x) = \begin{cases} x^2 & \text{if } -2 \leqslant x < 1 \\ x^3 & \text{if } \ \ 1 \leqslant x \leqslant 2 \end{cases}$

$f' = \begin{cases} 2x & \text{if } -2 < x < 1 \\ 3x^2 & \text{if } \ \ 1 < x < 2 \end{cases}$

CN $= 0$ and 1 (since f' is undefined at $x = 1$)

$f(-2) = 4$ $\qquad$ $f(0) = 0$ min

$f(1) = 1$ $\qquad$ $f(2) = 8$ max

51. If $f(x) = Ax^2 + Bx + C$ has a local minimum at 0,

$f'(x) = 2Ax + B \Rightarrow$ CN at $x = \dfrac{-B}{2A} = 0 \Rightarrow B = 0$

Hence, $f(x) = Ax^2 + C$. If $f(x)$ passes through $(0,2),(1,8)$, then

$f(0) = 2 = C \qquad f(1) = 8 = A + C \Rightarrow A = 8 - C$

$\Rightarrow A = 8 - 2 = 6$

Thus, $f(x) = 6x^2 + 2$

Exercise 4, pp. 215-216

3. $f(x) = x^2 - 2x - 2$ on $[0,2]$

$f' = 2x - 2; \quad f'(c) = 0 \Rightarrow c = 1$

7. $f(x) = x^3 - x + 2$ on $[-1,1]$

$f' = 3x^2 - 1; \quad f'(c) = 0 \Rightarrow c = \pm\, 1/\sqrt{3}$

11. $f(x) = x^2 - 2x + 1$ on $[-2,1]$

$f(-2) = 4 + 4 + 1 = 9 \qquad f(1) = 0$

$f(a) \neq f(b)$

15. $f(x) = x^2$ on $[-1,2]$

$f'(x) = 2x \qquad f(-1) = 1 \qquad f(2) = 4$

$2c = \dfrac{4 - 1}{2 + 1} = 1 \Rightarrow c = \dfrac{1}{2}$

19. $f(x) = x^3 - 5x^2 + 4x - 2$ on $[1,3]$

$f' = 3x^2 - 10x + 4 \qquad f(1) = -2 \qquad f(3) = -8$

$3c^2 - 10c + 4 = \dfrac{-8 + 2}{3 - 1} = \dfrac{-6}{2} = -3$

$3c^2 - 10c + 7 = 0$

$(3c-7)(c-1) = 0$

$c = 7/3 \quad \varepsilon(1,3)$

23. $f(x) = x^{2/3}$ on $[1,8]$

$f' = \dfrac{2}{3x^{1/3}} \qquad f(1) = 1 \qquad f(8) = 4$

$\dfrac{2}{3c^{1/3}} = \dfrac{4 - 1}{8 - 1} = \dfrac{3}{7}$

$$9c^{1/3} = 14$$
$$c^{1/3} = 14/9$$
$$c = (\frac{14}{9})^3 = \frac{2744}{729} \approx 3.7641$$

27.

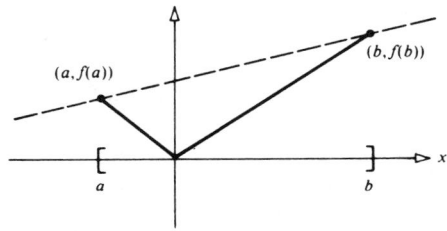

f' does not exist at 0.

31. $f(x) = (x-1)\sin x \Rightarrow f'(x) = (x-1)\cos x + \sin x$
f continuous on [0,1] and $f(0) = f(1) = 0$.
Then by Rolle's Theorem, there exists a $c\varepsilon(0,1)$ such
that $f'(c) = 0$. That is, $(c-1)\cos(c) + \sin(c) = 0$.
$$\Rightarrow \sin c = (1-c)\cos c$$

$$\Rightarrow \frac{\sin c}{\cos c} = 1 - c$$

$$\Rightarrow \tan c + c = 1$$
for some $c\varepsilon(0,1)$.

35. Let $f(x) = x^n + ax^2 + b$, n a positive odd integer.
Suppose that $f(x) = 0$ has more than 3 distinct
real roots. As a minimal case, suppose there exists
$x_1 < x_2 < x_3 < x_4$ such that $f(x_1) = f(x_2) = f(x_3)$
$= f(x_4) = 0$. Then by Rolle's Theorem, there must
exist three numbers, $c_1\varepsilon(x_1,x_2)$, $c_2\varepsilon(x_2,x_3)$,
$c_3\varepsilon(x_3,x_4)$ such that
$f'(c_1) = f'(c_2) = f'(c_3) = 0$.
Now, $f'(x) = nx^{n-1} + 2ax$ and $n - 1$ is EVEN means
$nx^{n-1} + 2ax = x(nx^{n-2} + 2a) = 0$
must have 3 distinct real solutions.
One solution is $x = 0$, the other two must come from
$nx^{n-2} + 2a = 0$, for $n - 2$ ODD. Unfortunately,
this odd power has only one real solution.

This contradiction then shows that f(x) cannot
have more than 3 distinct real roots.

<u>Exercise 5</u>, pp. 224-225

3. $f(x) = \dfrac{x}{x + 1}$ dom $f = \{x : x \neq -1\}$

   $f' = \dfrac{(x+1) - x}{(x+1)^2} = \dfrac{1}{(x+1)^2}$

   $f' > 0$ for $x \in (-\infty, -1) \cup (-1, +\infty)$

   f is increasing $x \in (-\infty, -1) \cup (-1, +\infty)$

7. $f(x) = -2x^2 + 4x - 5$

   $f'(x) = -4x + 4 = -4(x-1) \Rightarrow$ CN is 1

   CN is 1 $\underset{1}{\overset{\;\; + + + +\; |\; - - - -}{\rule{4cm}{0.4pt}}}$ f'

   f increasing on $(-\infty, 1]$, decreasing on $[1, +\infty)$

   $f(1) = -2 + 4 - 5 = -3 \Rightarrow$ Local max at $(1, -3)$

11. $f(x) = x^3 + 6x^2 + 12x + 1$

   $f'(x) = 3x^2 + 12x + 12 = 3(x+2)^2 \Rightarrow$ CN is -2

   $\underset{-2}{\overset{+ + + +\; |\; + + + +}{\rule{4cm}{0.4pt}}}$ f'

   f increasing on $(-\infty, +\infty) \Rightarrow$ No local extrema

15. $f(x) = 3x^4 - 4x^3$

   $f'(x) = 12x^3 - 12x^2 = 12x^2(x-1)$

   CN are 0,1

   $\underset{\;0 \qquad\quad 1}{\overset{- - -\; |\; - - - -\; |\; + + +}{\rule{5cm}{0.4pt}}}$ f'

   f increasing on $[1, +\infty)$, decreasing on $(-\infty, 1]$

   $f(1) = 3 - 4 = -1 \Rightarrow$ Local min at $(1, -1)$

19. $f(x) = x^2(x+1) = x^3 + x^2$

$f'(x) = 3x^2 + 2x = x(3x+2)$

CN are $0, -2/3$

$$\underset{}{\frac{+\,+\,+\ \ \big|\ \ -\,-\,-\ \ \big|\ \ +\,+\,+\,+}{}}\quad f'$$
$$-2/301$$

f increasing on $(-\infty, -2/3] \cup [0, +\infty)$, decreasing on $[-2/3, 0]$

$f(0) = 0;\quad f(\frac{-2}{3}) = (-\frac{2}{3})^3 + (\frac{-2}{3})^2 = \frac{4}{27}$

Local max at $(-2/3\ ,\ 4/27)$

Local min at $(0,0)$

23. $f(x) = x^{2/3} + x^{1/3}$

$f'(x) = \frac{2}{3x^{1/3}} + \frac{1}{3x^{2/3}} = \frac{2x^{1/3} + 1}{3x^{2/3}}$

CN are $0, -1/8$

$$\frac{-\,-\,-\ \ \big|\ \ +\,+\ \ \big|\ \ +\,+\,+}{}\quad f'$$
$$-1/80$$

f increasing on $[-1/8, +\infty)$, decreasing on $(-\infty, -1/8]$

$f(-\frac{1}{8}) = \frac{1}{4} - \frac{1}{2} = \frac{-1}{4}$

Local min at $(-1/8,\ -1/4)$

27. $f(x) = x^{2/3}(x^2-4) = x^{8/3} - 4x^{2/3}$

$f'(x) = \frac{8}{3}(x^{5/3}) - \frac{8}{3x^{1/3}} = \frac{8x^2 - 8}{3x^{1/3}}$

CN are $-1, 0, 1$

$$\frac{-\,-\,-\ \ \big|\ \ +\,+\,+\ \ \big|\ \ -\,-\,-\ \ \big|\ \ +\,+\,+}{}\quad f'$$
$$-101$$

f increasing on $[-1, 0] \cup [1, +\infty)$, decreasing on $(-\infty, -1] \cup [0, 1]$

$f(-1) = (1) - 4(1) = -3;$ $f(0) = 0$

$f(1) = 1 - 4(1) = -3$

Local minima at $(-1,-3)$ and $(1,-3)$

Local max at $(0,0)$

31. $f(x) = 3 \sin x$

$f'(x) = 3 \cos x$

CN are $\dfrac{(2k+1)\pi}{2}$, $k = 0, \pm1, \pm2 \ldots$

f increasing on $[(4k+3)\pi/2], (4k+5)\pi/2],$
decreasing on $[(4k+1)\pi/2], (4k+3)\pi/2],$ $k = 0, \pm 1, \ldots$

Local min at $[(4k+3)\pi/2,-1], k = 0, \pm1, \pm2, \ldots$

Local max at $[(4k+1)\pi/2,1], k = 0, \pm1, \pm2, \ldots$

35. $s = 2t^3 + 6t^2 - 18t + 1$

$v = 6t^2 + 12t - 18$

$ = 6(t+3)(t-1)$

$a = 12t + 12$

Particle moves right if
$t \in (-\infty,-3] \cup [1,+\infty);$
moves left if $t \in [-3,1];$
reverses direction at
$s(-3) = 55$ and $s(1) = -9$
v increasing if $t \in [-1,+\infty);$
v decreasing if $t \in (-\infty,-1].$

39. $s = 2 \sin 3t;$ $0 \leqslant t \leqslant \dfrac{2\pi}{3}$

$\dfrac{ds}{dt} = 6 \cos 3t,$

$\dfrac{d^2s}{dt^2} = -18 \sin 3t$

The particle moves to right if $t \in [0,\pi/6] \cup [\pi/2,$
$2\pi/3];$ moves to left if $t \in [\pi/6,\pi/2];$ reverses

direction at $s(\pi/6) = 2$ and $s(\pi/2) = -2$: decreasing if $t \; \varepsilon \; [0,\pi/3]$; v increasing if $t \; \varepsilon \; [\pi/3,2\pi/3]$ and v decreasing if $t \; \varepsilon \; [0,\pi/3]$

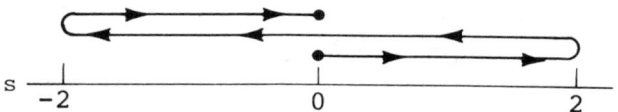

43. $f(x) = ax^3 + bx^2 + cx + d$, and we want local minimum at 0, local maximum at 4, and the graph through (0,5) and (4,33). We have $f'(x) = 3ax^2 + 2bx + c$, $f'(0) = 0$ $\Rightarrow c = 0$ and $f'(4) = 0 \Rightarrow 48a + 8b = 0$. Also, $f(0) = 5 \Rightarrow d = 5$, and $f(4) = 33 \Rightarrow 64a + 16b + 4c + 5 = 33 \Rightarrow 64a + 16b = 28$ (since $c = 0$) $\Rightarrow 32a + 8b = 14$.

Then

$48a + 8b = 0$
$\underline{32a + 8b = 14}$
$16a = -14 \Rightarrow a = -7/8 \Rightarrow 8b = -48a \Rightarrow b = -6a$

$\Rightarrow b = 21/4$

This gives $f(x) = -\dfrac{7}{8} x^3 + \dfrac{21}{4} x^2 + 5$.

47. $f(x) = x^{1/3}$ dom $f = (-\infty, +\infty)$

$f'(x) = \dfrac{1}{3x^{2/3}} \Rightarrow$ CN at $x = 0$

$$\begin{array}{c} 0 \\ \hline + + + + \mid + + + + + \\ \hline \end{array} \quad f'$$

f is increasing if $x \; \varepsilon \; (-\infty,0] \cup [0,+\infty) = (-\infty, +\infty)$ so $x = 0$ does not give f a local maximum or minimum.

51. To show that $1 - \dfrac{x^2}{2} \leqslant \cos x$, $0 \leqslant x \leqslant 2\pi$, let

$f(x) = \cos x + \dfrac{x^2}{2} - 1$. Then $f'(x) = -\sin x + x$.

From Problem 50, we know that $f'(x) \geqslant 0$; therefore, f is increasing on $[0,2\pi]$. Thus, $f(0) = 0$ is the minimum value and $\cos x + \frac{x^2}{2} - 1 \geqslant 0$ on $[0, 2\pi]$,

or $\cos x \geqslant 1 - \frac{x^2}{2}$ on $[0, 2\pi]$.

<u>Exercise 6</u>, pp. 230-232

3. $f(x) = x^3 - 9x^2 + 2$

   $f' = 3x^2 - 18x \Rightarrow f'' = 6x - 18 = 6(x-3)$

   $$\begin{array}{c} \underline{- - - - \;\Big|\; + + + +} \quad f'' \\ 3 \end{array}$$

   f concave down on $(-\infty, 3]$ and concave up on $[3, +\infty)$.
   Inflection point at $(3, -52)$

7. $f(x) = x + \frac{1}{x}$ dom $f = \{x : x \neq 0\}$

   $f' = 1 - \frac{1}{x^2}$

   $f'' = \frac{2}{x^3}$

   $$\begin{array}{c} 0 \\ \underline{- - - \;\Big|\; + + +} \quad f'' \end{array}$$

   f concave down on $(-\infty, 0)$ and concave up on $(0, +\infty)$.
   No inflection points since $0 \notin$ dom f.

11. $f(x) = 3 - \frac{4}{x} + \frac{4}{x^2}$ dom $f = \{x : x \neq 0\}$

    $f' = \frac{4}{x^2} - \frac{8}{x^3}$

    $f'' = -\frac{8}{x^3} + \frac{24}{x^4} = \frac{24 - 8x}{x^4}$ $\quad \begin{array}{c} \underline{+ + + \;\Big|\; + + + \;\Big|\; - -} \quad f'' \\ 0 3 \end{array}$

    f concave down on $[3, +\infty)$ and concave up on
    $(-\infty, 0) \cup (0, 3]$. Inflection point at $(3, 19/9)$.

15. $f(x) = -2x^3 + 15x^2 - 36x + 7$

    $f' = -6x^2 + 30x - 36 = -6(x^2 - 5x + 6) = -6(x-3)(x-2)$;
    CN at 2,3

    $f'' = -12x + 30 = -6(2x-5)$; $f''(5/2) = 0$

    $f''(3) < 0 \Rightarrow$ Local maximum at $(3, -20)$

    $f''(2) > 0 \Rightarrow$ Local minimum at $(2, -21)$

Concave up on $(-\infty, 5/2]$

Concave down on $[5/2, +\infty)$

Inflection point at $(5/2, -41/2)$

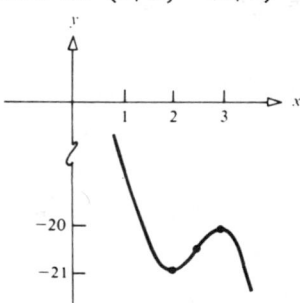

19. $f(x) = 5x^4 - x^5$

$f' = 20x^3 - 5x^4 = 5x^3(4-x)$; CN at 0,4

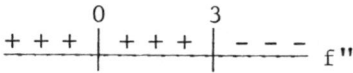

$f'' = 60x^2 - 20x^3 = 20x^2(3-x)$; $f''(0) = 0$; $f''(4) < 0$

Local maximum at $(4, 256)$

Local minimum at $(0,0)$ by first derivative test

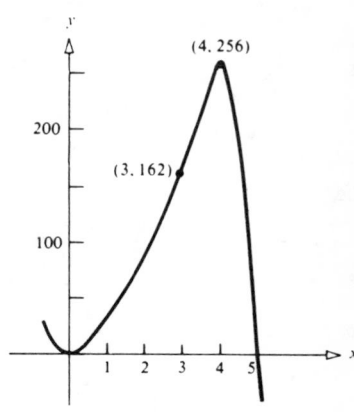

Concave up on $(-\infty, 3]$

Concave down on $[3, +\infty)$

Inflection point at $(3, 162)$

23. $f(x) = 6x^{4/3} - 3x^{1/3} = 3x^{1/3}(2x-1)$

$f' = 8x^{1/3} - x^{-2/3} = (8x-1)/x^{2/3}$; CN at 0 and 1/8

$$f'' = \frac{8}{3}\, x^{-2/3} + \frac{2}{3}\, x^{-5/3} = \frac{8x + 2}{3x^{5/3}} \; ; \; f''(1/8) > 0$$

$$\begin{array}{ccccc} & & -1/4 & & 0 \\ + \, + \, + & | & - \; - & | & + \, + \, + \, + \\ \hline -1 & & & & 1 \end{array} \quad f''$$

Local minimum at $(1/8, -9/8)$

Concave up on $(-\infty, -1/4] \cup [0, +\infty)$

Concave down on $[-1/4, 0]$

Inflection points at $(-1/4, 9/2^3\sqrt{4})$ and $(0,0)$

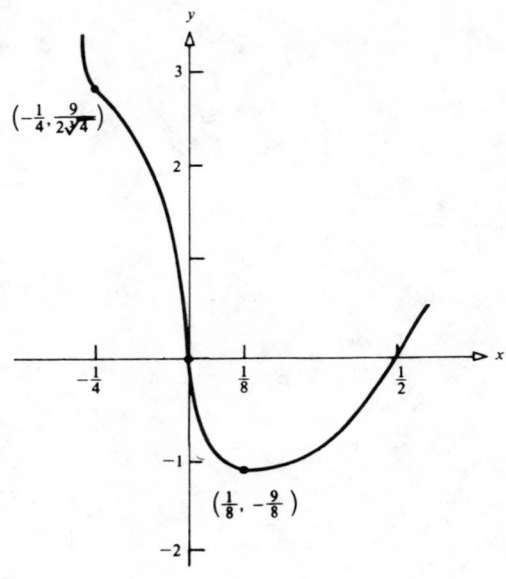

27. $f(x) = x^{2/3}(x^2 - 8) = x^{8/3} - 8x^{2/3}$

$$f' = \frac{8}{3}\, x^{5/3} - \frac{16}{3x^{1/3}} = \frac{8(x^2 - 2)}{3x^{1/3}} \; ; \; \text{CN at } 0 \text{ and } \pm\sqrt{2}$$

$$\begin{array}{ccccc} & -\sqrt{2} & & 0 & & \sqrt{2} \\ - \; - & | & + \, + & | & - \; - & | & + \, + \\ \hline \end{array} \quad f'$$

$$f'' = \frac{40}{9}\, x^{2/3} + \frac{16}{9x^{4/3}} = \frac{8(5x^2 + 2)}{9x^{4/3}} \; ; \; f''(-\sqrt{2}) > 0;$$

$f''(\sqrt{2}) > 0$

$$\begin{array}{c} 0 \\ \underline{+ + + \;\big|\; + + +} \\ f'' \end{array}$$

Local maximum (cusp) at $(0,0)$

Local minima at $(-\sqrt{2}, -6\sqrt[3]{2})$ and $(\sqrt{2}, -6\sqrt[3]{2})$

Concave up for all x

No inflection points

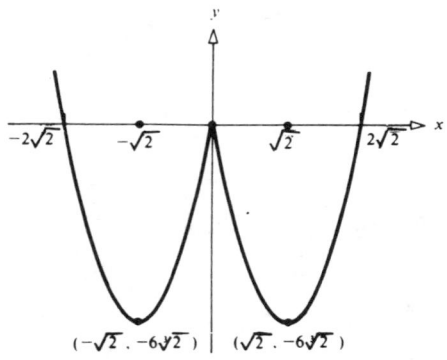

$(-\sqrt{2}, -6\sqrt[3]{2})$ $(\sqrt{2}, -6\sqrt[3]{2})$

31. $f(x) = \dfrac{\sqrt{x}}{1 + x}$ dom $f = [0, +\infty)$

$f' = \dfrac{(1+x)(1/2\sqrt{x}) - \sqrt{x}}{(1+x)^2} = \dfrac{1 + x - 2x}{2\sqrt{x}(1+x)^2} = \dfrac{1 - x}{2\sqrt{x}(1+x)^2}$;

CN at 0 and 1

$$\begin{array}{c} 0 1 \\ \big|+ + +\; \big|\; - - - \\ f' \end{array}$$

$f'' = \dfrac{2\sqrt{x}(1+x)^2(-1) - (1-x)[4\sqrt{x}(1+x) + (1+x)^2/\sqrt{x}]}{4x(1+x)^4}$

$= \dfrac{-2\sqrt{x}(x+1) + (x-1)(\dfrac{4x + (x+1)}{\sqrt{x}})}{4x(1+x)^3}$

$= \dfrac{-2x(x+1) + (x-1)(5x+1)}{4x^{3/2}(1+x)^3} = \dfrac{3x^2 - 6x - 1}{4x^{3/2}(1+x)^3}$;

$3x^2 - 6x - 1 = 0 \Rightarrow x = \dfrac{6 \pm \sqrt{48}}{6} \Rightarrow x = \dfrac{3 \pm 2\sqrt{3}}{3}$

$$0 \qquad (3+2\sqrt{3})/3 \approx 2.15$$

$$\vdash - - - \vert + + + \quad f''$$

Local maximum at $(1, 1/2)$, since $f''(1) < 0$

Concave down on $[0, \dfrac{3 + 2\sqrt{3}}{3}]$

Concave up on $[\dfrac{3 + 2\sqrt{3}}{3}, +\infty)$

Inflection point at $[\dfrac{3 + 2\sqrt{3}}{3}, f(\dfrac{3 + 2\sqrt{3}}{3})]$

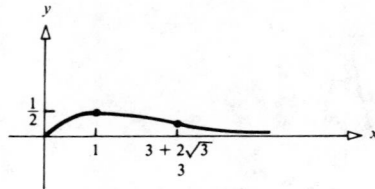

35. $f(x) = \dfrac{x^2}{3x + 1} \qquad \text{dom } f = \{x : x \neq -1/3\}$

$f' = \dfrac{(3x+1)(2x) - x^2(3)}{(3x+1)^2} = \dfrac{3x^2 + 2x}{(3x+1)^2} = \dfrac{x(3x+2)}{(3x+1)^2}$;

CN at 0 and $-2/3$

$$\begin{array}{c} -2/3 \quad -1/3 \quad \ 0 \\ + \, + \, + \ \vert \ - \ - \ \vert \ - \ - \ \vert \ + \, + \, + \quad f' \\ -1 \qquad\qquad\qquad 1 \end{array}$$

$f'' = \dfrac{(3x+1)^2(6x+2) - (3x^2+2x)(2)(3x+1)(3)}{(3x+1)^4}$

$\quad = \dfrac{(3x+1)(6x+2) - (18x^2+12x)}{(3x+1)^3}$

$\quad = \dfrac{18x^2 + 12x + 2 - 18x^2 - 12x}{(3x+1)^3} = \dfrac{2}{(3x+1)^3}$;

$f''(\dfrac{-2}{3}) < 0; \quad f''(0) > 0$

```
      -1/3
- - - | + + +
            f"
```

Local maximum at $(-2/3, -4/9)$

Local minimum at $(0,0)$

Concave down on $(-\infty, -1/3)$

Concave up on $(-1/3, +\infty)$

No inflection points

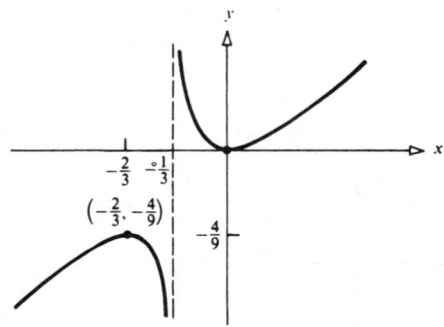

39. $(-\infty, 0)$ f concave down, decreasing

(0, 1) f concave down, increasing

$(1, +\infty)$ f concave up, increasing

$x = 0 \rightarrow f(0) = 1$

$x = 1 \rightarrow f(1) = 2$

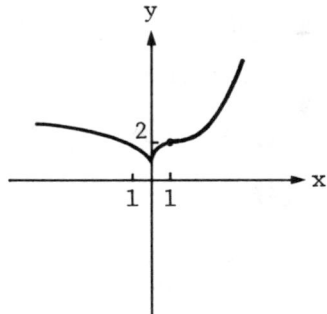

43. $x < 0 \rightarrow f''(x) > 0 \rightarrow$ f concave up

$x > 0 \rightarrow f''(x) > 0 \rightarrow$ f concave up

$x = 0 \rightarrow f(0) = 1, f'(0) = 0, f''(0) = 0$

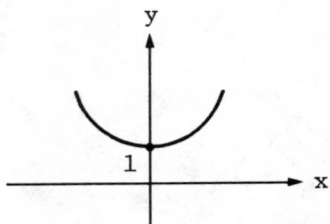

47. x < 0 → f"(x) > 0 → f concave up
    x > 0 → f"(x) > 0 → f concave up
    x = 0 → f(0) = 1, f'(0) does not exist

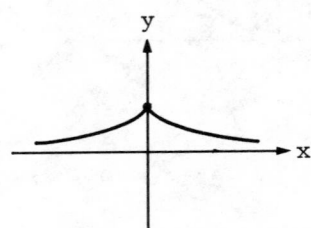

51. Let $f(x) = x^2 - 8x + 21$.

    $f'(x) = 2x - 8 \Rightarrow$ CN at $x = 4$

    $f''(x) = 2 \Rightarrow f$ is concave up for all x

    $f(4)$ is minimal value of f, and

    $f(4) = 16 - 32 + 21 = 5$. Hence,

    $x^2 - 8x + 21 \geqslant 5 > 0$ for all x.

55. $f(x) = \sqrt{3} \sin x + \cos x$ on $(0, 2\pi)$

    $f' = \sqrt{3} \cos x - \sin x$, and $f' = 0 \Rightarrow \tan x = \sqrt{3}$

    $\Rightarrow x = \pi/3, \dfrac{4\pi}{3}$

    $f'' = -\sqrt{3} \sin x - \cos x$, and $f'' = 0 \Rightarrow$

    $\tan x = -\dfrac{1}{\sqrt{3}} \Rightarrow x = \dfrac{5\pi}{6}, \dfrac{11\pi}{6}$

```
 0        5π              11π
          6               6      2π
|- - - - -|+ + + + +|- - -|   f"
|                    
          π
```

From the sign graph, $f''(\pi/3) < 0$ and

$f''(4\pi/3) > 0$ so that $(\frac{\pi}{3}, 2)$ is a local maximum

point and $(\frac{4\pi}{3}, -2)$ is a local minimum point.
There are inflection points at $(\frac{5\pi}{6}, 0)$ and
$(\frac{11\pi}{6}, 0)$.

59. $f(x) = (x-a)^n$ $n = 2k$

$f' = n(x-a)^{n-1}$

$f'' = n(n-1)(x-a)^{n-2} = n(n-1)(x-a)^{2k-2}$

```
                a
+ + + + + | + + + +    f"
          |
```

$f''(x)$ is positive for all $x \neq a$. Hence, there is
no change in concavity so there are no inflection
points.

Exercise 7, pp. 243-244

3. $\lim\limits_{x \to +\infty} \dfrac{2x + 4}{x - 1} = 2$

7. $\lim\limits_{x \to -\infty} \dfrac{5x^3 - 1}{x^2 + 1} = \lim\limits_{x \to -\infty} 5x = -\infty$

11. $\lim\limits_{x \to +\infty} \dfrac{\sqrt{x^2 + 4}}{3x - 1} = \lim\limits_{x \to +\infty} \dfrac{|x|}{3x} = \dfrac{1}{3}$

15. $\lim\limits_{x \to -1^+} \dfrac{5x + 3}{x(x+1)} = +\infty$ $\left[\begin{array}{l}\text{numerator} \to -2 \\ \text{denominator} \to 0^-\end{array}\right]$

19. $\lim\limits_{t \to -\infty} \dfrac{3 - t}{\sqrt{4+5t^2}} = \lim\limits_{t \to -\infty} \dfrac{-t}{\sqrt{5}|t|} = \dfrac{1}{\sqrt{5}}$

23. $\lim\limits_{x\to+\infty} (x - \sqrt{x^2+4}) \cdot \dfrac{x+\sqrt{x^2+4}}{x+\sqrt{x^2+4}} = \lim\limits_{x\to+\infty} \dfrac{x^2-(x^2+4)}{x+\sqrt{x^2+4}}$

$= \lim\limits_{x\to+\infty} \dfrac{-4}{x+|x|} = \lim\limits_{x\to+\infty} \dfrac{-4}{2x} = 0$

27. $\lim\limits_{x\to+\infty} \dfrac{\sin x}{x} = 0$, since $|\sin x| \le 1$

31. $f(x) = 3 + \dfrac{1}{x}$

Vertical asymptote (VA): $x = 0$

$\lim\limits_{x\to+\infty} f(x) = 3 \Rightarrow$ Horizontal asymptote (HA): $y = 3$

35. $f(x) = \dfrac{3x - 1}{x + 1}$

VA: $x = -1$

$\lim\limits_{x\to+\infty} f(x) = 3 \Rightarrow$ HA: $y = 3$

39. $f(x) = \dfrac{x^2 + 4}{x^2 + 1}$

VA: None   HA: $y = 1$

43. $f(x) = \dfrac{x^5}{x^2 + 1}$ VA: None   HA: None

47. $f(x) = x^{1/3}$ at 0

$f'(x) = \dfrac{1}{3x^{2/3}} \Rightarrow$ CN at 0

$\underset{\qquad\;0}{\underline{++++\;|\;++++}}\; f$

$\lim\limits_{x\to 0^+ \text{or } 0^-} \left[\dfrac{1}{3x^{2/3}} \right] = +\infty$

Vertical tangent at $x=0$

51. $f(x) = (x-3)^{2/3}+2$ at 3

$$f'(x) = \frac{2}{3(x-3)^{1/3}}$$

$$\lim_{x \to 3^-} \frac{2}{3(x-3)^{1/3}} = -\infty$$

$$\lim_{x \to 3^+} \frac{2}{3(x-3)^{1/3}} = +\infty$$

$$\underset{3}{\underline{\quad\quad\; \overset{+++++}{\big|}\quad}} \; f'$$

Vertical tangent, (cusp) at x=3

55. $f(x) = \dfrac{2x-1}{x+1}$

$$f'(x) = \frac{(x+1)2-(2x-1)}{(x+1)^2} = \frac{3}{(x+1)^2} \qquad \underset{-1}{\underline{\quad \overset{+++}{}\,\overset{+++}{\big|}\quad}} \; f'$$

$$f''(x) = \frac{-6}{(x+1)^3} \qquad \underset{-1}{\underline{\quad \overset{++++}{}\,\overset{----}{\big|}\quad}} \; f''$$

(1) Intercepts: $(\frac{1}{2}, 0)$, $(0, -1)$

(2) Symmetry: none
(3) Asymptotes: VA: x = -1; HA: y = 2
(4) Critical number: none
(5) Local extrema: none
(6) Concavity: down on $(-1, +\infty)$; up on $(-\infty, -1)$
(7) Inflection points: none

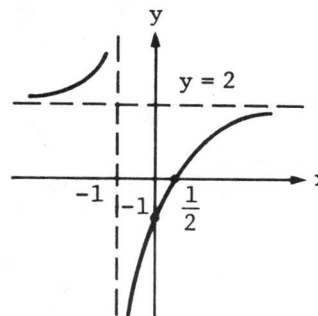

59. $f(x) = \dfrac{8}{x^2-16}$

$f'(x) = \dfrac{-8(2x)}{(x^2-16)^2} = \dfrac{-16x}{(x^2-16)^2}$

$$\begin{array}{ccccc} +++ & +++ & --- & --- \\ \hline & -4 & 0 & 4 \end{array} \quad f'$$

$f''(x) = \dfrac{(x^2-16)^2(-16)+(16x)\cdot 2(x^2-16)(2x)}{(x^2-16)^4}$

$= \dfrac{-16x^2+256+64x^2}{(x^2-16)^3}$

$= \dfrac{48x^2+256}{(x^2-16)^3}$

$$\begin{array}{ccc} +++ & ---- & +++ \\ \hline -4 & 4 \end{array} \quad f''$$

(1) Intercept: $(0, -\frac{1}{2})$
(2) Symmetric about y-axis: $f(-x) = f(x)$
(3) Asymptotes: VA: x = -4, x=4; HA: y = 0
(4) Critical number: x=0
(5) Local extrema: maximum at (0, -1/2)
(6) Concavity: down on (-4, 4); up on (-∞, -4)
    ∪ (4, +∞)
(7) Inflection points: none

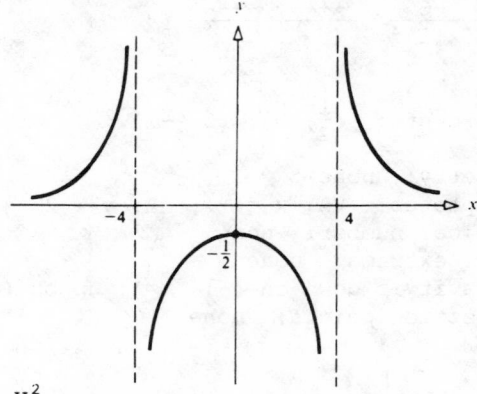

63. $f(x) = \dfrac{x^2}{x+3}$

$f'(x) = \dfrac{(x+3)(2x)-x^2(1)}{(x+3)^2} = \dfrac{x^2+6x}{(x+3)^2} = \dfrac{x(x+6)}{(x+3)^2}$

$$\begin{array}{cccc} +++ & --- & --- & +++ \\ \hline -6 & -3 & 0 \end{array} \quad f'$$

$$f''(x) = \frac{(x+3)^2(2x+6)-(x^2+6x)2(x+3)}{(x+3)^4} = \frac{2x^2+12x+18-2x^2-12x}{(x+3)^3}$$

$$= \frac{18}{(x+3)^3}$$

$$\begin{array}{c} \dfrac{\ \ ---\ \ |\ \ ++++}{-3}\ \ f'' \end{array}$$

(1) Intercepts: (0, 0)
(2) Symmetry: none
(3) Asymptotes: VA: x = -3; HA: none
(4) Critical numbers: x = -6, x=0
(5) Local extrema: maximum at (-6, -12); minimum at (0, 0)
(6) Concavity: down on (-∞, -3); up on (-3, +∞)
(7) Inflection points: none

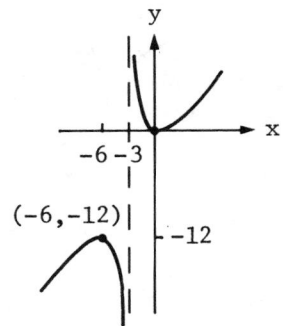

67. $f(x) = 1 + \dfrac{1}{x} + \dfrac{1}{x^2}$

$$f'(x) = -\frac{1}{x^2} - \frac{2}{x^3} = \frac{-(x+2)}{x^3} \qquad \begin{array}{c} \dfrac{\ ---\ |\ +++\ |\ ----}{\quad -2 \qquad 0}\ f' \end{array}$$

$$f''(x) = \frac{2}{x^3} + \frac{6}{x^4} = \frac{2(x+3)}{x^4} \qquad \begin{array}{c} \dfrac{\ ---\ |\ +++\ |\ +++++}{\quad -3 \qquad 0}\ f'' \end{array}$$

(1) Intercepts: none
(2) Symmetry: none
(3) Asymptotes: VA at x=0; HA at y=1
(4) Critical number: x = -2
(5) Local extrema: minimum at (-2, 3/4)
(6) Concavity: down on (-∞, -3]; up on [-3, 0) ∪ (0, +∞)

(7) Inflection ponts: (-3, 7/9)

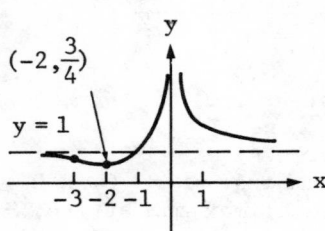

71. If $x^2+y^2 = 1$, then $2x+2yy' = 0 \Rightarrow y' = -\dfrac{x}{y} = \dfrac{-x}{\sqrt{1-x^2}}$

$\Rightarrow$ vertical tangent at -1 and +1, since $\lim\limits_{x \to \pm 1} |y'| = +\infty$

Exercise 8, pp. 252-256

3.

x

y

Given L meters of fence, then
$2x + 2y = L$, or $x + y = L/2$.

$A = xy = x(\dfrac{L}{2} - x) = \dfrac{L}{2}x - x^2$

$A' = \dfrac{L}{2} - 2x \Rightarrow$ CN is $x = L/4$

$A'' = -2$ for all $x \Rightarrow x = L/4$ gives A a maximum value.

Dimensions are $x = L/4$m by $y = L/4$ m.

7.

ℓ ℓ

h

$\dfrac{x}{2}$

x

For a perimeter of fixed length L,
$P = 2\ell + x = L \Rightarrow 2\ell = L - x$
$\Rightarrow \ell = (L-x)/2 \Rightarrow \ell^2 = (L-x)^2/4$.
From the figure, $h^2 + (x^2/4) = \ell^2$.

$h = \sqrt{\ell^2 - \dfrac{x^2}{4}} = \sqrt{\dfrac{(L-x)^2}{4} - \dfrac{x^2}{4}}$

$= \sqrt{\dfrac{L^2 - 2xL}{4}}$

$$A = \frac{1}{2} xh = \frac{1}{2} x \sqrt{\frac{L^2 - 2xL}{4}} = \frac{1}{4} x(L^2-2Lx)^{1/2} \quad \text{for}$$

$$0 < x < L/2$$

$$A' = \frac{1}{4} x \cdot \frac{1}{2}(L^2-2Lx)^{-1/2} (-2L) + \frac{1}{4}(L^2-2Lx)^{1/2}$$

$$= \frac{-Lx}{4\sqrt{L^2 - 2Lx}} + \frac{L^2 - 2Lx}{4\sqrt{L^2 - 2Lx}}$$

$$= \frac{L^2 - 3Lx}{4\sqrt{L^2 - 2Lx}} \Longrightarrow \text{CN at} \quad x = \frac{L}{3}$$

```
       L/3        L/2
 |+ + +|- - -|————|
 0                L   A'
```

$x = L/3$ gives A a maximum value. Solving for ℓ

in $\ell = \dfrac{L - x}{2}$ gives $\ell = \dfrac{L - (L/3)}{2} = \dfrac{2L/3}{2} = \dfrac{L}{3}$.

Thus, an equilateral triangle with each side $L/3$ gives the maximum area.

11.

$$V = x^2h = 2000 \text{ cm}^3 \Longrightarrow h = \frac{2000}{x^2}$$

For a box closed on top, the amount of material $A = 2x^2 + 4xh$

$$= 2x^2 + 4x(\frac{2000}{x^2}) = 2x^2 + \frac{8000}{x}$$

for $x > 0$

$$A' = 4x - \frac{8000}{x^2} = \frac{4x^3 - 8000}{x^2} \Longrightarrow \text{CN at} \quad x^3 = 2000$$

or $x = \sqrt[3]{2000}$

$$A'' = 4 + \frac{16,000}{x^3} \Longrightarrow A'' (\sqrt[3]{2000}) > 0 \Longrightarrow x = \sqrt[3]{2000} \text{ gives}$$

A a minimum value

$$h = \frac{2000}{x^2} = \frac{2000}{(\sqrt[3]{2000})^2} = \frac{2000}{2000^{2/3}} = \sqrt[3]{2000}$$

The box is a cube with each side $= \sqrt[3]{2000} = 10\sqrt[3]{2}$ cm.

15. Let $10 = base price per day, P = price over $10, and 24 - P = number of cars rented.

Income $I = (10+P)(24-P) = 240 + 14P - P^2$
Income $I' = 14 - 2P \Rightarrow$ CN at $P = 7$
Income $I'' = -2 \Rightarrow P = 7$ gives a maximum value

Maximum income at $P = 7 \Rightarrow \$10 + P = \$17$ per day should be charged, for an income of $I = \$289$.

19. If $C(x) = (\dfrac{1600}{x} + x)a + \dfrac{200b}{x} + c,$

then $C(x) = \dfrac{1600a + 200b}{x} + ax + c,$

so that $C' = \dfrac{-(1600a+200b)}{x^2} + a = \dfrac{ax^2 - (1600a+200b)}{x^2}$

$\Rightarrow$ CN if $x^2 = \sqrt{\dfrac{1600a + 200b}{a}}$

$C'' = \dfrac{2(1600a+200b)}{x^3} > 0 \Rightarrow x = \sqrt{\dfrac{1600a + 200b}{a}}$ gives

C a minimum value

(a) $a = \$1.50$, $b = 0$, $c = 0$: Economical speed is $x = \sqrt{1600} = 40$ mph

(b) $a = \$1.50$, $b = \$8$, $c = \$500$: Economical speed is $x = \sqrt{\dfrac{1600(1.50) + 200(8)}{1.50}} = \sqrt{2666.666}$

≈ 51.64 mph

(c) $a = \$1.60$, $b = \$10$, $c = 0$: Economical speed is $x = \sqrt{\dfrac{1600(1.60) + 200(10)}{1.60}} = \sqrt{2850}$

≈ 53.39 mph

23. (a) Let R denote the point on the road and suppose for convenience that $r > q$. Let R be x units from D on the road and let E be s units from D. Then, using Figure II, we have $s = \sqrt{p^2 - (r-q)^2}$. Letting d_1 and d_2 denote the distances $\overline{AR}$ and $\overline{BR}$, respectively, we have $d_1 = \sqrt{x^2 + q^2}$, and $d_2 = \sqrt{(s-x)^2 + r^2}$, and we

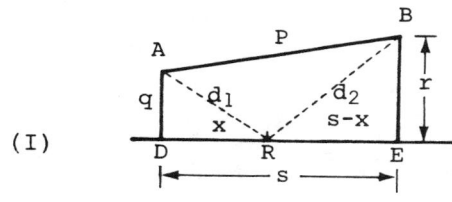

(I)

(II) A ‗‗‗‗ P ‗‗‗ B r-q

wish to minimize $d(x) = d_1 + d_2 = \sqrt{x^2 + q^2}$
$+ \sqrt{(s-x)^2 + r^2}$. This gives $d'(x) = \dfrac{x}{\sqrt{x^2 + q^2}}$

$+ \dfrac{-(s-x)}{\sqrt{(s-x)^2 + r^2}} = \dfrac{x\sqrt{(s-x)^2 + r^2} - (s-x)\sqrt{x^2 + q^2}}{\sqrt{x^2 + q^2}\ \sqrt{(s-x)^2 + r^2}}$,

and $d' = 0 \Rightarrow x\sqrt{(s-x)^2 + r^2} = (s-x)\sqrt{x^2 + q^2}$
$$x^2[s^2 - 2sx + x^2 + r^2] = (s^2 - 2sx + x^2)(x^2 + q^2)$$
$$(r^2 - q^2)x^2 + 2sq^2 x - s^2 q^2 = 0$$
$$[(r-q)x + sq][(r+q)x - sq] = 0,$$

which gives critical numbers for $x = \dfrac{-sq}{r - q}$,
or $x = \dfrac{sq}{r + q}$. We exclude the negative critical
number $\dfrac{-sq}{r - q}$, and conclude that $x = \dfrac{sq}{r + q}$
gives the desired local extrema. Then

$d(\dfrac{sq}{r+q}) = \sqrt{(\dfrac{sq}{r+q})^2 + q^2} + \sqrt{(s - \dfrac{sq}{r+q})^2 + r^2}$

$= \dfrac{\sqrt{q^2[s^2 + (r+q)^2]} + \sqrt{s^2(r+q-q)^2 + r^2(r+q)^2}}{r + q}$

$= \dfrac{q\sqrt{p^2 - (r-q)^2 + (r+q)^2} + r\sqrt{[p^2 - (r-q)^2] + (r+q)^2}}{r + q}$

[Since $s^2 = p^2 - (r-q)^2$]

$= \dfrac{q\sqrt{p^2 + 4rq} + r\sqrt{p^2 + 4rq}}{r + q}$

$$= \frac{(r+q)\sqrt{p^2 + 4rq}}{r + q} = \sqrt{p^2 + 4rq}$$

The shortest path from A to the road and then to B is $\sqrt{p^2 + 4rq}$ units.

(b) We can geometrically reason that if we reflect A and B about the road, we have the trapezoid AA'B'B as shown in Figure III. By the symmetry we see that the shortest path from A to the road to B will intersect the road at the point R where the line from A to B' intersects the road. Hence, the shortest distance has length equal to the length of the diagonal of the trapezoid AA'B'B. If we form a coordinate system placing the origin at A', then the length of the diagonal is the distance from $(0,0)$ to $(s,2q+r-q))$ $= (s,r+q)$, where

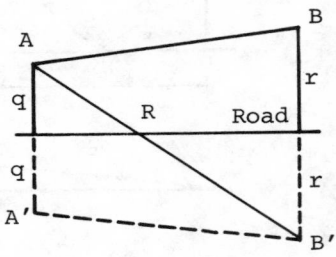

(III)

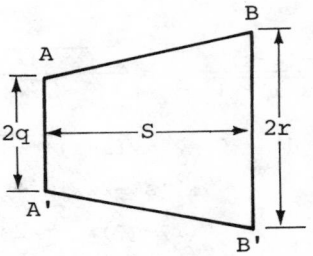

$s = \sqrt{p^2 + (r-q)^2}$. The distance is

$d = \sqrt{s^2 + (r+q)^2}$

$= \sqrt{p^2 + (r-q)^2 + (r+q)^2}$

since $s^2 = p^2 + (r-q)^2$

$= \sqrt{p^2 + 4rq}.$

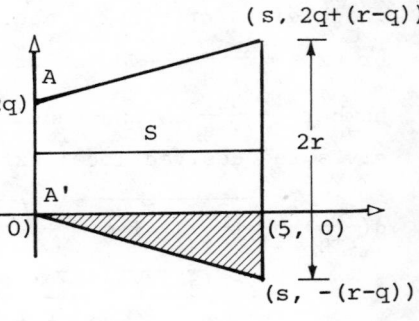

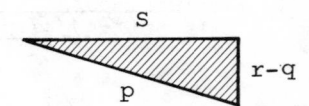

27. Let $\ell = \sqrt{(y+8)^2 + (x+1)^2}$,
   $x > 0$ similar triangles

   $\Rightarrow \dfrac{8}{x} = \dfrac{y + 8}{x + 1} \Rightarrow (y+8)^2$

   $= \dfrac{64}{x^2}(x+1)^2$. Hence,

   $\ell = \sqrt{\dfrac{64}{x^2}(x+1)^2 + (x+1)^2}$

   $= \dfrac{x + 1}{x} \sqrt{64 + x^2}$

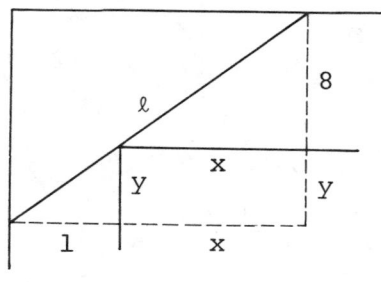

   $\ell' = (\dfrac{x+1}{x}) \dfrac{x}{\sqrt{64 + x^2}} + (\dfrac{-1}{x^2})\sqrt{64 + x^2} = \dfrac{x^3 - 64}{x^2\sqrt{64 + x^2}}$

   ```
   0        4
   |- - - | + + +
                    ℓ'
   ```

   Hence $x = 4$ minimizes ℓ.
   That is, $\ell(4) = (5/4)\sqrt{80}$
   $= 5\sqrt{5}$ m is the

   "clearance" distance for a rigid pipe being carried around the corner. Thus, any pipe less than $5\sqrt{5}$ m long will get around the corner.

31. Minimize $P = y + x + y = x + 2y$, given that
   $xy = C$ (constant), or $y = C/x$.

   $P(x) = x + \dfrac{2C}{x}$

   $P' = 1 - \dfrac{2C}{x^2} = \dfrac{x^2 - 2C}{x^2} \Rightarrow x = \sqrt{2C}$ is a CN

   $P'' = \dfrac{4C}{x^3} \Rightarrow P''(\sqrt{2C}) > 0 \Rightarrow x = \sqrt{2C}$ gives P a
   minimum value.

   The dimensions are x by y or $\sqrt{2C}$ by $\sqrt{\dfrac{C}{2}}$.

   This is a base-to-height ratio of 2 to 1.

35.

   All in square $\Rightarrow x^2 = 64$
   $\Rightarrow x = 8 \Rightarrow L = 32$
   All in circle $\Rightarrow \pi R^2 = 64$

   $R = \dfrac{8}{\sqrt{\pi}} \Rightarrow L = 16\sqrt{\pi} \sim 28.36$

$$A_s = x^2 \qquad\qquad A_c = \pi R^2$$

$$A_s + A_c = 64 \text{ cm}^2$$

$$P_s = 4x \qquad\qquad P_c = 2\pi R$$

Total area $A = x^2 + \pi R^2 = 64 \implies x^2 = 64 - \pi R^2$

$$\implies x = \sqrt{64 - \pi R^2}$$

when $x = 0$, $R = 8/\sqrt{\pi}$.

Total length $L = 4x + 2\pi R$

$$L(R) = 4(64 - \pi R^2)^{1/2} + 2\pi R \quad \text{for} \quad 0 \leqslant R \leqslant \frac{8}{\sqrt{\pi}}$$

$$L' = \frac{2(-2\pi R)}{\sqrt{64 - \pi R^2}} + 2\pi = \frac{2\pi[-2R + \sqrt{64 - \pi R^2}]}{\sqrt{64 - \pi R^2}}$$

$\implies$ CN when $2R = \sqrt{64 - \pi R^2}$

$$4R^2 + \pi R^2 = 64 \implies R^2 = \frac{64}{4 + \pi} \implies$$

$$R = \frac{8}{\sqrt{4 + \pi}} \sim 2.99 \text{ cm}$$

Test $L(R)$ at $R = 0$, $\dfrac{8}{\sqrt{4 + \pi}}$, $\dfrac{8}{\sqrt{\pi}}$:

$L(0) = 4\sqrt{64} + 0 = 4\cdot 8 = 32$ (all in square)

$L(\frac{8}{\sqrt{\pi}}) = 0 + 2\pi(\frac{8}{\sqrt{\pi}}) = 16\sqrt{\pi} \approx 28.36$ (all in circle)

$$L(\frac{8}{\sqrt{4 + \pi}}) = 4\sqrt{64 - \frac{64\pi}{4+\pi}} + 2\pi(\frac{8}{\sqrt{4 + \pi}})$$

$$= 32\sqrt{1 - \frac{\pi}{4+\pi}} + \frac{16\pi}{\sqrt{4 + \pi}} = \frac{64 + 16\pi}{\sqrt{4 + \pi}}$$

$$= 16\sqrt{4 + \pi} \approx 42.76$$

Maximum length of wire is $16\sqrt{4 + \pi} \approx 42.76$ cm.

Minimum length of wire (all in circle) is $16\sqrt{\pi} \approx 28.36$ cm.

39.

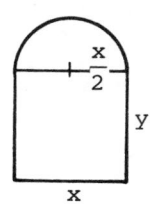

Circumference of a semi-circle is $\frac{1}{2}(\pi d)$. Here $d = x$. $P = x + 2y + \frac{1}{2}\pi x = 10$

$$y = 5 - \frac{2 + \pi}{4} x$$

$A = xy + \frac{1}{2}\pi R^2 = xy + \frac{\pi}{2}(\frac{x}{2})^2$, since $R = \frac{x}{2}$

$A = x(5 - \frac{2+\pi}{4}x) + \frac{\pi}{8}x^2 = 5x + (\frac{\pi}{8} - \frac{2 + \pi}{4})x^2$

$\quad = 5x - \frac{4 + \pi}{8}x^2$

$A' = 5 - \frac{4 + \pi}{4}x$

CN at $x = \frac{20}{4 + \pi}$ gives A a maximum value

```
0           CN
|+ + + + +  |  - - - - -
|           |              A'
```

$y = 5 - \frac{2 + \pi}{4}(\frac{20}{4 + \pi}) = \frac{20 + 5\pi - 10 - 5\pi}{4 + \pi} = \frac{10}{4 + \pi}$

Thus, for a maximum area, the diameter of the semi-circle is $\frac{20}{4 + \pi}$ m and the height of the rectangle is $\frac{10}{4 + \pi}$ m.

Exercise 9, pp. 266-268

3. $f(x) = 5x^{3/2}$

$F(x) = 2x^{5/2} + C$

7. $f(x) = \sqrt{x} = x^{1/2}$

$F(x) = \frac{2}{3}x^{3/2} + C$

11. $f(x) = (2-3x)^2$ or $f(x) = 4 - 12x + 9x^2$
$F(x) = -\frac{1}{9}(2-3x)^3 + C$ $F(x) = 4x - 6x^2 + 3x^3 + K$

15. $f(x) = \dfrac{x^2 + 10x + 21}{3x + 9} = \dfrac{(x+3)(x+7)}{3(x+3)} = \dfrac{1}{3}x + \dfrac{7}{3}$

$F(x) = \dfrac{1}{6}x^2 + \dfrac{7}{3}x + C$

19. $\dfrac{dy}{dx} = 3x^2 - 2x + 1; \quad y(0) = 1$

$y = x^3 - x^2 + x + C$

$1 = 0 - 0 + 0 + C \Rightarrow C = 1$

$y = x^3 - x^2 + x + 1$

23. $\dfrac{ds}{dt} = t^3 + t^{-2} \quad s(1) = 2$

$s = \dfrac{1}{4}t^4 - \dfrac{1}{t} + C$

$2 = \dfrac{1}{4} - 1 + C \Rightarrow C = 3 - \dfrac{1}{4} = \dfrac{11}{4}$

$s = \dfrac{1}{4}t^4 - \dfrac{1}{t} + \dfrac{11}{4}$

27. $a(t) = -32, \quad s(0) = 0, \quad v(0) = 128$

$a(t) = -32 \Rightarrow v(t) = -? \; c + C_1$

$v(0) = 128 \Rightarrow C_1 = 128 \Rightarrow v(t) = -32t + 128$

$s(t) = -16t^2 + 128t + C_2$

$s(0) = 0 \Rightarrow C_2 = 0$

$s(t) = -16t^2 + 128t$

31. $\dfrac{dF}{dx} = -x \sin x + 2 \cos x; \quad F(0) = 1$

$F(x) = x \cos x + \sin x + C$

$1 = 0 + 0 + C$

$F(x) = x \cos x + \sin x + 1$

35. $s_e = s_{earth} = -4.9t^2 + v_o t$ (Jump from ground
    $\Rightarrow s_o = 0$)

$v_e = -9.8t + v_o \Rightarrow$ CN at $t = \dfrac{v_o}{9.8}$

$s_m = s_{moon} = -1.6(1/2)t^2 + v_o t \ (s_o = 0)$

Suppose that $s_e(\dfrac{v_o}{9.8}) = 2 = -4.9(\dfrac{v_o}{9.8})^2 + \dfrac{v_o}{9.8}^2$,

then, $\dfrac{-v_o^2}{19.6} + \dfrac{v_o^2}{9.8} = 2 \Rightarrow v_o^2 = 2(19.6) = 39.2$

and this gives $v_o = \sqrt{39.2} \approx 6.26$ m/sec.

On the moon, $s_m = -0.8t^2 + v_o t$

$v_m = -1.6t + v_o \Rightarrow$ CN at $t = \dfrac{v_o}{1.6}$.

If $s_m = 0.8t^2 + \sqrt{39.2} \, t$, then the maximum height

occurs when $t = \dfrac{v_o}{1.6} = \dfrac{\sqrt{39.2}}{1.6}$ sec.

$s_m(\dfrac{v_o}{1.6}) = -0.8(\dfrac{v_o}{1.6})^2 + v_o(\dfrac{v_o}{1.6}) = (\dfrac{-0.8}{(1.6)^2} + \dfrac{1}{1.6})v_o^2$

$= \dfrac{0.8}{(1.6)^2}(39.2) = 12.25$ m

Therefore, the maximum high jump would be 12.25 m.

39. Force = F (constant), mass = 4 grams, $v(0) = 0$,
    $v(6) = 12$ cm/sec

$F = ma = 4a \qquad a = dv/dt$

$a(t) = \dfrac{F}{4} = \dfrac{dv}{dt} \Rightarrow v(t) = \dfrac{F}{4}t + C_1$

$v(0) = 0 \Rightarrow C_1 = 0 \Rightarrow v(t) = \dfrac{F}{4}t$

$v(6) = 12 = \dfrac{F}{4}(6) \Rightarrow F = \dfrac{48}{6} = 8 \ \dfrac{gm - cm}{sec^2}$

3. If the store sells x units at a price of p ¢ per
unit, where $p = 20 - 0.03x$, and the cost is C ¢
per unit, where $C = 3 + 0.02x$, we want to find the
number of units x that will maximize profit.

Revenue = $R = xp = 20x - 0.03x^2$

Profit = $P = R - C = (20x-0.03x^2) - (3+0.02x)$
$$= -0.03x^2 + 19.98x - 3$$

$P = -0.03(x^2-666x+100)$

$P' = -0.03(2x-666) \Rightarrow$ CN at $x = 333$

$P'' = -0.06 \Rightarrow x = 333$ units gives a maximum P

7. We are given that 500 articles can be sold weekly if
$p = \$20$; for each 50¢ decrease in p, sales will
rise 50 per week; and the cost function is
$C(x) = 4200 + 5.10x + 0.0001x^2$.

Revenue = R = (Number sold)(Price charged)

$R = xp$ where $p = 20 - 0.50(\frac{x-500}{50}) = 20 - 0.01(x-500)$
$$= 20 + 5 - 0.01x = 25 - 0.01x$$

$R = xp = 25x - 0.01x^2$

Profit = $P = R - C = -0.01x^2 + 25x - 4200 - 5.10x$
$$- 0.0001x^2$$

$P = -0.0101x^2 + 19.9x - 4200$

$P' = -0.0202x + 19.9 \Rightarrow$ CN at $x = 985.15$

(a) $p(x) = 25 - 0.01x$

(b) Weekly level for maximum profit is $x = 985$
articles

(c) $p(985) = 25.00 - 9.85 = \$15.15$ per article

Miscellaneous Exercises, pp. 274-277

3. $f(x) = \sqrt{x}$ on $[0,9]$

f is continuous for $x \in [0,9]$ and $f'(x) = \frac{1}{2\sqrt{x}}$
is continuous for $x \in (0,9)$. Therefore, there is a
number c such that $f'(c) = \frac{f(9) - f(0)}{9 - 0}$

or $\dfrac{1}{2\sqrt{c}} = \dfrac{3-0}{9-0} = \dfrac{1}{3}$

$2\sqrt{c} = 3 \implies c = \left(\dfrac{3}{2}\right)^2 = \dfrac{9}{4}$

7. $f(x) = x - 2x^{1/2} \qquad \text{dom } f = [0,\infty)$

$f' = 1 - x^{-1/2} = 1 - \dfrac{1}{\sqrt{x}} = \dfrac{\sqrt{x} - 1}{\sqrt{x}}$

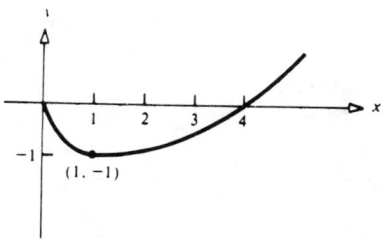

$f'' = \dfrac{1}{2} x^{-3/2} = \dfrac{1}{2x^{3/2}}$

(1) Intercepts: $(0,0), (4,0)$

(2) Symmetry: None

(3) Asymptotes: VA: None; HA: None

(4) Critical numbers: $x = 0$, $x = 1$

(5) Local extrema: Minimum at $(1,-1)$

(6) Concavity: Up on $[0, +\infty)$

(7) Inflection points: None

11. If $y = f(x)$, such that $y' > 0$ for all x and $y'' < 0$ for all x, then f is increasing and concave downward. Therefore, (b) could be part of graph of $y = f(x)$.

15. $f(x) = x(x^2-4)^{1/3}$

$f' = x \cdot \dfrac{2x}{3(x^2-4)^{2/3}} + (x^2-4)^{1/3} = \dfrac{5x^2 - 12}{3(x^2-4)^{2/3}}$

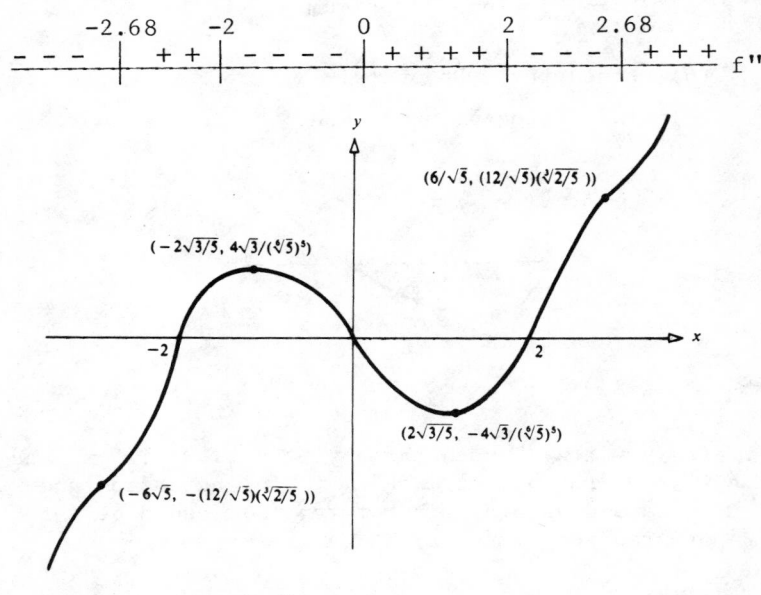

$$f'' = \frac{3(x^2-4)^{2/3}(10x) - (5x^2-12)\left[2 \cdot \dfrac{2x}{(x^2-4)^{1/3}}\right]}{9(x^2-4)^{4/3}}$$

$$= \frac{30x(x^2-4) - 4x(5x^2-12)}{9(x^2-4)^{5/3}}$$

$$= \frac{10x^3 - 72x}{9(x^2-4)^{5/3}} = \frac{2x(5x^2-36)}{9(x^2-4)^{5/3}}$$ Possible points of

inflection are $x = 0, \pm \dfrac{6}{\sqrt{5}}$ and $x = \pm 2$.

(1)   Intercepts: $(0,0)$ and $(\pm 2, 0)$

(2)   Symmetry: None

(3)   Asymptotes: None

(4) Critical numbers: $x = \pm 2$ (vertical tangent)

and $x = \pm\sqrt{12/5}$ (horizontal tangent)

(5) Local extrema: Maximum at $(-\sqrt{12/5}, -1.812)$;

Minimum at $(\sqrt{12/5}, 1.812)$

(6) Concavity: Down on $(-\infty, -\frac{6}{\sqrt{5}}] \cup [-2,0] \cup [2, \frac{6}{\sqrt{5}}]$;

Up on $[-\frac{6}{\sqrt{5}}, -2] \cup [0,2] \cup [\frac{6}{\sqrt{5}}, +\infty)$

(7) Inflection points: $(\frac{-6}{\sqrt{5}}, \frac{-12}{\sqrt{5}} \sqrt[3]{\frac{2}{5}})$, $(-2,0)$,

$(0,0)$, $(2,0)$, $(\frac{6}{\sqrt{5}}, \frac{12}{\sqrt{5}} \sqrt[3]{\frac{2}{5}})$

19. If the motorcycle accelerates 0 to 72 km/hr in 10 sec and acceleration is constant, then

$$\text{ave. acc.} = \frac{72 \text{ km/hr} - 0 \text{ km/hr}}{10 \text{ sec} - 0 \text{ sec}} = \frac{72 \text{ km/hr}}{10 \text{ sec} \cdot \frac{1 \text{ min}}{60 \text{ sec}} \cdot \frac{1 \text{ hr}}{60 \text{ min}}}$$

$$= 7.2(3600) \text{ km/hr}^2 = 25{,}920 \text{ km/hr}^2$$

$a(t) = 25{,}920 \text{ km/hr}^2$

$v(t) = 25{,}920t + v_o = 25{,}920t$, since $v_o = 0$

$s(t) = 12{,}960t^2 + s_o = 12{,}960t^2$, since $s_o = 0$

$s(10 \text{ sec}) = s(\frac{10}{3600} \text{ hr}) = s(\frac{1}{360}) = \frac{12{,}960}{(360)^2} = 0.1 \text{ km}$

23. (a)

x	1	10	100	10,000	guess
$(x+1/x)^x$	2	2.5937	2.7048	2.7181	2.7181

(b)

x	1	10	100	guess
$(1+1/x)^{x^2}$	2	13,780.6	1.6358×10^{43}	$+\infty$

(c)

x	1	10	100	10,000	guess
$(1-1/x)^x$	0	0.3487	0.3660	0.3679	0.3679

(d)

x	1	10	100	guess
$\frac{\sin x}{x}$	0.8415	-0.0544	-0.00506	0

(e)

x	1	10	100	guess
$\left\|\sin x\right\|^{x}$	0.8415	0.0023	2.795×10^{-30}	0

<u>Note</u>: For $x = \pi/2 + k\pi$, $k = 1,2,3,\ldots$, $\left\|\sin x\right\|^{x} = 1$, so in (e), the guess appears doubtful.

27. Let $s(t)$ be the height of the ball t seconds after it was thrown upward with $v_o = 19.6$ m/sec from a height $s_o = 1$ m. Then, using

$a(t) = -9.8$ m/sec^2, we have $s(t) = \frac{1}{2}at^2 + v_ot + s_o$

$= -4.9t^2 + 19.6t + 1$. Then $v(t) = s'(t)$

$= -9.8t + 19.6$ and $v(t) = 0$ when $t = 2$. Thus, the ball reaches its maximum height when $t = 2$, so that at $t = 3$ the ball is moving downward.

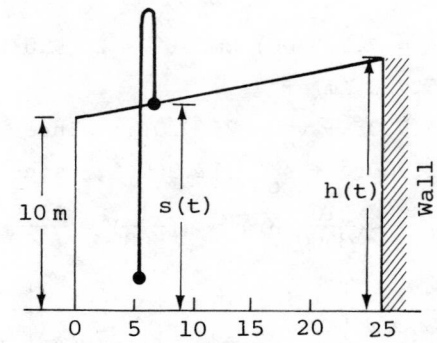

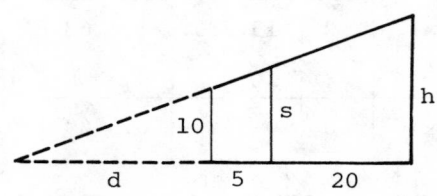

(a) The height of the ball at t = 3 sec is
s(3) = 15.7 m.

(b) Since the ball is moving downward at t = 3 sec,
the shadow on the wall is moving downward.

A relationship between the height of the shadow on
the wall h(t) and the height of the ball s(t)
can be formed by extending the line joining h,s,10
until it hits ground level. By similar triangles we
conclude that $\dfrac{h}{25 + d} = \dfrac{s}{5 + d} = \dfrac{10}{d}$. This gives

sd = 10(5+d),   or $d = \dfrac{50}{s - 10}$,   and   h(5+d)

= s(25+d). Substituting $d = \dfrac{50}{s - 10}$ in this last
result and simplifying gives

$$h = s\left[\frac{25 + d}{5 + d}\right] = s\left[\frac{25 + \dfrac{50}{s - 10}}{5 + \dfrac{50}{s - 10}}\right] = s\left[\frac{25(s-10) + 50}{5(s-10) + 50}\right]$$

$$= s\left[\frac{25s - 200}{5s}\right] ,$$

or h = 5s − 40. That is, h(t) = s(t) − 40.
Then, h'(t) = 5s'(t), and we use this to find that
at t = 3 the shadow is moving on the wall at a rate
of $|h'(3)| = |5s'(3)| = |5[-9.8(3) + 19.6]| = |-49|$
= 49 m/sec, downward.

31. If f'(x) > g'(x) for all real x, then the graphs
of f and g can intersect no more than once
because f is increasing at a faster rate than g.
Thus, statement (b) must be true.

35. Let A and B denote the two bodies. If we let
$s_A(t)$ and $s_B(t)$ denote the height of the bodies

t seconds after the first body, say A, is released,
then we have $s_A(t) = -4.9t^2 + s_o$, since the initial
velocity v_o is zero, and $s_B(t) = -4.9(t-1)^2 + s_o$,
t ⩾ 1, since $v_o = 0$ and since B began to fall
one second after A. We want to find t such that

$|s_B(t) - s_A(t)| = 10.$ This gives

$[-4.9(t^2-2t+1) + s_o] - [-4.9t^2 + S_o] = 10,$

$-4.9t^2 + 9.8t - 4.9 + 4.9t^2 = 10$

$$9.8t = 10 + 4.9 = 14.9$$

$$t = \frac{14.9}{9.8} \approx 1.52 \text{ seconds}$$

39.   If we let b_n denote the length of each side of the n-sided regular polygon inscribed in the circle of radius R, then the perimeter P_n of the polygon is $P_n = n \cdot b_n$. Since the angle $\theta = 2\pi/n$, we have that

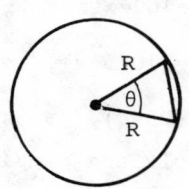

$\sin \dfrac{\theta}{2} = \dfrac{\frac{1}{2} b_n}{R}$, or

$\sin \dfrac{\theta}{2} = \dfrac{b_n}{2R}$.

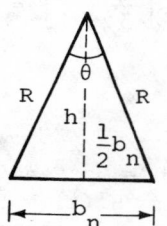

(a)   Hence, $P_n = nb_n = n(2R \sin \dfrac{\theta}{2}) = 2nR \sin \dfrac{\pi}{n}$, since $\theta = \dfrac{2\pi}{n}$.

(b)   We have

$$\lim_{n \to +\infty} P_n = \lim_{n \to +\infty} 2nR \sin \frac{\pi}{n} = 2\pi R \lim_{n \to +\infty} \left(\frac{\sin \frac{\pi}{n}}{\pi/n}\right)$$

$$= 2\pi R \lim_{K \to 0^+} \frac{\sin K}{K} , \text{ letting } K = \frac{\pi}{n} .$$

Thus, $\lim\limits_{n \to +\infty} P_n = 2\pi R(1) = 2\pi R.$

Chapter Five

Exercise 1, p. 284

3. $[-1,4]$; $n = 10$

$P = \{-1, -\frac{1}{2}, 0, \frac{1}{2}, 1, \frac{3}{2}, 2, \frac{5}{2}, 3, \frac{7}{2}, 4\}$

$[-1, -1/2], [-1/2, 0], [0, \frac{1}{2}], [\frac{1}{2}, 1], [1, 3/2], [\frac{3}{2}, 2],$

$[2, 5/2], [5/2, 3], [3, 7/2], [7/2, 4]$

7. $f(x) = x^2$ on $[0,2]$;

$f' = 2x > 0$ on $[0,2] \Rightarrow f$ is increasing on $[0,2]$.

x	x^2
0	0
$\frac{1}{4}$	1/16
$\frac{1}{2}$	$\frac{1}{4}$
$\frac{3}{4}$	9/16
1	1
$\frac{5}{4}$	25/16
$\frac{3}{2}$	9/4
$\frac{7}{4}$	49/16
2	4

(a) $n = 4 \Rightarrow P = \{0, \frac{1}{2}, 1, \frac{3}{2}, 2\}$;

$\Delta x = 1/2$

c_i = left end pt.:

$s_4 = [f(0) + f(\frac{1}{2}) + f(1) + f(\frac{3}{2})](\frac{1}{2})$

$= [0 + \frac{1}{4} + 1 + \frac{9}{4}]\frac{1}{2} = \frac{7}{4}$

C_i = right end pt.:

$S_4 = [f(\frac{1}{2}) + f(1) + f(\frac{3}{2}) + f(2)](\frac{1}{2})$

$= [\frac{1}{4} + 1 + \frac{9}{4} + 4]\frac{1}{2} = \frac{15}{4}$

(b) $n = 8 \Rightarrow P = \{0, \frac{1}{4}, \frac{1}{2}, \frac{3}{4}, 1, \frac{5}{4}, \frac{3}{2}, \frac{7}{4}, 2\}$; $\Delta x = 1/4$

$s_8 = [f(0) + f(\frac{1}{4}) + \cdots + f(\frac{7}{4})]\frac{1}{4}$

$= [0 + \frac{1}{16} + \frac{4}{16} + \frac{9}{16} + \frac{16}{16} + \frac{25}{16} + \frac{36}{16} + \frac{49}{16}](\frac{1}{4})$

$= \frac{35}{16}$

$$S_8 = [f(\tfrac{1}{4}) + f(\tfrac{1}{2}) + \cdots + f(\tfrac{7}{4}) + f(2)](\tfrac{1}{4})$$

$$= [\tfrac{1}{16} + \tfrac{4}{16} + \tfrac{9}{16} + \cdots + \tfrac{49}{16} + \tfrac{64}{16}](\tfrac{1}{4}) = \tfrac{51}{16}$$

Exercise 2, pp. 291-293

3. $\displaystyle\sum_{i=1}^{n} (i^3 - 2i + 1) = \frac{n^2(n+1)^2}{4} - 2[\frac{n(n+1)}{2}] + n$

$$= n[\frac{n(n^2+2n+1)}{4}] - (n+1) + 1]$$

$$= n[\frac{n^3}{4} + \frac{n^2}{2} + \frac{n}{4} - n] = \frac{n^2}{4}(n^2 + 2n - 3)$$

$$= \frac{n^2}{4}(n+3)(n-1)$$

7. $\displaystyle\sum_{i=1}^{n} [(\frac{i}{n})^2 - (\frac{i}{n})](\frac{1}{n}) = \sum_{i=1}^{n} \frac{i^2 - ni}{n^3} = \frac{1}{n^3}\sum_{i=1}^{n} i^2 - \frac{1}{n^2}\sum_{i=1}^{n} i$

$$= \frac{n(n+1)(2n+1)}{6n^3} - \frac{n(n+1)}{2n^2}$$

$$= \frac{n+1}{6n^2}[(2n+1) - 3n] = \frac{(n+1)[(2n+1) - 3n]}{6n^2}$$

$$= \frac{(n+1)(1-n)}{6n^2} = -\frac{n^2 - 1}{6n^2} \quad \text{or} \quad \frac{1 - n^2}{6n^2}$$

11. $1 + 2 + 4 + 8 + \ldots + 2^n = \displaystyle\sum_{i=0}^{n} 2^i = \sum_{i=1}^{n+1} 2^{i-1}$

15. $1\cdot 2 + 2\cdot 3 + 3\cdot 4 + \ldots + n(n+1) = \displaystyle\sum_{i=0}^{n-1} (i+1)(i+2)$

$$= \displaystyle\sum_{i=1}^{n} i(i+1)$$

19. $f(x) = 2x + 1, \quad x \in [0,4]; \quad f' = 2 > 0 \Rightarrow f$ is increasing; therefore, use x_{i-1} for inscribed rectangles.

$\Delta x = 4/n; \quad x_{i-1} = 0 + (i-1)\Delta x = 4(i-1)/n$

$$f(x_{i-1}) = 2x_{i-1} + 1 = \frac{8(i-1)}{n} + 1$$

$$s_n = \sum_{i=1}^{n} f(x_{i-1})\Delta x = \sum_{i=1}^{n} [\frac{8(i-1)}{n} + 1]\frac{4}{n}$$

$$= \frac{32}{n^2} \sum_{i=1}^{n} (i-1) + \frac{4}{n} \sum_{i=1}^{n} 1$$

$$= \frac{32}{n^2} [\frac{n(n+1)}{2} - n] + \frac{4}{n}(n) = \frac{16(n+1)}{n} - \frac{32}{n} + 4$$

$$= \frac{20n - 16}{n} = 20 - \frac{16}{n}$$

$$\text{Area} = \lim_{n \to +\infty} s_n = \lim_{n \to +\infty} [20 - \frac{16}{n}] = 20$$

23. $f(x) = 4 - x^2$, $x \in [0,2]$; $f' = -2x < 0$ if
$x \in [0,2] \Rightarrow f$ is decreasing; therefore, on $[0,2]$
use x_i for inscribed rectangles

$$\Delta x = \frac{2}{n} ; \quad x_i = 0 + i(\frac{2}{n}) ; \quad f(x_i) = 4 - \frac{4i^2}{n^2} = \frac{4}{n^2}(n^2 - i^2)$$

$$s_n = \sum_{i=1}^{n} f(x_i)\Delta x = \frac{8}{n^3} \sum_{i=1}^{n} (n^2 - i^2) = \frac{8}{n} \sum_{i=1}^{n} 1 - \frac{8}{n^3} \sum_{i=1}^{n} i^2$$

$$= 8 - \frac{8}{n^3} \cdot \frac{n(n+1)(2n+1)}{6} = 8 - \frac{16}{6} - \frac{24}{6n} - \frac{8}{6n^2}$$

$$\text{Area} = \lim_{n \to +\infty} s_n = 8 - \frac{8}{3} = \frac{16}{3}$$

27. $f(x) = 2x + 1$, $x \in [1,3]$; $f' = 2 > 0 \Rightarrow f$ is
increasing on $[1,3]$; therefore, use x_{i-1} for
inscribed rectangles.

$$\Delta x = \frac{3-1}{n} = \frac{2}{n} ; \quad x_{i-1} = 1 + (i-1)\Delta x = 1 + \frac{2(i-1)}{n} ;$$

$$f(x_{i-1}) = 2x_{i-1} + 1 = 2 + \frac{4(i-1)}{n} + 1 = 3 + \frac{4i}{n} - \frac{4}{n} ,$$

$$s_n = \sum_{i=1}^{n} f(x_{i-1})\Delta x = \sum_{i=1}^{n} (3 + \frac{4i}{n} - \frac{4}{n}) \frac{2}{n}$$

$$= \frac{6}{n} \sum_{i=1}^{n} 1 + \frac{8}{n^2} \sum_{i=1}^{n} i - \frac{8}{n^2} \sum_{i=1}^{n} 1$$

$$= \frac{6}{n}(n) + \frac{8}{n^2} \cdot \frac{n(n+1)}{2} - \frac{8}{n^2}(n)$$

$$= 6 - \frac{8}{n} + \frac{4(n+1)}{n}$$

$$\text{Area} = \lim_{n \to +\infty} s_n = 6 - 0 + 4 = 10$$

31. $f(x) = x^2$, $x \in [0,2]$; $f'(x) = 2x > 0 \Rightarrow f$ is increasing on $[0,2]$; therefore, use x_i for circumscribed rectangles.

$$\Delta x = \frac{2}{n} \; ; \; x_i = 0 + i(\frac{2}{n}); \; f(x_i) = (\frac{2i}{n})^2 = \frac{4i^2}{n^2}$$

$$S_n = \sum_{i=1}^{n} f(x_i)\Delta x = \sum_{i=1}^{n} (\frac{4i^2}{n^2})(\frac{2}{n}) = \frac{8}{n^3} \sum_{i=1}^{n} i^2$$

$$= \frac{8}{n^3} \cdot \frac{n(n+1)(2n+1)}{6} = \frac{8}{3} + \frac{4}{n} + \frac{4}{3n^2}$$

$$\text{Area} = \lim_{n \to +\infty} S_n = \frac{8}{3}$$

35. $f(x) = x^3$, $x \in [0,2]$; $f' = 3x^2 > 0 \Rightarrow f$ is increasing on $[0,2]$; therefore, use x_i for circumscribed rectangles.

$$\Delta x = \frac{2}{n} \; ; \; x_i = 0 + i(\frac{2}{n}) \; ; \; f(x_i) = (\frac{2i}{n})^3 = \frac{8i^3}{n^3}$$

$$S_n = \sum_{i=1}^{n} f(x_i)\Delta x = \sum_{i=1}^{n} (\frac{8i^3}{n^3}) \frac{2}{n} = \frac{16}{n^4} \sum_{i=1}^{n} i^3$$

$$= \frac{16}{n^4} \frac{n^2(n+1)^2}{4} = \frac{4(n+1)^2}{n^2} = 4 + \frac{8}{n} + \frac{4}{n^2}$$

Area $= \lim\limits_{n \to +\infty} S_n = 4$

39. $\sum\limits_{i=1}^{n} [(i+1)^2 - i^2] = (2^2-1^2) + (3^2-2^2) + (4^2-3^2) + \cdots$

$$+ (n^2-(n-1)^2) + ((n+1)^2-n^2)$$

$$= (n+1)^2 - 1^2 = n^2 + 2n$$

43.

y

(0, H)

H

(B, 0)

B

x

The slope of the line determined by (0,H) and (B,0) is $m = -\dfrac{H}{B}$ so the equation of the line is $y - 0 = -\dfrac{H}{B}(x-B)$.

Thus, $f(x) = -\dfrac{H}{B}x + H$ for $x \in [0,B]$.

$\Delta x = \dfrac{B}{n}$; $x_i = \dfrac{Bi}{n}$; $f(x_i) = \dfrac{-H}{B}\left(\dfrac{Bi}{n}\right) + H = -\dfrac{Hi}{n} + H$

$s_n = \sum\limits_{i=1}^{n} \left(\dfrac{-Hi}{n} + H\right)\dfrac{B}{n} = \dfrac{-BH}{n^2} \sum\limits_{i=1}^{n} i + \dfrac{BH}{n} \sum\limits_{i=1}^{n} 1$

$= \dfrac{-BH}{n^2} \cdot \dfrac{n(n+1)}{2} + \dfrac{BHn}{n} = \dfrac{-BH}{2}\left(\dfrac{n+1}{n}\right) + BH$

$A = \lim\limits_{n \to +\infty} s_n = \dfrac{-BH}{2} + BH = \dfrac{1}{2}BH$

47. $f(x) = x$, $x \in [a,b]$; $f' = 1 > 0 \Rightarrow f$ is increasing on $[a,b]$; therefore, use x_{i-1} for s_n and x_i for S_n.

$\Delta x = \dfrac{b-a}{n}$; $x_{i-1} = a + (i-1)\left(\dfrac{b-a}{n}\right)$

$$= a + \dfrac{i(b-a)}{n} - \dfrac{(b-a)}{n} ;$$

$x_i = a + i\left(\dfrac{b-a}{n}\right) = a + \dfrac{i(b-a)}{n}$

$f(x_{i-1}) = a + \dfrac{i(b-a)}{n} - \dfrac{(b-a)}{n}$

$$s_n = \sum_{i=1}^{n} [a + \frac{i(b-a)}{n} - \frac{(b-a)}{n}](\frac{b-a}{n})$$

$$= \frac{a(b-a)}{n} \sum_{i=1}^{n} 1 + \frac{(b-a)^2}{n^2} \sum_{i=1}^{n} i - \frac{(b-a)^2}{n^2} \sum_{i=1}^{n} 1$$

$$= a(b-a) + \frac{(b-a)^2}{n^2} \cdot \frac{n(n+1)}{2} - \frac{(b-a)^2}{n}$$

$$= a(b-a) + \frac{(b-a)^2}{2} \cdot \frac{n+1}{n} - \frac{(b-a)^2}{n}$$

$$= (b-a)[a + \frac{b-a}{2}(1 + \frac{1}{n}) - \frac{b-a}{n}]$$

$$= (b-a)[a + \frac{b}{2} - \frac{a}{2} + \frac{b-a}{2n} - \frac{2(b-a)}{2n}]$$

$$= (b-a)[\frac{b+a}{2} - \frac{b-a}{2n}]$$

$$= \frac{b^2-a^2}{2} - \frac{(b-a)^2}{2n}$$

$$f(x_i) = a + \frac{i(b-a)}{n}$$

$$S_n = \sum_{i=1}^{n} [a + \frac{i(b-a)}{n}](\frac{b-a}{n}) = \frac{a(b-a)}{n} \sum_{i=1}^{n} 1 + \frac{(b-a)^2}{n^2} \sum_{i=1}^{n} i$$

$$= a(b-a) + \frac{(b-a)^2}{n^2} \cdot \frac{n(n+1)}{2}$$

$$= a(b-a) + \frac{(b-a)^2}{2} \cdot \frac{n+1}{n}$$

$$= (b-a)[a + \frac{b-a}{2}(1 + \frac{1}{n})]$$

$$= (b-a)[a + \frac{b}{2} - \frac{a}{2} + \frac{b-a}{2n}]$$

$$S_n = (b-a)[\frac{b+a}{2} + \frac{b-a}{2n}] = \frac{b^2-a^2}{2} + \frac{(b-a)^2}{2n}$$

Thus, $s_n < \dfrac{b^2-a^2}{2} < S_n$.

3. $\int_0^{-4} (2x^2)\,dx = -\int_{-4}^0 (2x^2)\,dx$ Let $\Delta x = \dfrac{0 - (-4)}{n} = \dfrac{4}{n}$;

   $u_i = x_i = \dfrac{4i}{n}$; $f(x_i) = 2\left(\dfrac{4i}{n}\right)^2 = \dfrac{32i^2}{n^2}$

   $R_n = \sum_{i=1}^n \left(\dfrac{32i^2}{n^2}\right)\left(\dfrac{4}{n}\right) = \dfrac{128}{n^3} \sum_{i=1}^n i^2 = \dfrac{128}{n^3} \cdot \dfrac{n(n+1)(2n+1)}{6}$

   $= \dfrac{64(n+1)(2n+1)}{3n^2}$ and $\lim_{n \to +\infty} R_n = \dfrac{128}{3}$

   $\int_0^{-4} (2x^2)\,dx = -\dfrac{128}{3}$

7. $f(x) = x$, $[0,2]$; $\Delta x_i = \dfrac{1}{4}$; $f(u_i) = u_i$

   $R_8 = \dfrac{1}{4}\left(\dfrac{1}{8} + \dfrac{3}{8} + \dfrac{5}{8} + \dfrac{7}{8} + \dfrac{9}{8} + \dfrac{11}{8} + \dfrac{13}{8} + \dfrac{15}{8}\right) = 2$

11. $f(x) = \sqrt{x}$, $[0,1]$; $\Delta x = \dfrac{1}{n}$; $x_i = \dfrac{i}{n}$

   $f(x_i) = \sqrt{\dfrac{i}{n}}$ $R_n = \sum_{i=1}^n \sqrt{\dfrac{i}{n}}\left(\dfrac{1}{n}\right) = \dfrac{1}{n^{3/2}} \sum_{i=1}^n \sqrt{i}$

15. $f(x) = \dfrac{2}{x^2}$, $[1,4]$; $\Delta x = \dfrac{3}{n}$; $x_i = 1 + \dfrac{3i}{n}$

   $f(x_i) = 2 / \left(\dfrac{n+3i}{n}\right)^2 = \dfrac{2n^2}{(n+3i)^2}$

   $R_n = \sum_{i=1}^n \dfrac{2n^2}{(n+3i)^2}\left(\dfrac{3}{n}\right) = 6n \sum_{i=1}^n \dfrac{1}{(n+3i)^2}$

19. $\int_a^b k\,dx$ Then $f(x) = k$ on $[a,b]$. $x_0 = a$, $x_n = b$,

   $\Delta x_i = \Delta x = \dfrac{b - a}{n}$.   For any $u_i \in [x_{i-1}, x_i]$, $f(u_i) = k$

   Hence, $R_n = \sum_{i=1}^n f(u_i)\,\Delta x_i = \sum_{i=1}^n k\,\dfrac{(b-a)}{n} = \dfrac{k(b-a)}{n} \sum_{i=1}^n 1$

   $= \dfrac{k(b-a)}{n} \cdot n = k(b-a)$.

   Thus, $\int_a^b k\,dx = \lim_{n \to +\infty} R_n = k(b-a)$.

<u>Exercise 4</u>, pp. 305-306

3. $\int_0^1 3x^2 dx = x^3 \Big|_0^1 = 1$

7. $\int_0^{\pi/3} \sin x \, dx = -\cos x \Big|_0^{\pi/3}$

$= -\cos \pi/3 + \cos 0 = -\frac{1}{2} + 1 = \frac{1}{2}$

11. $\int_{-2}^3 (x-1)(x+3) dx = \int_{-2}^3 (x^2+2x-3) dx$

$= \frac{x^3}{3} + x^2 - 3x \Big|_{-2}^3 = (9+9-9) - (-\frac{8}{3} + 4 + 6)$

$= 9 - 10 + \frac{8}{3} = \frac{8}{3} - \frac{3}{3} = \frac{5}{3}$

15. $\int_0^1 (t^{2/5}+1) dt = \frac{5}{7} t^{7/5} + t \Big|_0^1 = \frac{5}{7} + 1 = \frac{12}{7}$

19. $\int_0^1 (ax^4+b) dx = \frac{a}{5} x^5 + bx \Big|_0^1 = \frac{a + 5b}{5}$

23. $A = \int_{-2}^2 (x^2+9) dx = \int_{-2}^2 (x +9) dx$

$= [\frac{x^3}{3} + 9x] \Big|_{-2}^2$

$= [\frac{8}{3} + 18] - [\frac{-8}{3} - 18]$

$= 36 + \frac{16}{3} = \frac{124}{3}$

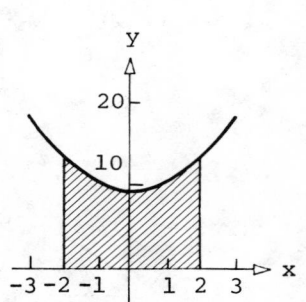

27. $\frac{d}{dt}[\int_0^t (3+x^2)^{3/2} dx] = (3+t^2)^{3/2}$

31. $\frac{d}{dx}[\int_1^{2x^3} \sqrt{t^2 + 1} \, dt] = \sqrt{4x^6 + 1} \, (6x^2)$

35. $\int_0^3 6x^2 dx = f(u)(3-0); \ u \in [0,3]$

$\int_0^3 6x^2 dx = 2x^3 \Big|_0^3 = 54, \quad \text{and} \quad f(x) = 6x^2$

$\Rightarrow 3f(u) = 18u^2$

Thus, $18u^2 = 54 \Rightarrow u^2 = 3 \Rightarrow u = \sqrt{3} \in [0,3]$

112   CHAPTER FIVE

39. $\int_1^3 (5x+1)\,dx$; $1 \leqslant x \leqslant 3 \Rightarrow 5 \leqslant 5x \leqslant 15$

$\Rightarrow 6 \leqslant 5x + 1 \leqslant 16$; $b - a = 2$

$12 \leqslant \int_1^3 (5x+1)\,dx \leqslant 32$

43. $\int_0^1 \sqrt{1 + x^2}\,dx$; $0 \leqslant x \leqslant 1 \Rightarrow 1 \leqslant \sqrt{1 + x^2} \leqslant \sqrt{2}$;

$b - a = 1$

$1 \leqslant \int_0^1 \sqrt{1 + x^2} \leqslant \sqrt{2}$

47, For f' continuous on [a, b], we also have f
continuous and differentiable.

Since $\frac{d}{dx}\left(\frac{1}{2}[f(x)]^2\right) = \frac{1}{2} \cdot 2 \cdot f(x) \cdot f'(x) = f(x)f'(x)$,

we can apply Theorem (5. 36):

$\int_a^b f(x)f'(x)\,dx = \frac{1}{2}[f(x)]^2 \Big|_a^b = \frac{1}{2}[f(b)]^2 - \frac{1}{2}[f(a)]^2$

Exercise 5, pp. 308-309

3. $\int_0^4 f(x)\,dx - \int_6^4 f(x)\,dx = \int_0^4 f(x)\,dx + \int_4^6 f(x)\,dx = \int_0^6 f(x)\,dx$

Property (5.39) and Definition (5.22)

7. $\int_1^3 [5f(x) - 3g(x)]\,dx = 5\int_1^3 f(x)\,dx - 3\int_1^3 g(x)\,dx$

$= 5(5) - 3(-2) = 25 + 6 = 31$

11. If $x \in [0,1]$, then $x \geqslant x^3$. Hence, $\int_0^1 x\,dx \geqslant \int_0^1 x^3\,dx$
(using Property 5.44).

15. If f and g are continuous on [a,b] and $f \leqslant g$
on [a,b], then

$\int_a^b f(x)\,dx \leqslant \int_a^b g(x)\,dx$

Proof: $f \leqslant g$ on [a,b] $\Rightarrow h(x) = g(x) - f(x) \geqslant 0$

on [a,b]. Hence, $\int_a^b h(x)\,dx \geqslant 0 \Rightarrow \int_a^b [g(x) - f(x)]\,dx$

≥ 0, by (5.43); thus, $\int_a^b g(x)dx - \int_a^b f(x) \geq 0$, so

that $\int_a^b g(x)dx \geq \int_a^b f(x)dx$ or $\int_a^b f(x)dx \leq \int_a^b g(x)dx$.

Exercise 6, pp. 314-316

3. $\int 3xdx = \frac{3}{2} x^2 + C$ 7. $\int(x^2+2)dx = \frac{1}{3} x^3 + 2x + C$

11. $\int(4x^3-3x^2+5x-2)dx = x^4 - x^3 + \frac{5}{2} x^2 - 2x + C$

15. $\int(x-x^{-2})dx = \frac{1}{2} x^2 + \frac{1}{x} + C$

19. $\int \frac{3x^5 + 1}{x^2} dx = \int(3x^3+x^{-2})dx = \frac{3}{4} x^4 - \frac{1}{x} + C$

23. Let $u = 2x + 1$, then $du = 2dx$ so $\frac{1}{2} du = dx$.

$\int(2x+1)^5 dx = \int u^5 \cdot \frac{1}{2} du = \frac{1}{2} \cdot \frac{u^6}{6} + C = \frac{1}{12}(2x+1)^6 + C$

27. Let $u = 1 + x^2$, then $du = 2xdx$ so $\frac{1}{2} du = xdx$.

$\int \frac{x}{\sqrt{1 + x^2}} dx = \frac{1}{2} \int \frac{du}{u^{1/2}} = \frac{1}{2} \int u^{-1/2} du = \frac{1}{2} \cdot \frac{u^{1/2}}{1/2} + C$

$= u^{1/2} + C = \sqrt{1 + x^2} + C$

31. Let $u = 3x$, then $du = 3dx$ so $\frac{1}{3} du = dx$.

$\int \sin 3x\, dx = \int \sin u(\frac{1}{3} du) = \frac{1}{3} \int \sin u\, du$

$= -1/3 \cos u + C = -\frac{1}{3} \cos 3x + C$

35. Let $u = x + 1$, then $x = u - 1$ so $dx = du$.

$$\int x^2 \sqrt{x + 1} \, dx = \int (u-1)^2 \sqrt{u} \, du = \int (u^{5/2} - 2u^{3/2} + u^{1/2}) \, du$$

$$= \frac{2}{7} u^{7/2} - \frac{4}{5} u^{5/2} + \frac{2}{3} u^{3/2} + C$$

$$= \frac{2}{7}(x+1)^{7/2} - \frac{4}{5}(x+1)^{5/2} + \frac{2}{3}(x+1)^{3/2} + C$$

39. Let $u = s - 5$ so $du = ds$.

$$\int (s-5)^{1/2} \, ds = \int u^{1/2} \, du = \frac{2}{3} u^{3/2} + C = \frac{2}{3}(s-5)^{3/2} + C$$

43. Let $u = x + 1$, then $x = u - 1$ so $dx = du$.

$$\int (x-5)\sqrt{x + 1} \, dx = \int (u-6)u^{1/2} \, du = \int (u^{3/2} - 6u^{1/2}) \, du$$

$$= \frac{2}{5} u^{5/2} - (6)\frac{2}{3} u^{3/2} + C$$

$$= \frac{2}{5}(x+1)^{5/2} - 4(x+1)^{3/2} + C$$

47. Let $u = x^2 + 3$ so $du = 2xdx$: $\frac{1}{2} du = xdx$;
$x = 0 \Rightarrow u = 3$; $x = -2 \Rightarrow u = 7$

$$\int_{-2}^{0} \frac{x \, dx}{(x^2+3)^2} = \frac{1}{2} \int_{7}^{3} \frac{1}{u^2} \, du = \frac{1}{2}(\frac{-1}{u})\Big|_{7}^{3} = \frac{1}{2}[-\frac{1}{3} + \frac{1}{7}]$$

$$= \frac{-7 + 3}{42} = \frac{-2}{21}$$

51. Let $u = x + 3$, then $x = u - 3$ so $dx = du$;
$x = 6 \Rightarrow u = 9$; $x = 1 \Rightarrow u = 4$.

$$\int_{6}^{1} x\sqrt{x+3} \, dx = \int_{9}^{4} (u-3)\sqrt{u} \, du = \int_{9}^{4} (u^{3/2} - 3u^{1/2}) \, du$$

$$= (\frac{2}{5} u^{5/2} - 2u^{3/2})\Big|_{9}^{4}$$

$$= [\frac{2}{5}(2^5) - 2(2^3)] - [\frac{2}{5}(3^5) - 2(3^3)]$$

$$= \frac{64}{5} - 16 - \frac{486}{5} + 54 = 38 - \frac{422}{5} = \frac{-232}{5}$$

55. Let $u = x + 1$, then $x = u - 1$ so $dx = du$;
$x = -1 \Rightarrow u = 0$; $x = 7 \Rightarrow u = 8$.

$$\int_{-1}^{7} x(x+1)^{1/3} dx = \int_{0}^{8} (u-1)u^{1/3} du = \int_{0}^{8} (u^{4/3} - u^{1/3}) du$$

$$= (\frac{3}{7} u^{7/3} - \frac{3}{4} u^{4/3})\Big|_{0}^{8} = \frac{3(2^7)}{7} - \frac{3(2^4)}{4}$$

$$= 3(16)[\frac{8}{7} - \frac{1}{4}]$$

$$= \frac{48}{1}[\frac{32-7}{28}] = \frac{(12)(25)}{7} = \frac{300}{7}$$

59. Area $= \int_{0}^{\pi/2} \sin x \, dx = -\cos x \Big|_{0}^{\pi/2} = 0 + 1 = 1$

63. Let $u = x - 1$, then $du = dx$; $x = 1 \Rightarrow u = 0$;
$x = 2 \Rightarrow u = 1$.

$$A = \int_{1}^{2} \sqrt{x-1} \, dx = \int_{0}^{1} u^{1/2} du = \frac{2}{3} u^{3/2} \Big|_{0}^{1} = \frac{2}{3}$$

67. $\int x\sqrt{x} \, dx = \int x^{3/2} dx = \frac{2}{5} x^{5/2} + C$

$$(\int x \, dx)(\int x^{1/2} dx) = (\frac{x^2}{2} + C_2)(\frac{2}{3} x^{3/2} + C_3)$$

$$= \frac{1}{3} x^{7/2} + \text{other terms}$$

Thus, $\int x\sqrt{x} \, dx \neq (\int x \, dx)(\int x^{1/2} dx)$,

since the degrees are not equal.

71. $\int_{-1}^{1} f(x) dx = \int_{-1}^{0} (x+1) dx + \int_{0}^{1} \cos \pi x \, dx$

$$= (\frac{x^2}{2} + x)\Big|_{-1}^{0} + \frac{1}{\pi} \sin \pi x \Big|_{0}^{1} = 0 - (\frac{1}{2} - 1) + [0-0] = \frac{1}{2}$$

75. Let $\int_0^2 f(x-3)dx = 8$. Using the substitution
$w = x - 3$ with $dw = dx$ and $x = 0 \Rightarrow w = -3$, $x = 2$
$\Rightarrow w = -1$, we have $\int_0^2 f(x-3)dx = \int_{-3}^{-1} f(w)dw$
and hence $\int_{-3}^{-1} f(w)dw = 8$.

Historical Exercises, pp. 317-320

3. (a) The area of $\triangle ABE = \frac{1}{2}(t_0)(kt_0) = \frac{1}{2}kt_0^2$, since
kt_0 is the height of the triangle. The area of
rectangle $ABCD = t_0 \cdot v(\frac{t_0}{2}) = t_0(k\frac{t_0}{2}) = \frac{1}{2}kt_0^2$.

   (b) Area of $\triangle ABE = \int_0^{t_0}(kt)dt = k\frac{t^2}{2}\Big|_0^{t_0} = \frac{1}{2}kt_0^2$.

7. (a) We have Area $R_1 = ab$, Area $R_2 = AB$.
If the rectangles are similar, then
we have $\frac{a}{A} = \frac{b}{B}$, so that $a = \frac{Ab}{B}$

and $b = \frac{aB}{A}$. Thus,

$$\frac{\text{Area } R_1}{\text{Area } R_2} = \frac{ab}{AB} = \frac{a(\frac{aB}{A})}{AB} = \frac{a^2B}{A^2B} = \frac{a^2}{A^2}$$

and

$$\frac{\text{Area } R_1}{\text{Area } R_2} = \frac{ab}{AB} = \frac{(\frac{Ab}{B})b}{AB} = \frac{Ab^2}{AB^2} = \frac{b^2}{B^2}.$$

This gives the conclusion.

(b) If the triangles are similar,

then $\frac{b}{h} = \frac{B}{H}$, which gives the

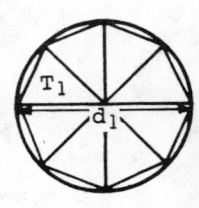

statements $h = \frac{bH}{B}$ and $b = \frac{Bh}{H}$.

Also, Area $T_1 = \frac{1}{2} bh$ and

Area $T_2 = \frac{1}{2} BH$. Hence, we have

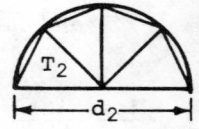

$$\frac{\text{Area } T_1}{\text{Area } T_2} = \frac{\frac{1}{2} bh}{\frac{1}{2} BH} = \frac{b(\frac{bH}{B})}{BH} = \frac{b^2 H}{B^2 H} = \frac{b^2}{B^2}$$

and

$$\frac{\text{Area } T_1}{\text{Area } T_2} = \frac{\frac{1}{2} bh}{\frac{1}{2} BH} = \frac{(\frac{Bh}{H})h}{BH} = \frac{Bh^2}{BH^2} = \frac{h^2}{H^2}$$. This gives the

conclusion.

(c) Let B and b be the lengths of the sides of the polygons in the circles of diameter d_1 and d_2, respectively. Then $P_1 = n(\text{Area } T_1)$ and $P_2 = n(\text{Area } T_2)$, where n is the number of sides of the regular polygons, and T_1 and T_2 are the triangles indicated. T_1 and T_2 are similar

triangles so that $\frac{H}{h} = \frac{B}{b}$, or $H = \frac{Bh}{b}$. Then

$$\frac{d_1^2}{d_2^2} = \frac{(2r_1)^2}{(2r_2)^2} = \frac{r_1^2}{r_2^2} = \frac{H^2 + \frac{1}{4}B^2}{h^2 + \frac{1}{4}b^2} = \frac{4H^2 + B^2}{4h^2 + b^2} = \frac{4(\frac{B^2 h^2}{b^2}) + B^2}{4h^2 + b^2}$$

$$= \frac{B^2(4h^2 + b^2)}{b^2(4h^2 + b^2)} = \frac{B^2}{b^2} = \frac{\text{Area } T_1}{\text{Area } T_2}$$

$$= \frac{n(\text{Area } T_1)}{n(\text{Area } T_2)} = \frac{P_1}{P_2}, \quad \text{using (b) above.}$$

3. Let $u = 2\sqrt{x^2 + 3} - \dfrac{4}{x} + 9$, then

$du = (\dfrac{2x}{\sqrt{x^2 + 3}} + \dfrac{4}{x^2})dx$ or $\dfrac{1}{2} du = (\dfrac{x}{\sqrt{x^2 + 3}} + \dfrac{2}{x^2})dx$.

$\int(2\sqrt{x^2 + 3} - \dfrac{4}{x} + 9)^6 (\dfrac{x}{\sqrt{x^2 + 3}} + \dfrac{2}{x^2})dx$

$= \dfrac{1}{2} \int u^6 du = (\dfrac{1}{2}) \dfrac{u^7}{7} + C = \dfrac{1}{14}(2\sqrt{x^2 + 3} - \dfrac{4}{x} + 9)^7 + C$

7. Let $u = 5t^3 - 1$, $du = 15t^2 dt$, $\dfrac{1}{15} du = t^2 dt$,

$t = 1 \Rightarrow u = 4$, $t = b \Rightarrow u = 5b^3 - 1$. Then

$\int_1^b t^2(5t^3-1)^{1/2}dt = \dfrac{1}{15} \int_4^{5b^3-1} u^{1/2} du = \dfrac{2}{45} u^{3/2}\Big|_4^{5b^3-1} = \dfrac{38}{45}$

$(5b^3-1)^{3/2} - (4)^{3/2} = 19 \Rightarrow (5b^3-1)^{3/2} = 19 + 8 = 27$

Therefore, $(5b^3-1) = (27)^{2/3} = 9 \Rightarrow 5b^3 = 10 \Rightarrow b^3$

$= 2 \Rightarrow b = \sqrt[3]{2}$

11. Let $u = 1 - \dfrac{1}{t^3} = 1 - t^{-3}$, $du = 3t^{-4}dt$, $\dfrac{1}{3} du = \dfrac{1}{t^4} dt$

$t = 1 \Rightarrow u = 1 - 1 = 0$; $t = \sqrt[3]{\dfrac{1}{2}} \Rightarrow t^3 = \dfrac{1}{2} \Rightarrow u = -1$.

$\int_1^{\sqrt[3]{\frac{1}{2}}} \dfrac{[1 - (\frac{1}{t^3})]^3}{t^4} dt = \int_1^{\sqrt[3]{\frac{1}{2}}}[1 - (\dfrac{1}{t^3})]^3(\dfrac{1}{t^4} dt)$

$= \dfrac{1}{3} \int_0^{-1} u^3 du = \dfrac{1}{12} u^4\Big|_0^{-1} = \dfrac{1}{12}$

15. Because $\displaystyle\lim_{\|P\|\to 0}\sum_{i=1}^{n}\cos(u_i)\Delta x_i = \int_{0}^{\pi/2}\cos x\,dx$

$$= \sin x\Big|_{0}^{\pi/2} = 1.$$

19. Find n such that $\displaystyle\int_{0}^{1}x^n dx = \int_{0}^{1}(1-x)^n dx$ if $n \geqslant 0$.

$$\int_{0}^{1}x^n dx = \frac{x^{n+1}}{n+1}\Big|_{0}^{1} = \frac{1}{n+1}$$

If $u = 1 - x$, then $du = -dx$ so that

$$\int_{0}^{1}(1-x)^n dx = -\frac{(1-x)^{n+1}}{n+1}\Big|_{0}^{1} = -0 + \frac{1}{n+1} = \frac{1}{n+1}$$

Thus, the given integrals are equal for all $n = 0,1,2,\cdots$.

23. $\displaystyle\int_{0}^{3}|x-1|\,dx = \int_{0}^{1}(1-x)\,dx + \int_{1}^{3}(x-1)\,dx$

$$= (x-\frac{x^2}{2})\Big|_{0}^{1} + (\frac{x^2}{2}-x)\Big|_{1}^{3}$$

$$= (1-\frac{1}{2}) + (\frac{9}{2}-3) - (\frac{1}{2}-1)$$

$$= \frac{1}{2} + \frac{3}{2} + \frac{1}{2} = \frac{5}{2}$$

27. $\displaystyle\frac{d}{dx}\int_{x}^{1}(t-1)^2 dt = -\frac{d}{dx}\int_{1}^{x}(t-1)^2 dt = -(x-1)^2$

Also, note that $\displaystyle\int_{x}^{1}(t-1)^2 dt = \frac{(t-1)^3}{3}\Big|_{x}^{1}$

$$= 0 - \frac{(x-1)^3}{3} = -\frac{1}{3}(x-1)^3.$$

Then $\displaystyle\frac{d}{dx}(\int_{x}^{1}(t-1)^2 dt) = \frac{d}{dx}(\frac{-1}{3}(x-1)^3) = -(x-1)^2.$

$$\implies \int_{a+2c}^{b+2c} f(u)\,du$$

Thus, integrals I and II have the same value.

39. $\int_{-4}^{-1} (2x^3+9x^2+12x+32)\,dx$

Let $f(x) = 2x^3 + 9x^2 + 12x + 32$,

then $f'(x) = 6x^2 + 18x + 12 = 6(x+2)(x+1)$

$\implies$ CN at $x = -1,-2$.

$f(-4) = 0$; $f(-1) = 27$; $f(-2) = 28$

Thus, by (4.23), Chapter 4, Section 3, on the internal $[-4,-1]$, the maximum value for f is 28 and the minimum value for f is 0. Now, by Theorem (5.26) with $b - a = -1 -(-4) = 3$:

$$0 \le \int_{-1}^{-4} (2x^3+9x^2+12x+32)\,dx \le 3.28 = 84$$

43. The average value of a function g on $[a,b]$ is

$\dfrac{1}{b - a} \int_a^b g(x)\,dx$, so the average slope is the average

of f' on $[a,b]$. This is

$$\frac{1}{b - a} \int_a^b f'(x)\,dx = \frac{1}{b - a}[f(x)]\Big|_a^b = \frac{f(b) - f(a)}{b - a}.$$

Geometrically, this is the slope of the secant line joining $(a,f(a))$ and $(b,f(b))$.

47. If $f(x) = \begin{cases} 1 & \text{if } 0 \le x \le 2 \\ -1 & \text{if } 2 < x \le 4 \end{cases}$, then

$$\int_0^4 f(x)\,dx = \int_0^2 (+1)\,dx + \int_2^4 (-1)\,dx = x\Big|_0^2 - x\Big|_2^4$$
$$= 2 - 2 = 0$$

51. (a) $R - \int_a^b f(x)\,dx = R - \sum_{i=1}^n \int_{x_{i-1}}^{x_i} f(x)\,dx$, (extend (5.39))

$$\int_a^b f\,dx = \int_a^{x_1} f\,dx + \int_{x_1}^{x_2} f\,dx + \cdots + \int_{x_{n-1}}^b f\,dx = \sum_{i=1}^n \int_{x_{i-1}}^{x_i} f\,dx$$

(b) By the mean value theorem for integrals, there is some $t_i \in [x_{i-1},x_i]$ such that

$$f(t_i)\,\Delta x_i = \int_{x_{i-1}}^{x_i} f(x)\,dx, \text{ where } \Delta x_i = x_i - x_{i-1}.$$

(c) Since $R = \sum\limits_{i=1}^{n} f(t_i)\Delta x_i = \sum\limits_{i=1}^{n} f(u_i)\Delta x_i - \sum\limits_{i=1}^{n} f(t_i)\Delta x_i$

$= \sum\limits_{i=1}^{n} (f(u_i) - f(t_i)]\Delta x_i$, we apply the mean value

theorem on f on the interval from u_i to t_i
to conclude there is some c_i between u_i and
t_i such that
$$f'(c_i) = \frac{f(u_i) - f(t_i)}{u_i - t_i} , \quad \text{or}$$

$$f'(c_i)(u_i - t_i) = f(u_i) - f(t_i).$$

Thus, $R - \sum\limits_{i=1}^{n} f(t_i)\Delta x_i = \sum\limits_{i=1}^{n} f'(c_i)(u_i - t_i)\Delta x_i$

(d) Then
$$\left| R - \sum\limits_{i=1}^{n} f(t_i)\Delta x_i \right| = \left| \sum\limits_{i=1}^{n} f'(c_i)(u_i - t_i)\Delta x_i \right|$$

$$\leqslant \sum\limits_{i=1}^{n} |f'(c_i)| |u_i - t_i| |\Delta x_i|, \quad \text{since}$$

$|\Sigma a_i| \leqslant \Sigma |a_i|$ by the triangle inequality.

(e) Since $|f'(x)| \leqslant M$ for all $x \in [a,b]$, then
for each $i = 1,2,\cdots,n$ we have $|f'(c_i)| \leqslant M$.
Also, $|u_i - t_i| < |x_i - x_{i-1}| = \Delta x_i$, since
u_i and t_i are in (x_{i-1}, x_i). Hence,
$|f'(c_i)| |u_i - t_i| \leqslant M\Delta x_i$, which gives
$$\sum\limits_{i=1}^{n} |f'(c_i)| |u_i - t_i| \Delta x_i \leqslant M \sum\limits_{i=1}^{n} (\Delta x_i)^2.$$

(f) However, $\Delta x_i = x_i - x_{i-1} = \dfrac{b - a}{n}$ for each i,

so that $\sum\limits_{i=1}^{n} (\Delta x_i)^2 = n(\dfrac{b-a}{n})^2$. Thus,

$$M \sum\limits_{i=1}^{n} (\Delta x_i)^2 = Mn(\dfrac{b-a}{n})^2 = \dfrac{M}{n}(b-a)^2.$$

Chapter Six

Exercise 1, pp. 331-332

3. If $f(x) = x^2$, $g(x) = x$, then $x^2 = x \Rightarrow x(x-1) = 0$
   $\Rightarrow x = 0, x = 1$.

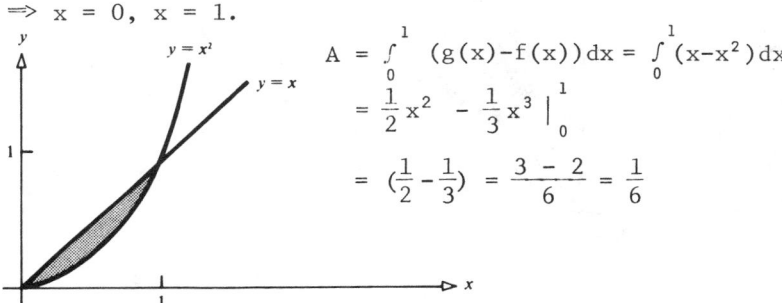

$$A = \int_0^1 (g(x)-f(x))dx = \int_0^1 (x-x^2)dx$$

$$= \frac{1}{2}x^2 - \frac{1}{3}x^3 \Big|_0^1$$

$$= (\frac{1}{2} - \frac{1}{3}) = \frac{3-2}{6} = \frac{1}{6}$$

7. If $f(x) = \sqrt{x}$, $g(x) = x^3$, then $x^3 = \sqrt{x} \Rightarrow x^6 - x = 0$
   $\Rightarrow x(x^5-1) = 0 \Rightarrow x = 0, 1$.

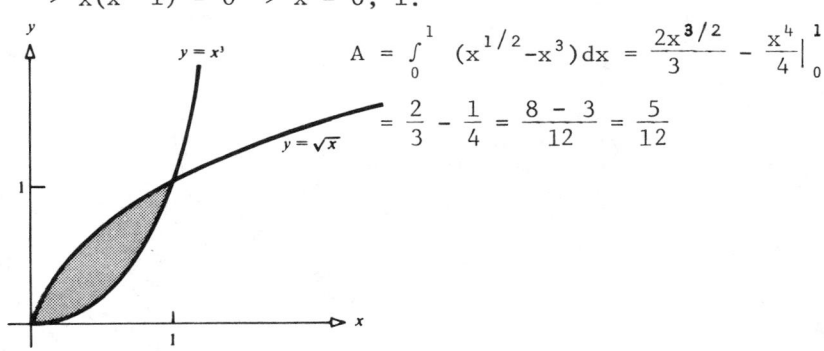

$$A = \int_0^1 (x^{1/2}-x^3)dx = \frac{2x^{3/2}}{3} - \frac{x^4}{4} \Big|_0^1$$

$$= \frac{2}{3} - \frac{1}{4} = \frac{8-3}{12} = \frac{5}{12}$$

11. If $f(x) = x^2 - 4x$, $g(x) = -x^2$, then $x^2 - 4x = -x^2$
    $\Rightarrow 2x^2 - 4x = 0 \Rightarrow 2x(x-2) = 0 \Rightarrow x = 0, 2$.

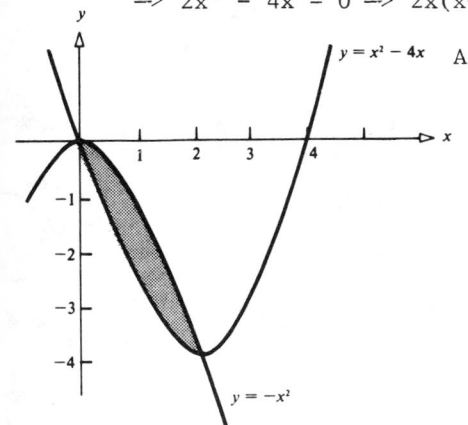

$$A = \int_0^2 (g(x)-f(x))dx$$

$$= \int_0^2 [-x^2-(x^2-4x)]dx$$

$$= \int_0^2 (4x-2x^2)dx$$

$$= 2x^2 - \frac{2x^3}{3} \Big|_0^2 = 8 - \frac{16}{3}$$

$$= \frac{24-16}{3} = \frac{8}{3}$$

15. If $f(x) = x^3$, $g(x) = 4x$ then $x^3 = 4x \Rightarrow x^3 - 4x = 0$

$\Rightarrow x(x^2-4) = 0 \Rightarrow x = 0, -2, 2$.

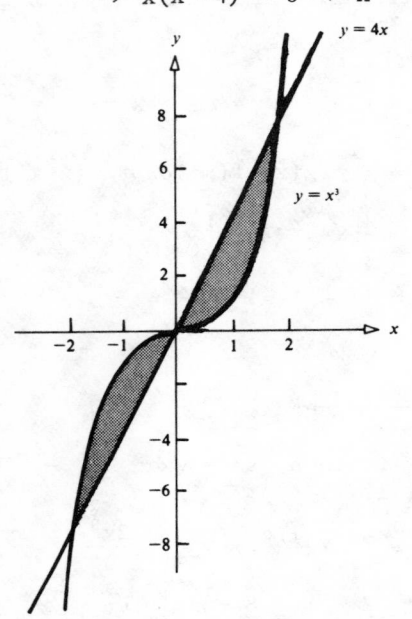

$A = \int_{-2}^{0} (x^3-4x)\,dx + \int_{0}^{2} (4x-x^3)\,dx$

$= 2 \int_{0}^{2} (4x-x^3) = 2(2x^2- \frac{x^4}{4})\Big|_{0}^{2}$

$= 2(8-4) = 2(4) = 8$

19. If $y = \sqrt{9 - x}$, then $y^2 = 9 - x$ and $f(y) = x = 9 - y^2$; if $y = \sqrt{9 - 3x}$, then $y^2 = 9 - 3x$ and $g(y) = x = \frac{9 - y^2}{3}$.

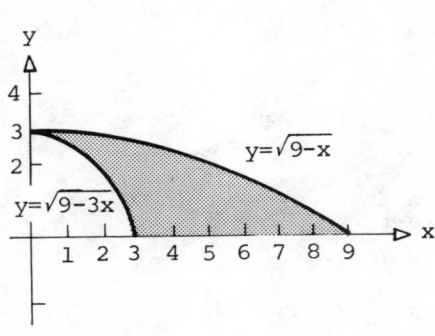

$A = \int_{0}^{3} (f-g)\,dy$

$= \int_{0}^{3} [(9-y^2) - (\frac{9 - y^2}{3})]\,dy$

$= \frac{2}{3} \int_{0}^{3} (9-y^2)\,dy = \frac{2}{3}[9y-\frac{y^3}{3}]\Big|_{0}^{3}$

$= \frac{2}{3}[27-9] = \frac{2}{3}(18) = 12$

23. The region bounded by $y = \cos x$, $y = 1$, $x = \pi/6$ is shown in the figure.

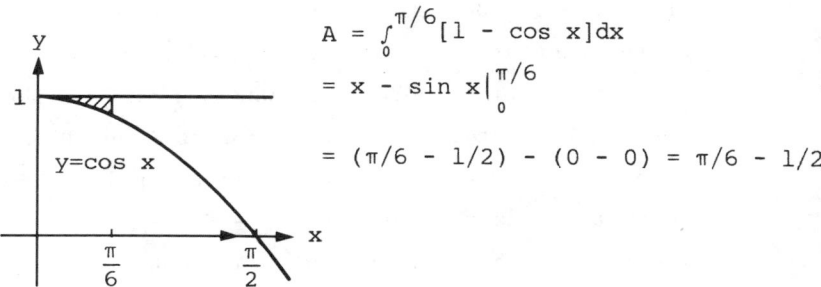

$$A = \int_0^{\pi/6}[1 - \cos x]dx$$

$$= x - \sin x\Big|_0^{\pi/6}$$

$$= (\pi/6 - 1/2) - (0 - 0) = \pi/6 - 1/2$$

27. If $y^2 = x$, $x + y = 2$, then $y^2 = 2 - y$
$\Rightarrow$ $y^2 + y - 2 = 0 \Rightarrow (y+2)(y-1) = 0 \Rightarrow y = -2, 1.$

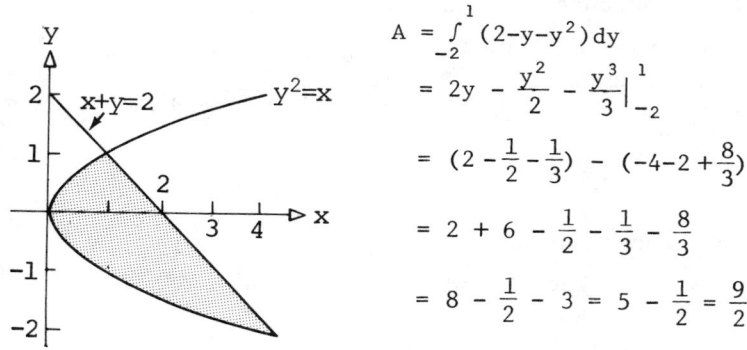

$$A = \int_{-2}^{1}(2-y-y^2)dy$$

$$= 2y - \frac{y^2}{2} - \frac{y^3}{3}\Big|_{-2}^{1}$$

$$= (2 - \frac{1}{2} - \frac{1}{3}) - (-4 - 2 + \frac{8}{3})$$

$$= 2 + 6 - \frac{1}{2} - \frac{1}{3} - \frac{8}{3}$$

$$= 8 - \frac{1}{2} - 3 = 5 - \frac{1}{2} = \frac{9}{2}$$

31. The region bounded by $x^{1/2} + y^{1/2} = 1$, $x \geqslant 0$, $y \geqslant 0$, the x-axis and the y-axis are shown in the figure. We have $\sqrt{y} = 1 - \sqrt{x} \Rightarrow y = (1-\sqrt{x})^2$, so that

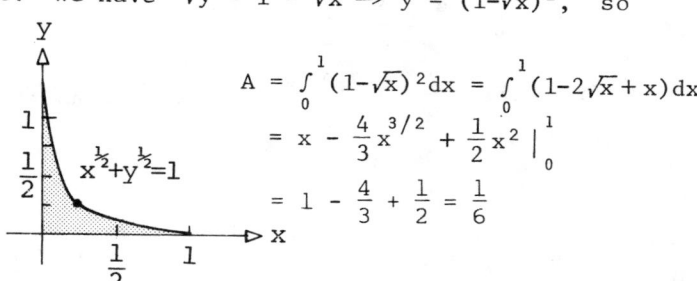

$$A = \int_0^1 (1-\sqrt{x})^2 dx = \int_0^1 (1 - 2\sqrt{x} + x)dx$$

$$= x - \frac{4}{3}x^{3/2} + \frac{1}{2}x^2 \Big|_0^1$$

$$= 1 - \frac{4}{3} + \frac{1}{2} = \frac{1}{6}$$

35. The slope of $\overline{BC}$ equals $m = \dfrac{4 - 1}{-2 - 1} = -1$. The slope of the tangent line to $f(x) = x^2$ at $x = a$ is $f'(a) = 2a$. Then, $2a = -1$ when $a = -\dfrac{1}{2}$. The desired tangent line (through points A and D) has slope $m = -1$ and passes through the point $(-\dfrac{1}{2}, \dfrac{1}{4})$, so that it has the equation $y - \dfrac{1}{4} = -1(x + \dfrac{1}{2}) \Rightarrow 4x + 4y = -1$. The coordinates of A and D are not necessary to find the areas required.

The area of the parallelogram ABCD is found by $A = bh$ where $b = d(B,C) = \sqrt{(1+2)^2 + (1-4)^2} = \sqrt{18} = 3\sqrt{2}$, and h is the distance from C to the point of intersection of $4x + 4y = -1$ and the line through C perpendicular to $4x + 4y = -1$, which is $y = x + 6$. This point is $(\dfrac{-25}{8}, \dfrac{23}{8})$, so that

$$h = \sqrt{(\dfrac{-25}{8} + 2)^2 + (\dfrac{23}{8} - 4)^2} = \sqrt{\dfrac{81}{64} + \dfrac{81}{64}} = \dfrac{9}{8}\sqrt{2}.$$

Thus, $A = 3\sqrt{2}(\dfrac{9}{8}\sqrt{2}) = \dfrac{27}{4}$. The equation of the line through B and C is $y - 1 = -1(x-1)$, or $y = -x + 2$. The shaded area is given by

$$A_s = \int_{-2}^{1} [(-x+2) - x^2]\,dx = \int_{-2}^{1} (2-x-x^2)\,dx$$

$$= 2x - \dfrac{1}{2}x^2 - \dfrac{1}{3}x^3 \Big|_{-2}^{2} = (2 - \dfrac{1}{2} - \dfrac{1}{3}) - (-4 - 2 + \dfrac{8}{3})$$

$$= 8 - \dfrac{1}{2} - \dfrac{9}{3} = \dfrac{9}{2}.$$

We note that $\dfrac{2}{3}A = \dfrac{2}{3}(\dfrac{27}{4}) = \dfrac{9}{2} = A_s$, as we wished to show.

3. $y = x^3$ for $0 \leqslant x \leqslant 2$

$$V = \pi \int_0^2 f^2(x)\,dx$$

$$= \pi \int_0^2 x^6\,dx$$

$$= \frac{\pi}{7} x^7 \Big|_0^2 = \frac{128\pi}{7}$$

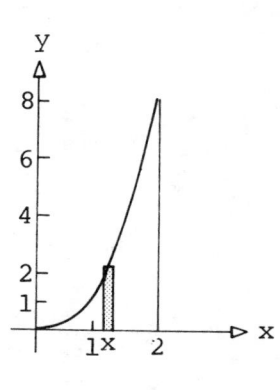

Region

Solid of
Revolution

7. $y = \dfrac{1}{x}$ for $1 \leqslant x \leqslant 2$

$$V = \pi \int_1^2 f^2(x)\,dx$$

$$= \pi \int_1^2 \frac{1}{x^2}\,dx$$

$$= \pi \int_1^2 x^{-2}\,dx$$

$$= \pi \left(\frac{-1}{x}\right) \Big|_1^2$$

$$= \pi \left[-\frac{1}{2} + 1\right] = \frac{\pi}{2}$$

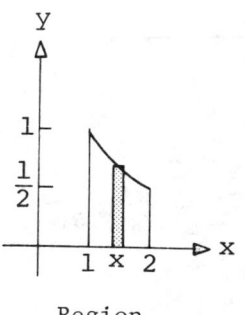

Region

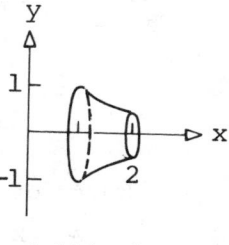

Solid of
Revolution

11. $y = \dfrac{1}{x}$ for $1 \leqslant y \leqslant 4$;

revolve about the y-axis.

$y = \dfrac{1}{x} \Rightarrow x = \dfrac{1}{y} = g(y)$

$V = \pi \int_1^4 g^2(y)\,dy$

$\quad = \pi \int_1^4 y^{-2}\,dy$

$\quad = \pi \left[\dfrac{-1}{y}\right]\Big|_1^4$

$\quad = \pi \left[\dfrac{-1}{4} + 1\right] = \dfrac{3\pi}{4}$

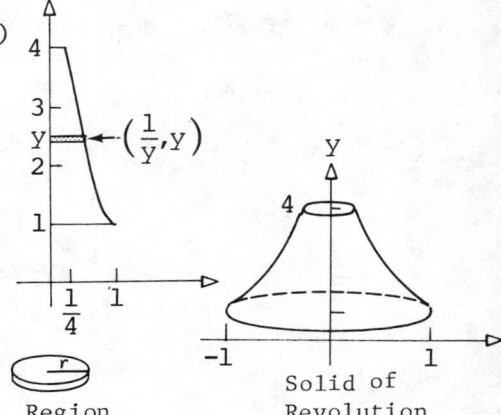

Region   Solid of
                Revolution

15.

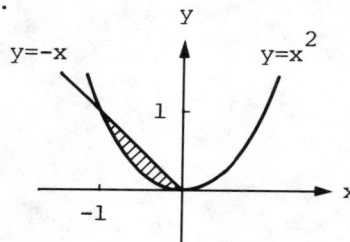

Revolve the region bounded by $y = -x$, $y = x^2$, $x = -1$, and $x = 0$ about the x-axis.

Volume of washer $= \pi[(-x)^2 - (x^2)^2]\Delta x$

$V = \pi \int_{-1}^0 (x^2 - x^4)\,dx = \pi\left(\dfrac{1}{3}x^3 - \dfrac{1}{5}x^5\right)\Big|_{-1}^0$

$\quad = \pi\left[(0-0) - \left(-\dfrac{1}{3} + \dfrac{1}{5}\right)\right] = \pi\left(\dfrac{1}{3} - \dfrac{1}{5}\right) = \dfrac{2\pi}{15}$

19. Revolve the region enclosed by
$y = x^2$, the y-axis, and
$y = 4$ about the x-axis. Note
that if $y = 4$, then $x = 2$.

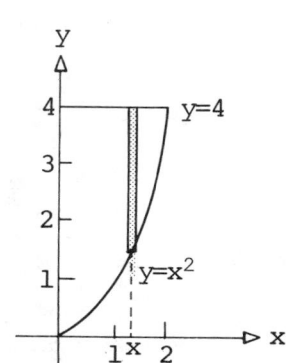

$V = \pi\int_0^2 [f^2(x) - g^2(x)]dx$,
where $f(x) = 4$ and $g(x) = x^2$.

$V = \pi\int_0^2 (16-x^4)dx$

$\quad = \pi(16x - \frac{1}{5}x^5)\Big|_0^2$

$\quad = \pi(32 - \frac{32}{5}) = \frac{128\pi}{5}$

23. Revolve the region enclosed
by $y = 2\sqrt{x}$, the y-axis,
and $y = 4$ about the
x-axis. Note that if
$y = 4$, then $x = 4$.

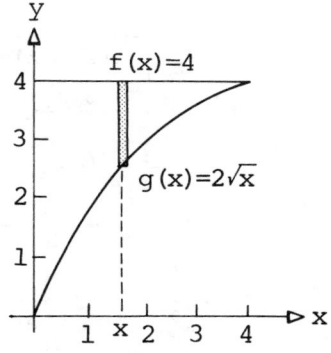

$V = \pi\int_0^4 [f^2(x) - g^2(x)]dx$,
where $f(x) = y = 4$ and
$g(x) = y = 2\sqrt{x}$

$V = \pi\int_0^4 (16-4x)dx$

$\quad = \pi(16x-2x^2)\Big|_0^4$

$\quad = \pi(64-32) = 32\pi$

27. Revolve the region under
$y = \sqrt{x}$, $0 \leqslant x \leqslant 4$, about
$x = 4$. Note that $y = \sqrt{x}$
implies $x = y^2$.

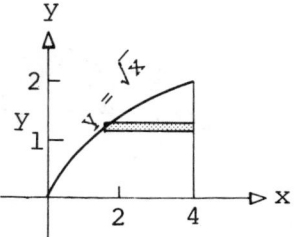

$V = \pi\int_0^2 r^2 dy$ and $r = 4 - x$

$= 4 - y^2$, since we want r
as a function of y.

$V = \pi\int_0^2 (4-y^2)^2 dy$

$$= \pi \int_0^2 (16-8y^2+y^4)\,dy$$

$$= \pi(16y - \frac{8}{3}y^3 + \frac{1}{5}y^5)\Big|_0^2$$

$$= \pi(32 - \frac{64}{3} + \frac{32}{5}) = \pi(\frac{480-320+96}{15}) = \frac{256\pi}{15}$$

<u>Exercise 3</u>, pp. 350-351

3. Revolve the region under
   $y = \sqrt{x} + x$, $1 \leqslant x \leqslant 4$,
   about the y-axis.

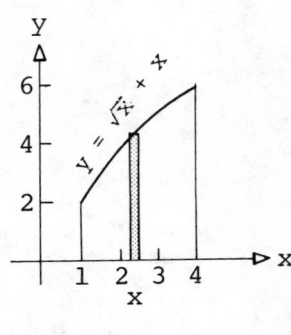

$V = 2\pi \int_1^4$ rhdx, where $r = x$

and $h = \sqrt{x} + x$

$= 2\pi \int_1^4 x(\sqrt{x} + x)\,dx$

$= 2\pi \int_1^4 (x^{3/2} + x^2)\,dx$

$= 2\pi(\frac{2}{5}x^{5/2} + \frac{1}{3}x^3)\Big|_1^4$

$= 2\pi[(\frac{64}{5} + \frac{64}{3}) - (\frac{2}{5} + \frac{1}{3})] = 2\pi(\frac{62}{5} + 21) = \frac{334\pi}{5}$

7. Revolve the region enclosed by
   $y = x^3$, the y-axis, and $y = 8$
   about the x-axis. If $x^3 = 8$,
   then $x = 2$. Also, $y = x^3$
   $\Rightarrow x = y^{1/3} = g(y)$.

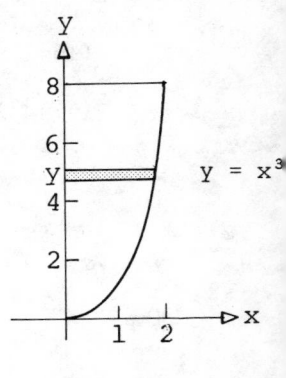

$V = 2\pi \int_0^8 y\, g(y)\,dy$

$= 2\pi \int_0^8 y\, y^{1/3}\,dy$

$= 2\pi \int_0^8 y^{4/3}\,dy = 2\pi(\frac{3}{7}y^{7/3})\Big|_0^8$

$= \frac{6\pi}{7} \cdot 2^7 = \frac{768\pi}{7}$

11. Revolve the region enclosed by
$y = x$ and $y = x^2$ about the
x axis. We have $x^2 = x$
$\Rightarrow x(x-1) = 0 \Rightarrow x = 0, 1 \Rightarrow$
$y = 0, 1$. Also, $y = x^2 \Rightarrow x = y^{1/2}$.

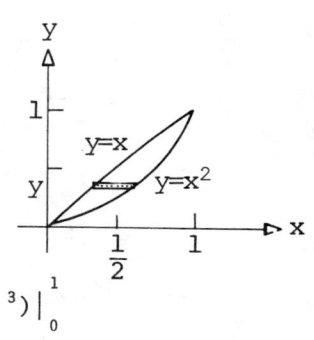

$$V = 2\pi \int_0^1 (\text{Average radius})(\text{Height})\,dy$$

$$= 2\pi \int_0^1 y(y^{1/2} - y)\,dy$$

$$= 2\pi \int_0^1 (y^{3/2} - y^2)\,dy = 2\pi(\tfrac{2}{5}y^{5/2} - \tfrac{1}{3}y^3)\Big|_0^1$$

$$= 2\pi(\tfrac{2}{5} - \tfrac{1}{3}) = \tfrac{2\pi}{15}$$

15. Revolve the region enclosed by
$y = x^3$ and $y = x$ in first
quadrant about the y-axis.
The graphs intersect when
$x^3 = x \Rightarrow x(x^2-1) = 0 \Rightarrow x = 0$
and $x = 1$ in the first quadrant.
The shell method gives

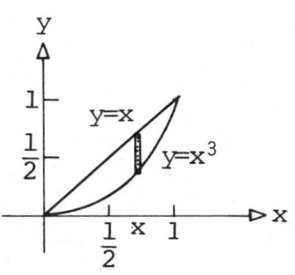

$$V = 2\pi \int_0^1 r\,h\,dx, \quad \text{where} \quad r = x$$

$$\text{and} \quad h = x - x^3$$

$$V = 2\pi \int_0^1 (x^2 - x^4)\,dx$$

$$= 2\pi(\tfrac{1}{3}x^3 - \tfrac{1}{5}x^5)\Big|_0^1$$

$$= 2\pi(\tfrac{1}{3} - \tfrac{1}{5}) = \tfrac{4\pi}{15}$$

19. Revolve the region enclosed by
$y = x^2$, $y = 8 - x^2$, to the
right of $x = 1$ about the
y-axis. The graphs intersect
when $x^2 = 8 - x^2 \Rightarrow x^2 = 4$
$\Rightarrow x = \pm2$. The shell method
is more efficient.

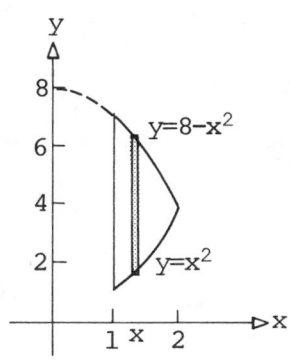

$$V = 2\pi \int_1^2 r\,h\,dx$$

$$= 2\pi \int_1^2 x[(8-x^2) - x^2]\,dx$$

$$= 2\pi \int_1^2 (8x - 2x^3)\,dx$$

$$= 2\pi (4x^2 - \frac{1}{2}x^4)\,\Big|_1^2$$

$$= 2\pi[(16-8) - (4 - \frac{1}{2})] = 9\pi$$

23. If $x = \sqrt{a^2 + by^2}$, then $x^2 = a^2 + by^2 \implies y^2$
$= \frac{1}{b}(x^2 - a^2) \implies y = \pm\sqrt{\dfrac{x^2 - a^2}{b}}$. Given $a = 147$, $b = 0.16$.
Using the shell method, we will find the volume for
$0 \leqslant x \leqslant 230$ (height $= 123 + 442 = 565$) and subtract
two other volumes: $h = \sqrt{\dfrac{x^2 - a^2}{b}} + \sqrt{\dfrac{x^2 - a^2}{b}}$ for
$147 \leqslant x \leqslant 155$ and $h = 123 + \sqrt{\dfrac{x^2 - a^2}{b}}$ for
$155 \leqslant x \leqslant 230$.

$$V = 2\pi \int_0^{230} x(565)\,dx - 2\pi \int_{147}^{155} x\left(\sqrt{\frac{x^2 - a^2}{b}} + \sqrt{\frac{x^2 - a^2}{b}}\right)dx$$

$$-2\pi \int_{155}^{230} x\left(123 + \sqrt{\frac{x^2 - a^2}{b}}\right)dx$$

$$V = 2\pi(565)\,\frac{x^2}{2}\Big|_0^{230}$$

$$-2\pi \int_{147}^{155} 2x\sqrt{\frac{x^2 - a^2}{b}}\,dx$$

$$-2\pi(123)\,\frac{x^2}{2}\Big|_{155}^{230}$$

$$-2\pi \int_{155}^{230} x\sqrt{\frac{x^2 - a^2}{b}}\,dx$$

Use $u = x^2 - a^2$, $du = 2x\,dx$ and $\dfrac{1}{\sqrt{b}}\int 2x\sqrt{x^2 - a^2}\,dx$

$$= \frac{1}{\sqrt{b}}\int u^{1/2}\,du = \frac{3}{2\sqrt{b}} u^{3/2} = \frac{3}{2\sqrt{b}}(x^2 - a^2)^{3/2}.$$

Evaluating at the limits gives $V \approx 53{,}135{,}993$ ft^3.

3. The base is bounded by $x^2 + y^2 = 1$
   and cross sections are squares.
   For each x with $-1 \leqslant x \leqslant 1$,
   we have $A(x) = (2y)^2 = 4y^2$
   $= 4(1-x^2)$. Hence,

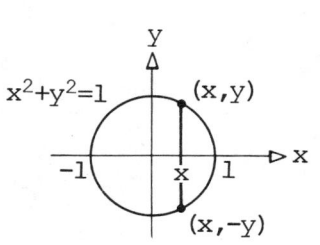

   $$V = \int_{-1}^{1} A(x)\,dx = 4\int_{-1}^{1}(1-x^2)\,dx$$

   $$= 4[x - \frac{1}{3}x^3]\Big|_{-1}^{1}$$

   $$= 4[(1-\frac{1}{3}) - (-1+\frac{1}{3})] = \frac{16}{3}$$

7. Choosing any h with $0 \leqslant h \leqslant 40$,
   we have $A(h) = (2h)^2 = 4h^2$. Then

   $$V = \int_{0}^{40} 4h^2\,dh = \frac{4}{3}h^3\Big|_{0}^{40}$$

   $$= \frac{4}{3}(64,000) = \frac{256,000}{3}\ m^3$$

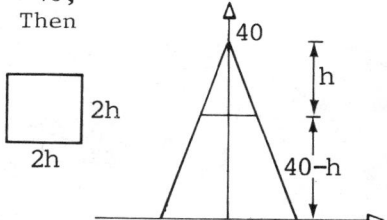

11. The slope of the line from $(0,0)$
    to (h,R) is $m = R/h$ and hence
    the equation of the line joining

    these points is $y - 0 = \frac{R}{h}(x-0)$,

    or $y = \frac{R}{h}x$. Revolving this

    line about the x-axis gives a
    right circular cone and the
    volume is

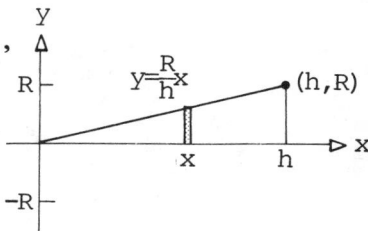

    $$V = \pi \int_{0}^{h} f^2(x)\,dx = \frac{R^2\pi}{h^2}\int_{0}^{h}x^2\,dx \quad \text{or}$$

    $$V = \frac{\pi R^2}{h^2}(\frac{x^3}{3})\Big|_{0}^{h} = \frac{\pi R^2 h}{3}$$

15.

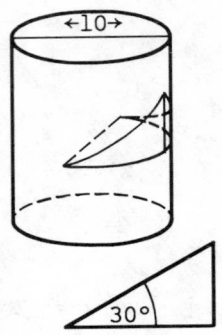

Wedge 30° angle
Wedge sides meet at a diameter.

Every vertical cross section
is a right triangle with a 30°
angle.

If we orient the bottom of this wedge on the x-axis
symmetrically about the y-axis, with the positive
y-axis pointing along the wedge, we get:

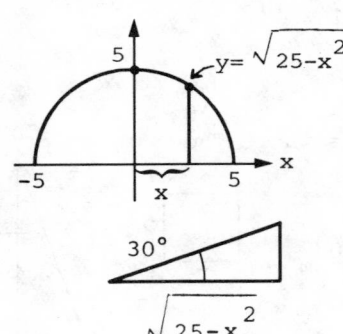

For a typical cross
section x units from the
middle, the adjacent leg
to the 30° is $y = \sqrt{25-x^2}$.
And the height is
$(\tan 30°) \sqrt{25-x^2}$.

Hence the cross sectional area is

$$A(x) = \frac{1}{2}\sqrt{25-x^2} \ (\tan 30° \ \sqrt{25-x^2}) = \frac{1}{2} \frac{\sqrt{3}}{3}(25-x^2)$$

Using symmetry we can double the volume on the
positive x-axis, hence

$$V = 2\int_0^5 \frac{1}{2} \frac{\sqrt{3}}{3}(25-x^2)\,dx = \frac{\sqrt{3}}{3}(25x-\frac{1}{3}x^3)\Big|_0^5$$

$$= \frac{\sqrt{3}}{3}(125-\frac{125}{3}) = \frac{250\sqrt{3}}{9}.$$

3. If $2x - 3y + 4 = 0$, then $y = \frac{2}{3}x + \frac{4}{3}$. For

    $P(1,2)$ and $Q(4,4)$, we have

    $$s = \int_1^4 \sqrt{1 + y'^2}\, dx = \int_1^4 \sqrt{1 + \frac{4}{9}}\, dx = \sqrt{\frac{13}{9}}\, x \Big|_1^4$$

    $$= \frac{\sqrt{13}}{3}(4-1) = \sqrt{13}$$

    Using the distance formula, we have

    $$d(P,Q) = \sqrt{(4-1)^2 + (4-2)^2} = \sqrt{9 + 4} = \sqrt{13}$$

7. If $y = x^{3/2}$ for $0 \leqslant x \leqslant 4$, then $y' = \frac{3}{2}x^{1/2}$

    $$s = \int_0^4 \sqrt{1 + y'^2}\, dx = \int_0^4 \sqrt{1 + \frac{9}{4}x}\, dx.$$ If $u = 1 + \frac{9}{4}x$,

    then $du = \frac{9}{4}\, dx$, or $dx = \frac{4}{9}\, du$, and $x = 0 \Rightarrow u = 1$,

    $x = 4 \Rightarrow u = 10$.

    $$s = \frac{4}{9} \int_1^{10} u^{1/2}\, du = \frac{4}{9} \cdot \frac{2}{3} u^{3/2} \Big|_1^{10} = \frac{8}{27}(10^{3/2} - 1)$$

    $$= \frac{1}{27}(80\sqrt{10} - 8)$$

11. If $y = \frac{2}{3}(x^2+1)^{3/2}$ for $1 \leqslant x \leqslant 4$, then

    $y' = 2x(x^2+1)^{1/2}$. $y'^2 = 4x^2(x^2+1)$ and $\sqrt{1 + y'^2}$

    $= \sqrt{4x^4 + 4x^2 + 1} = \sqrt{(2x^2+1)^2} = 2x^2 + 1$.

    $$s = \int_1^4 \sqrt{1 + y'^2}\, dx = \int_1^4 (2x^2+1)dx = \frac{2}{3}x^3 + x \Big|_1^4$$

    $$= (\frac{128}{3} + 4) - (\frac{2}{3} + 1) = \frac{126}{3} + 3 = 45$$

15. If $y = \frac{1}{8}x^4 + \frac{1}{4}x^{-2}$ for $1 \leqslant x \leqslant 2$, then

    $y' = \frac{1}{2}x^3 - \frac{1}{2}x^{-3}$. $y'^2 = \frac{1}{4}x^6 - \frac{1}{2} + \frac{1}{4}x^{-6}$ and

$$1 + y'^2 = \frac{1}{4}x^6 + \frac{1}{2} + \frac{1}{4}x^{-6} = (\frac{1}{2}x^3 + \frac{1}{2}x^{-3})^2$$

$$s = \int_1^2 \sqrt{1 + y'^2}\ dx = \int_1^2 (\frac{1}{2}x^3 + \frac{1}{2}x^{-3})dx = \frac{1}{8}x^4 - \frac{1}{4}x^{-2} \Big|_1^2$$

$$= (2 - \frac{1}{16}) - (\frac{1}{8} - \frac{1}{4}) = \frac{33}{16}$$

19. If $(x+1)^2 = 4y^3$ for $0 \leqslant y \leqslant 1$, then

$x = -1 + 2y^{3/2}$. $\frac{dx}{dy} = 3y^{1/2}$, so that $1 + (\frac{dx}{dy})^2$

$= 1 + 9y$.

$$s = \int_0^1 \sqrt{1 + (\frac{dx}{dy})^2}\ dy = \int_0^1 \sqrt{1 + 9y}\ dy$$

Let $u = 1 + 9y$, then $du = 9dy \Rightarrow \frac{1}{9}du = dy$, and

$y = 0 \Rightarrow u = 1$, $y = 1 \Rightarrow u = 10$.

$$s = \frac{1}{9}\int_1^{10} u^{1/2} du = \frac{1}{9} \cdot \frac{2}{3} u^{3/2} \Big|_1^{10} = \frac{2}{27}(10\sqrt{10} - 1)$$

23. Distance along $y^2 = x^3$ from $(1, 1)$ to $(3, 3\sqrt{3})$.

$y^2 = x^3 \Rightarrow y = x^{3/2} \Rightarrow y' = \frac{3}{2}x^{1/2}$

$$s = \int_1^3 \sqrt{1 + [\frac{3}{2}x^{1/2}]^2}\ dx = \int_1^3 \sqrt{1 + \frac{9}{4}x}\ dx$$

Let $u = 1 + \frac{9}{4}x$, $du = \frac{9}{4}dx$, $x = 1 \Rightarrow u = \frac{13}{4}$

and $x = 3 \Rightarrow u = \frac{31}{4}$.

$$s = \frac{4}{9}\int_{\frac{13}{4}}^{\frac{31}{4}} u^{1/2} du = \frac{4}{9} \cdot \frac{2}{3} u^{3/2} \Big|_{\frac{13}{4}}^{\frac{31}{4}}$$

$$= \frac{8}{27}(\sqrt{u})^3 \Big|_{\frac{13}{4}}^{\frac{31}{4}} = \frac{8}{27}(\frac{31\sqrt{31}}{8} - \frac{13\sqrt{13}}{8})$$

$$= \frac{31\sqrt{31} - 13\sqrt{13}}{27}$$

27. If $6xy = y^4 + 3$ and $1 \leqslant y \leqslant 2$, then

$x = \frac{1}{6}y^3 + \frac{1}{2}y^{-1}$. $\frac{dx}{dy} = \frac{1}{2}y^2 - \frac{1}{2}y^{-2}$ and $1 + (\frac{dx}{dy})^2$

$= 1 + \frac{1}{4}y^4 - \frac{1}{2} + \frac{1}{4}y^{-4} = (\frac{1}{2}y^2 + \frac{1}{2}y^{-2})^2$

$s = \int_1^2 \sqrt{1 + (\frac{dx}{dy})^2} \, dy = \int_1^2 (\frac{1}{2}y^2 + \frac{1}{2}y^{-2}) dy = \frac{1}{6}y^3 - \frac{1}{2}y^{-1}\Big|_1^2$

$= \frac{y^3}{6} - \frac{1}{2y}\Big|_1^2 = (\frac{8}{6} - \frac{1}{4}) - (\frac{1}{6} - \frac{1}{2}) = \frac{17}{12}$

31. Let $y = \sqrt{25 - x^2}$ for $0 \leqslant x \leqslant 4$. Then

$y' = \frac{1}{2}(25-x^2)^{-1/2}(-2x) = \frac{-x}{\sqrt{25 - x^2}}$, so that

$1 + y'^2 = 1 + \frac{x^2}{25 - x^2} = \frac{25}{25 - x^2}$.

$s = \int_0^4 \sqrt{1 + y'^2} \, dx = 5\int_0^4 \frac{dx}{\sqrt{25 - x^2}}$

35. (a) P_1 and P_2 both above the x-axis,

$$x^2 + y^2 = 1 \Rightarrow y = \sqrt{1 - x^2} \Rightarrow y' = \frac{-x}{\sqrt{1 - x^2}}$$

$$\Rightarrow \sqrt{1 + (y')^2} = \sqrt{1 + \frac{x^2}{1 - x^2}} = \sqrt{\frac{1}{1 - x^2}} = \frac{1}{\sqrt{1 - x^2}}.$$

Thus

$$\overline{P_1 P_2} = \int_{x_2}^{x_1} \frac{1}{\sqrt{1 - x^2}} \, dx. \quad (x_2, Y_2) \qquad (x_1, y_1)$$

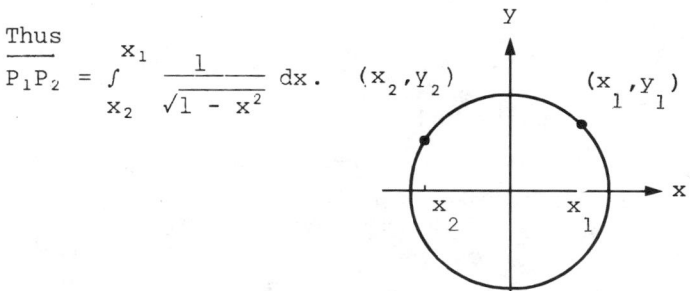

(b) P_1 and P_2 both below the x-axis (and $y_1 < y_2$).

$$x^2 + y^2 = 1 \Rightarrow y = -\sqrt{1 - x^2} \Rightarrow y' = \frac{x}{\sqrt{1 - x^2}}$$

$$\Rightarrow \sqrt{1 + [y']^2} = \frac{1}{\sqrt{1 - x^2}}$$

Thus,

$$\overline{P_1P_2} = \int_{x_1}^{x_2} \frac{1}{\sqrt{1-x^2}}\,dx$$

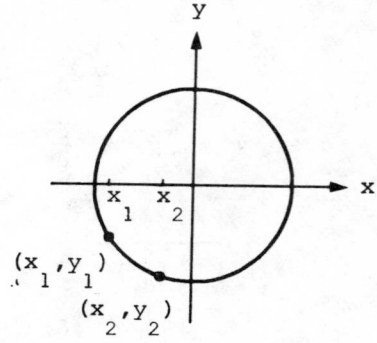

(c) P_1 in the second quadrant and P_2 in the fourth

In this case we must avoid using $x = -1$ with $1/\sqrt{1-x^2}$. We use both the symmetry of the original equation and the quarter circle circumference of $\pi/2$ (for the arc from $(-1,0)$ to $(0,-1)$. The counterclockwise arc length is then:

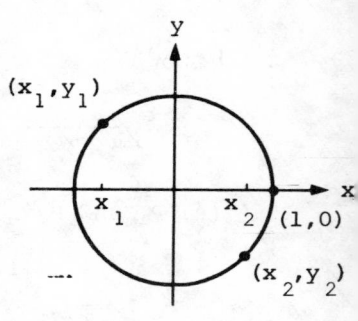

$$\overline{P_1P_2} = \int_0^{Y_1} \frac{1}{\sqrt{1-y^2}}\,dy + \frac{\pi}{2}$$

$$+ \int_0^{x_2} \frac{1}{\sqrt{1-x^2}}\,dx$$

Exercise 6, pp. 368-369

3. The unstretched length is 2m. Also $F(x) = kx$ and we are given that the length is compressed to $\frac{1}{2}$m when a force of 10 newtons is applied. Hence, $F(\frac{3}{2}) = \frac{3k}{2} = 10 \Rightarrow k = \frac{20}{3}$. The work required to compress the spring to a length of 1m is found by integrating between $x = 0$ and $x = 1$, since $x = 0$ indicates the unstretched length of 2m. Thus,

$$W = \int_0^1 F(x)\,dx = \frac{20}{3}\int_0^1 x\,dx = \frac{20}{6}x^2\Big|_0^1 = \frac{10}{3} \text{ joules}$$

7. The mass density of water is
$\rho = 1000$ kg/m^3, and $g \approx 9.8$m/sec^2.
Partition the depth interval $[0,6]$
into n subintervals of equal
thickness Δy. Then the volume
ΔV_i of the ith layer y_i units

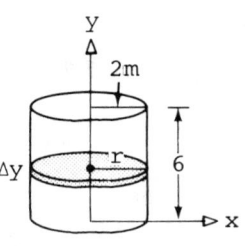

above the bottom is $\pi r^2 \Delta y = 4\pi \Delta y$,
since the radius r has a constant
value of $r = 2$. Thus, the weight
of this ith layer equals

ρg(Volume) = 4000 gπ. The distance this slice must
be lifted is $6-y_i$ meters, so that the work done in
lifting the ith layer is $4000 \text{ g}\pi(6-y_i)$. The work
required to pump all the water over the top is

$$W = \int_0^6 4000 \text{ g}\pi(6-y)dy = 4000 \text{ g}\pi(6y - \tfrac{1}{2}y^2)\Big|_0^6$$

$$= 4000 \text{ g}\pi(18) = 72,000 \text{ g}\pi \approx 2,216,708 \text{ joules.}$$

11. The depth of the water initially
is 2 meters so we subdivide the
interval $[-4,-2]$ on the
y-axis. Choosing the ith
layer, the radius is x_i,
which equals $\sqrt{16 - y_i^2}$. Thus,
the weight of the ith layer is

$\pi \rho g r^2 = \pi \rho g(16-y_i^2)$. This layer
must be lifted $|y_i| = -y_i$ meters, so that the work
required to empty the tank is

$$W = -\int_{-4}^{-2} \pi \rho g(16-y^2)y \; dy = -\pi \rho g \int_{-4}^{-2}(16y-y^3)dy$$

$$= \pi \rho g(\tfrac{1}{4}y^4 - 8y^2)\Big|_{-4}^{-2} = \pi \rho g[(4-32) - (64-128)]$$

$$= 36 \; \pi \rho g \approx 1,108,354 \text{ joules.}$$

15. Using 6.32, we have

$$W = \int_R^r \frac{GMm}{x^2} \; dx = GMm(\frac{1}{R} - \frac{1}{r}) = gRm(1 - \frac{R}{R+d}) = gRm \frac{d}{R+d}$$

where $gR \approx (9.8)(6.37 \times 10^6)$, $m = 30$ kg, $R \approx 6370$ km, and $d = 500$ km. Hence,

$$W = (9.8)(6.37 \times 10^6)(30)(\frac{500}{6870}) \approx 1.363 \times 10^8 \text{ joules}$$

19. The stiffness of the spring is 300,000 newtons per meter so that $F(x) = 300,000 \ x$. Hence, the work done compressing the spring 0.1 m is

$$W = \int_0^{0.1} 300,000 \ x \ dx = 150,000 \ x^2 \Big|_0^{0.1} = 150,000(0.01)$$
$$= 1500 \text{ joules.}$$

23. Using $pv^{1.4} = c$ and the fact that when $v = 2 \text{ ft}^3$ $= 3456 \text{ in}^3$, $p = 100 \text{ lb/in}^2$, we see that the constant $c = 100(3456)^{1.4} \approx 8,995,042.7$. This gives $p = cv^{-1.4} = 8,995,042.7 \ v^{-1.4}$.

If the volume is doubled, then

$$W = \int_a^b p(v)dv = 8,995,042.7 \int_{3456}^{6912} v^{-1.4} dv$$

$$= -\frac{8,995,042.7}{0.4 \ v^{0.4}} \Big|_{3456}^{6912}$$

$$= -22,487,607(\frac{1}{(6912)^{0.4}} - \frac{1}{(3456)^{0.4}})$$

$$\approx -22,487,607(0.02912 - 0.03842)$$

$$\approx +22,487,607(0.0093) = 209,135 \text{ in/lb.}$$

Exercise 7, pp. 373-374

3. Subdivide [0,6] and choose the ith subdivision $[y_{i-1}, y_i]$. We will use the approximation $\rho g = 62.5 \text{ lb/ft}^3$ as in the opening remarks of Section 7. Hence,

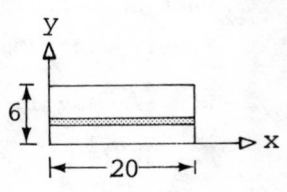

$$F = \int_0^6 \rho g(6-y)20dy = 20\rho g(6y - \frac{1}{2}y^2)\Big|_0^6 = 360 \ \rho g$$

$$\approx 360(62.5) = 22,500 \text{ lb.}$$

Note: Using $\rho g = (1.94)(32.2) = 62.468$, we have $F \approx 360(62.468) = 22,488 \text{ lb}$

140   CHAPTER SIX

7. Subdivide the interval $[-2,0]$ on the y-axis. The depth of the ith rectangle is $0 - y_i = -y_i$ and its area is $2x_i \Delta y = 2\sqrt{4 - y_i^2}\ \Delta y$. The total force is

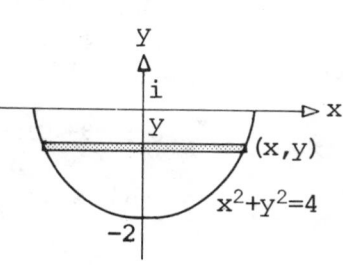

$$F = \int_{-2}^{0} 2\rho g\sqrt{4 - y^2}\,(-y)\,dy$$

$$= -2\rho g \int_{-2}^{0} y\sqrt{4 - y^2}\ dy.$$

Let $u = 4 - y^2$, then $du = -2y\,dy \implies y\,dy = -\frac{1}{2}\,du$ and $y = -2 \implies u = 0$ and $y = 0 \implies u = 4$. Then

$$F = \rho g \int_{0}^{4} u^{1/2}\,du = \frac{2\rho g}{3}\,u^{3/2}\Big|_{0}^{4} = \frac{16}{3}\,\rho g,$$

or, using $\rho g \approx 9800$, $\quad F \approx \frac{16}{3}(9800) = 52{,}267$ N

11. For y chosen with $-4 \leqslant y \leqslant 0$, the depth is $0 - y = -y$, and the area of the rectangle is $2x\,\Delta y = 2\sqrt{16 - y^2}\ \Delta y$. The total force is

$$F = \rho g \int_{-4}^{0} \sqrt{16 - y^2}\ (-2y\,dy)$$

Let $u = 16 - y^2$, so that $du = -2y\,dy$. Then,

$$\int u^{1/2}\,du = \frac{2}{3}\,u^{3/2} \implies$$

$$F = \rho g \cdot \frac{2}{3}(16-y^2)^{3/2}\Big|_{-4}^{0} = \frac{2}{3}\,\rho g[\,(16)^{3/2}-0\,] = \frac{128}{3}\,\rho g$$

Using $\rho g \approx 60(32.2)$, $\quad F \approx 128(20)(32.2) \approx 82{,}432$ lb.

Exercise 8, p. 377

3. If $f(x) = 1 - x^2$ over $[-1,1]$, then

$$\bar{y} = \frac{1}{1 + 1} \int_{-1}^{1} (1-x^2)\,dx = \frac{1}{2}\left(x - \frac{1}{3}x^3\right)\Big|_{-1}^{1}$$

$$= \frac{1}{2}\left[\left(1 - \frac{1}{3}\right) - \left(-1 + \frac{1}{3}\right)\right] = \frac{1}{2}\cdot\frac{4}{3} = \frac{2}{3}.$$

7. If $f(x) = -5x^4 + 4x - 10$ over $[-2,2]$, then

$$\bar{y} = \frac{1}{2+2} \int_{-2}^{2} (-5x^4+4x-10)\,dx = \frac{1}{4}(-x^5+2x^2-10x)\Big|_{-2}^{2}$$

$$= \frac{1}{4}[(-32+8-20) - (32+8+20)]$$

$$= \frac{1}{4}(-104) = -26$$

11. In free fall $a(t) = g$, so that $v = gt + C$. If the object falls from rest, then $v(0) = 0 \Rightarrow C = 0$. Thus, $v = gt$. Considering $v(t) = gt$ for $0 \le t \le 5$, we have

$$\bar{v}(t) = \frac{1}{5-0} \int_{0}^{5} v(t)\,dt = \frac{g}{5}\int_{0}^{5} t\,dt = \frac{g}{10} t^2 \Big|_{0}^{5}$$

$$= \frac{25g}{10} = \frac{5}{2} g \approx 24.5 \text{ m/sec}$$

We have the formula $v = gt$. Also, the distance s is $s = \frac{1}{2} gt^2 + k$, and since $s(0) = 0$, we have $k = 0$, so that $s = \frac{1}{2} gt^2$. Thus, $t = \sqrt{\frac{2s}{g}}$. This implies $v(s) = g\sqrt{\frac{2s}{g}} = \sqrt{2g}\, s^{1/2}$. Note that from $s = \frac{1}{2} gt^2$, $s(5) = \frac{25}{2} g$. The average value of $v(s)$ for $0 \le s \le \frac{25}{2} g$ is

$$\bar{v}(s) = \frac{1}{\frac{25}{2} g} \int_{0}^{(25g/2)} \sqrt{2g}\, s^{1/2}\,ds = \frac{2\sqrt{2g}}{25g}(\frac{2}{3} s^{3/2})\Big|_{0}^{(25g/2)}$$

$$= \frac{4\sqrt{2}}{75\sqrt{g}} (\frac{5\sqrt{g}}{\sqrt{2}})^3 = \frac{10}{3} g \approx 32.7 \text{ m/sec}$$

15. If $a(t) = 3$ m/sec^2, then $v(t) = 3t + k$. However, the car starts from rest so that $v(0) = 0 \Rightarrow k = 0$. Thus, $v(t) = 3t$. The average speed for t in the interval $[0,8]$ is

$$|\bar{v}| = \frac{1}{8} \int_{0}^{8} 3t\,dt = \frac{3}{16} t^2 \Big|_{0}^{8} = \frac{3(64)}{16} = 12 \text{ m/sec.}$$

3. The curves $x = 2(y-1)^2$ and $x = (y-1)^2 + 1$
   intersect when $2(y-1)^2 = (y-1)^2 + 1 \Rightarrow (y-1)^2 = 1$
   $\Rightarrow y - 1 = \pm 1 \Rightarrow y = 0$ or $y = 2$. Note that when
   $y = 1$, we have $x = 2(1-1)^2 = 0$ on one curve and
   $x = (1-1)^2 + 1 = 1$ on the other. Thus, $f(y)$
   $= (y-1)^2 + 1 > g(y) = 2(y-1)^2$ for $0 \leqslant y \leqslant 2$.
   The area is

   $$A = \int_0^2 [f(y) - g(y)]dy = \int_0^2 [(y-1)^2 + 1 - 2(y-1)^2]dy$$
   $$= \int_0^2 [1 - (y-1)^2]dy = \int_0^2 (2y-y^2)dy$$
   $$= (y^2 - \frac{1}{3}y^3)\Big|_0^2 = 4 - \frac{8}{3} = \frac{4}{3}.$$

7. $v = \pi \int_1^2 \frac{1}{x^2} dx = \pi \int_1^2 x^{-2} dx = \frac{-\pi}{x}\Big|_1^2 = \frac{\pi}{2}$

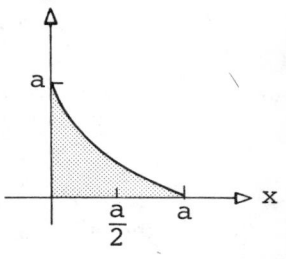

11. If $\sqrt{x} + \sqrt{y} = \sqrt{a}$, then we must
    have $x \geqslant 0$ and $y \geqslant 0$. Also,
    $\sqrt{y} = \sqrt{a} - \sqrt{x}$, and $y = (\sqrt{a} - \sqrt{x})^2$.
    We must have $0 \leqslant x \leqslant a$, as seen by
    $\sqrt{y} = \sqrt{a} - \sqrt{x}$. The area is

    $$A = \int_0^a (\sqrt{a} - \sqrt{x})^2 dx$$
    $$= \int_0^a (a - 2\sqrt{a}\, x^{1/2} + x) dx$$
    $$= (ax - \frac{4}{3}\sqrt{a}\, x^{3/2} + \frac{1}{2}x^2)\Big|_0^a$$
    $$= a^2(1 - \frac{4}{3} + \frac{1}{2}) = \frac{1}{6}a^2.$$

15. (a) The curve $y = -x^2 + 6x - 8$
        $= -(x-2)(x-4)$ intersects
        the x-axis at $x = 2$ and
        $x = 4$. Subdivide the
        interval $[2,4]$ on the
        x-axis. The maximum
        value of y occurs at the
        point $(3,1)$. The shell
        method gives

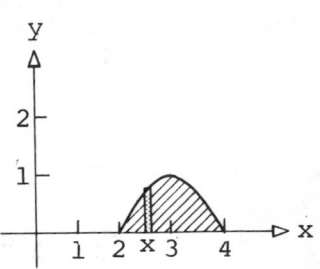

$$V = 2\pi \int_2^4 r\,h\ dx = 2\pi \int_2^4 x(-x^2+6x-8)\,dx$$

$$= 2\pi \int_2^4 (-x^3+6x^2-8x)\,dx = 2\pi(\frac{-x^4}{4} + 2x^3 - 4x^2)\Big|_2^4$$

$$= 2\pi[(-64+128-64) - (-4+16-16)] = 8\pi$$

(b) In order to use the shell method, subdivide [0,1] on the y-axis and resolve a horizontal rectangle about the y-axis. Given a $y \in [0,1]$, we have $x^2 - 6x + (8+y) = 0$,

so that $x = \dfrac{6 \pm \sqrt{36 - 4(8+y)}}{2} = 3\pm\sqrt{1 - y}$

Thus,

$$V = \pi \int_0^1 (R^2-r^2)\,dy = \pi \int_0^1 [(3+\sqrt{1-y})^2 - (3-\sqrt{1-y})^2]\,dy$$

$$= \pi \int_0^1 12\sqrt{1-y}\ dy = -8\pi(1-y)^{3/2}\Big|_0^1 = 8\pi$$

19. To determine the spring constant k we have $4 = k(0.8-0.6)$, or $k = 20$. Since the spring has length 0.6 m, if the upper end is attached to the point $(0,0.6)$ and the lower end is moved to $(0.8,0)$, which is $\sqrt{0.36 + 0.64} = 1$ m apart, then the spring has been stretched 0.4 m. The work done is $W = \int_0^{0.4} 20x\,dx = 10x^2\Big|_0^{0.4} = 1.6$ joules.

23. If $F = 100x$, then

(a) $F_1 = 100(0.1) = 10$ newtons

   $F_2 = 100(0.2) = 20$ newtons

   $F_3 = 100(0.4) = 40$ newtons

(b) $W_1 = \int_0^{0.1} 100x\ dx = 50x^2\Big|_0^{0.1} = 0.5$ joules

   $W_2 = \int_0^{0.2} 100x\ dx = 50x^2\Big|_0^{0.2} = 2.0$ joules

   $W_3 = \int_0^{0.4} 100x\ dx = 50x^2\Big|_0^{0.4} = 8.0$ joules

27. In Problem 26 we have $f(x) = \frac{1}{2}x$. The distribution function F for $f(x) = \frac{1}{2}x$, $0 \leqslant x \leqslant 2$, is

$$F(x) = \int_0^x f(t)\,dt = \int_0^x \frac{1}{2}t\ dt = \frac{1}{4}t^2\Big|_0^x$$

$$= \frac{1}{4}x^2,\ 0 \leqslant x \leqslant 2.$$

Chapter Seven

Exercise 2, pp. 391-393

3. $\ell n \sqrt[2]{5} = \ell n 5^{1/2} = \frac{1}{2}\ell n 5 = \frac{b}{2}$

7. $\ell n(\frac{5}{2}) = \ell n 5 - \ell n 2 = b - a$

11. $y = \ell n(3x)$ Domain: $x > 0$

   $y' = \frac{1}{3x} \cdot 3 = \frac{1}{x} > 0$ for all x in the domain

   => y always increasing

   $y'' = -\frac{1}{x^2} < 0$ for all x in the domain

   => y always concave down

   $\lim\limits_{x \to +\infty} \ell n(3x) = +\infty$ $\lim\limits_{x \to 0+} \ell n(3x) = -\infty$

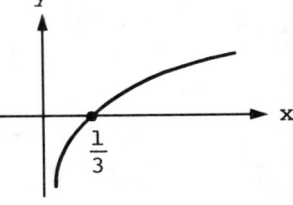

15. $y = x \; \ell n(x^2+4)$

   $y' = x(\frac{2x}{x^2 + 4}) + \ell n(x^2+4) = \frac{2x^2}{x^2 + 4} + \ell n(x^2+4)$

19. $y = \frac{1}{2} \ell n(\frac{1+x}{1-x}) = \frac{1}{2} \ell n(1+x) - \frac{1}{2} \ell n(1-x)$

   $y' = (\frac{1}{2})\frac{1}{1 + x} -(\frac{1}{2})\frac{-1}{1 - x} = \frac{(1-x) + (1+x)}{2(1+x)(1-x)} = \frac{1}{1 - x^2}$

23. $y = \ell n(\frac{x}{\sqrt{x^2 + 1}}) = \ell n \; x - \frac{1}{2} \ell n(x^2+1)$

   $y' = \frac{1}{x} -(\frac{1}{2})\frac{2x}{x^2 + 1} = \frac{1}{x} - \frac{x}{x^2 + 1} = \frac{x^2 + 1 - x^2}{x(x^2+1)}$

   $= \frac{1}{x(x^2+1)}$

27. $y = \ell n(\sin x)$

   $y' = \frac{\cos x}{\sin x} = \cot x$

31. $y = x \, \ln\sqrt{\cos 2x} = \frac{x}{2} \ln(\cos 2x)$

$y' = (\frac{x}{2})\frac{-2 \sin 2x}{\cos 2x} + \frac{1}{2} \ln(\cos 2x)$

$= -x \tan 2x + \frac{1}{2} \ln(\cos 2x)$

35. $y = \ln(x + \sqrt{x^2 + a^2})$

$y' = (\frac{1}{x + \sqrt{x^2 + a^2}})(1 + \frac{x}{\sqrt{x^2 + a^2}})$

$= (\frac{1}{x + \sqrt{x^2 + a^2}})(\frac{\sqrt{x^2 + a^2} + x}{\sqrt{x^2 + a^2}}) = \frac{1}{\sqrt{x^2 + a^2}}$

39. $y = (x^3+1)(x-1)(x^4+5)$

$\ln y = \ln(x^3+1) + \ln(x-1) + \ln(x^4+5)$

$\frac{1}{y} y' = [\frac{3x^2}{x^3 + 1} + \frac{1}{x - 1} + \frac{4x^3}{x^4 + 5}]$

$y' = (x^3+1)(x-1)(x^4+5)[\frac{3x^2}{x^3 + 1} + \frac{1}{x - 1} + \frac{4x^3}{x^4 + 5}]$

43. $y = \frac{x \cos x}{(x^2+1)^3 \sin x}$

$\ln y = \ln x + \ln \cos x - 3 \ln(x^2+1) - \ln \sin x$

$\frac{1}{y} y' = [\frac{1}{x} - \frac{\sin x}{\cos x} - \frac{6x}{x^2 + 1} - \frac{\cos x}{\sin x}]$

$y' = \frac{x \cos x}{(x^2+1)^3 \sin x}[\frac{1}{x} - \tan x - \frac{6x}{x^2 + 1} - \cot x]$

47. $\ln(\frac{y}{x}) - \ln(\frac{x}{y}) = 1$

$(\frac{x}{y})\frac{xy' - y}{x^2} - (\frac{y}{x})\frac{y - xy'}{y^2} = 0$

$\frac{xy' - y}{xy} - \frac{y - xy'}{xy} = 0$

$xy' - y - y + xy' = 0$

$y' = \frac{2y}{2x} = \frac{y}{x}$

51. $\ln x < 2(\sqrt{x}-1)$ for $x > 1$

Proof: Let $f(t) = \frac{1}{t}$, $g(t) = \frac{1}{\sqrt{t}}$ for $t > 1$.

We have $f(t) < g(t)$ for all $t > 1$, so

that $\int_1^x f(t)dt < \int_1^x g(t)dt$ for all $x > 1$.

Thus, $\int_1^x \frac{1}{t} dt < \int_1^x 1/\sqrt{t}\, dt$, which gives

$\ln|t|\Big|_1^x < 2\sqrt{t}\,\Big|_1^x$, or $\ln x - \ln 1 < 2\sqrt{x} - 2\sqrt{1}$

$\Rightarrow \ln x < 2(\sqrt{x}-1)$ for all $x > 1$.

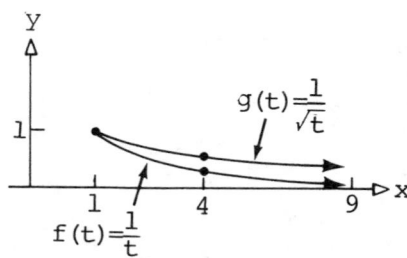

55. Consider

$f(x) = -x + \ln(1+x)^{1+x} = -x + (1+x)\ln(1+x)$

$f'(x) = -1 + 1 \cdot \ln(1+x) + (1+x) \cdot \frac{1}{1 + x} = \ln(1+x)$

$f'(0) = 0 \Rightarrow x = 0$ is a critical number.

$f''(x) = \frac{1}{1 + x} > 0$ for $x > -1$ (the domain of f)

$\Rightarrow$ f concave up on its domain.

Hence $f(0) = 0$ is a minimum value for f on its domain.

That is $f(x) = -x + \ln(1+x)^{1+x} \geqslant 0$ for $x > -1$

or $\ln(1+x)^{1+x} \geqslant x$ for $x > -1$.

59. $y = \ln(x^2+y^2)$

$y' = \frac{1}{x^2+y^2}[2x+2yy')$

At $(1, 0)$: $y' = \frac{1}{1+0}[2+0] = 2$

63. $v = kx^2\ln(\frac{1}{x}) = -kx^2\ln x$ dom $v = (0, +\infty)$

$v' = -kx^2(\frac{1}{x})-2kx \ln x = -kx - 2kx \ln x$

$v' = -kx(1 + 2 \ln x) \Rightarrow$ CN at

$x=e^{-1/2} \approx 0.607$

$$\begin{array}{c} 0 \quad\quad e^{-1/2} \\ \hline |\!+\!+|\!-\!-\!-\!-\! \\ 1 \end{array}$$

Thus, $x = e^{-1/2}$ gives a maximum value of v, assuming $k > 0$.

Exercise 3, pp. 398-401

3. $e^{\ln(\frac{1}{x})} = \frac{1}{x}$

7. $e^{(x+\ln x)} = e^x e^{\ln x} = e^x x$ or xe^x

11. $\ln(e^{-\ln x^2}) = 5 \Rightarrow \ln(e^{\ln(x^{-2})}) = 5$

$\Rightarrow \ln(x^{-2}) = 5$

$\Rightarrow x^{-2} = e^5 \Rightarrow x^2 = e^{-5}$

$\Rightarrow x = \pm \sqrt{e^{-5}}$

15. $\lim_{x\to+\infty} 3e^{3x} = 3 \lim_{x\to+\infty}(e^x)^3 = +\infty$

19. $\lim_{x\to+\infty}(1-e^{-x^3}) = \lim_{x\to+\infty} 1 - \lim_{x\to+\infty} (\frac{1}{e^x})^3$

$= 1 - (\lim_{x\to+\infty} \frac{1}{e^x})^3 = 1 - (0)^3 = 1$

23. $y = 5e^{3x}$

   $y' = 15e^{3x}$

27. $y = e^{-3x} \ln 2x$

   $y' = e^{-3x} \dfrac{2}{2x} + (-3)e^{-3x} \ln 2x$

   $= e^{-3x}[\dfrac{1}{x} - 3 \ln 2x]$

31. $y = e^{ax} \sin bx$

   $y' = e^{ax}(b \cos bx) + ae^{ax} \sin bx$

   $= e^{ax}(a \sin bx + b \cos bx)$

35. $y = e^{1/x}$

   $y' = e^{1/x}(-\dfrac{1}{x^2})$

   $= \dfrac{-1}{x^2} e^{1/x}$

39. $e^{x+y} = y$

   $e^{x+y}(1+y') = y'$

   $y' = \dfrac{-e^{x+y}}{e^{x+y}-1} = \dfrac{e^{x+y}}{1-e^{x+y}}$

43. $y = \dfrac{100}{1+99e^{-x}}$

   $y' = \dfrac{0-100(-99e^{-x})}{(1+99e^{-x})^2}$

   $= \dfrac{990e^{-x}}{(1+99e^{-x})^2}$

47. $y = \ln(\sin e^x)$, assuming $\sin e^x > 0$

$$y' = \frac{1}{\sin e^x}[\cos e^x(e^x)] = \frac{e^x \cos e^x}{\sin e^x} = e^x \cot e^x$$

51. $e^x \sin y + e^y \cos x = 4$

$e^x \cos y(y') + e^x \sin y + e^y(-\sin x) + y'e^y \cos x = 0$

$y'(e^x \cos y + e^y \cos x) = e^y \sin x - e^x \sin y$

$$y' = \frac{e^y \sin x - e^x \sin y}{e^x \cos y + e^y \cos x}$$

55. $y = e^{2x}$; $y' = 2e^{2x}$, $y'' = 4e^{2x}$

$y''-4y = 4e^{2x}-4e^{2x} = 0$

59. $y = Ae^{2x} + Be^{3x}$; $y' = 2Ae^{2x} + 3Be^{3x}$; $y'' = 4Ae^{2x} + 9Be^{3x}$

$y''-5y'+6y = e^{2x}(4A-10A+6A) + e^{3x}(9B-15B+6B) = 0$

63. $y = 3e^{3x} \Rightarrow y' = 9e^{3x} > 0$ for all x: y always
increasing, no critical numbers,
no local extrema

$\Rightarrow y'' = 27e^{3x} > 0$ for all x: y always
concave up, no points of inflection

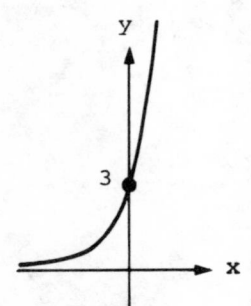

67. $y = e^{1/x}$

$$\lim_{x \to +\infty} e^{1/x} = \lim_{x \to -\infty} e^{1/x} = 1$$

$$\lim_{x \to 0+} e^{1/x} = +\infty \qquad \lim_{x \to 0-} e^{1/x} = 0$$

$$y' = -x^{-2}e^{1/x} = \frac{-e^{1/x}}{x^2}$$

y' undefined at $x = 0$ $y' \underline{---- \underset{\underset{0}{|}}{} ----}$

Therefore, y decreases for all x in its domain.

$$y'' = 2x^{-3}e^{1/x} + x^{-4}e^{1/x} = \frac{e^{1/x}(2x+1)}{x^4}$$

$y'' \underline{---- \underset{\underset{-\frac{1}{2}}{|}}{} +++ \underset{\underset{0}{|}}{} +++++}$ y concave down on $(-\infty, -\frac{1}{2})$; y concave up on $(-\frac{1}{2}, 0)$ and $(0, +\infty)$.

Point of inflection at $x = -\frac{1}{2}$.

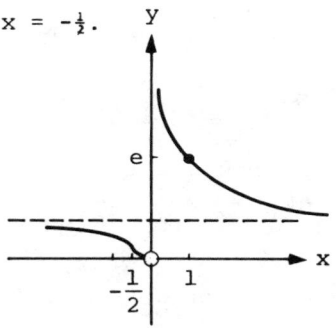

71. (a) $y = 2e^{\cos x} \Rightarrow y' = -2(\sin x)e^{\cos x}$

   $$y'' = -2 \sin x(-\sin x)e^{\cos x} - 2 \cos x \ e^{\cos x}$$

   $$= 2e^{\cos x}(\sin^2 x - \cos x)$$

   (b) $y'(\pi/2) = -2 \sin \pi/2 \ e^{\cos \pi/2} = -2(1)e^0 = -2$

   $$\frac{dy}{dx} = \frac{dy/dt}{dx/dt} \Rightarrow -2 = \frac{5}{dx/dt} \Rightarrow \frac{dx}{dt} = -\frac{5}{2}$$

75. Consider $f(x) = e^x - 1 - x$

   $f'(x) = e^x - 1 > 0$ for $x > 0$

   Thus f increases for $x > 0$. Since $f(0) = 0$, we have $e^x - 1 - x > 0$ for $x > 0$. That is, $e^x > 1 + x$ for $x > 0$.

79. $y = e^x \Rightarrow y' = e^x$

Slope of tangent at (x_0, y_0) is e^{x_0}

Equation of tangent line:

$(y-y_0) = e^{x_0}(x-x_0)$

And since $y_0 = e^{x_0}$ we get

$y = e^{x_0}x - e^{x_0}x_0 + e^{x_0}$

This line intersects the x-axis when $y = 0$, thus

$0 = e^{x_0}x - e^{x_0}x_0 + e^{x_0}$.

Dividing both sides by e^{x_0} yields

$0 = x - x_0 + 1$

or $x = x_0 - 1$.

Therefore, the tangent line intersects the x-axis at $x_0 - 1$, which is just 1 unit away from the original x_0 point on the x-axis.

83. We are given $\dfrac{dx}{dt} = 1$ unit/sec. Also, we have $A = xe^x$, so that $\dfrac{dA}{dt} = xe^x\dfrac{dx}{dt} + e^x\dfrac{dx}{dt}$

$= (x+1)e^x\dfrac{dx}{dt}$. When $x=10$, $\dfrac{dA}{dt} = (10+1)e^{10}(1)$,

since $\dfrac{dx}{dt} = 1$. Hence, at $x=10$, $\dfrac{dA}{dt} = 11e^{10}$

square units per second.

87. $f(x) = \dfrac{5000}{1+5e^{-x}}$

$f'(x) = \dfrac{-5000(-5e^{-x})}{(1+5e^{-x})^2} = \dfrac{25,000e^{-x}}{(1+5e^{-x})^2}$

$f''(x) = \dfrac{(1+5e^{-x})^2[-25,000e^{-x}]-(25,000e^{-x})2(1+5e^{-x})(-5e^{-x}}{(1+5e^{-x})^4}$

$$= \frac{-25,000e^{-x}(1+5e^{-x}-10e^{-x})}{(1+5e^{-x})^3}$$

$$= \frac{-25,000e^{-x}(1-5e^{-x})}{(1+5e^{-x})^3}$$

$f''(x) = 0 \Rightarrow e^{-x} = \frac{1}{5} \Rightarrow -x = \ell n \frac{1}{5} = -\ell n \ 5$

$\Rightarrow x = \ell n \ 5 \approx 1.61$ years

Sales are changing at a maximum rate during second year and are increasing in every year.

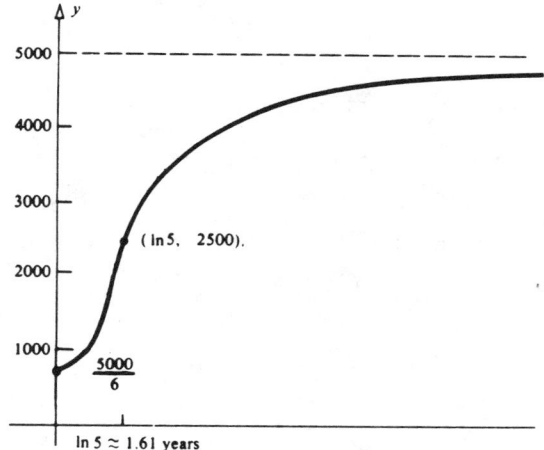

(ln 5, 2500).

$\frac{5000}{6}$

ln 5 ≈ 1.61 years

Exercise 4, pp. 404-405

3. $\int e^{3x+1} dx = \frac{1}{3} e^{3x+1} + C$

Let $u = 3x + 1$; $du = 3dx$;
$dx = \frac{1}{3} du;$ $\frac{1}{3} \int e^u du = \frac{1}{3} e^u + C$

7. $\int \frac{dx}{3x - 1} = \frac{1}{3} \int \frac{3x}{3x - 1} dx = \frac{1}{3} \ell n \ |3x-1| + C$

11. $\int_0^2 \frac{e^{2x}}{e^{2x} + 1} dx = \frac{1}{2} \int_2^{e^4+1} \frac{du}{u} = \frac{1}{2} \ell n \ |u| \Big|_2^{e^4+1}$

$$= \frac{1}{2} \ell n(e^4+1) - \frac{1}{2} \ell n \ 2$$

Let $u = e^{2x} + 1$; $du = 2e^{2x}dx$; $\frac{1}{2} du = e^{2x}dx$

$x = 0 \Rightarrow u = e^0 + 1 = 2$; $x = 2 \Rightarrow u = e^4 + 1$

15. $\int \dfrac{e^{1/x}}{x^2} dx = -\int e^u du = -e^u + C = -e^{-1/x} + C$

Let $u = \dfrac{1}{x}$; $du = -\dfrac{1}{x^2} dx$; $-du = \dfrac{1}{x^2} dx$

19. $\int \dfrac{3e^x}{(e^x-1)^{1/4}} dx = 3\int u^{-1/4} du = 4(e^x-1)^{3/4} + C$

Let $u = e^x - 1$; $du = e^x dx$

23. $\int \dfrac{dx}{x(\ell n\ x)^n} dx = \dfrac{(\ell n\ x)^{1-n}}{1-n} + C$ if $n \neq 1$, $n > 0$ or

$= \ell n|\ell n\ x| + C$ if $n = 1$

Let $u = \ell n\ x$, $du - \dfrac{1}{x} dx$

$\int u^{-n} du = \dfrac{u^{-n+1}}{-n+1} + C = \dfrac{u^{1-n}}{1-n} + C$ if $n \neq 1$ or $n > 0$

$= \ell n|u| + C$ if $n = 1$

27. $\int \dfrac{4}{1+e^x} dx = \int \dfrac{4e^{-x}}{e^{-x}+1} dx = -4\ \ell n|e^{-x}+1| + C$

Let $u = e^{-x} + 1$; $du = -e^{-x}dx$

31. $\int \dfrac{dx}{\sqrt{x}(1+\sqrt{x})} = 2\int \dfrac{du}{u} = 2\ \ell n(1+\sqrt{x}) + C$

Let $u = 1 + \sqrt{x}$; $du = \dfrac{1}{2\sqrt{x}} dx$; $2\ du = \dfrac{1}{\sqrt{x}} dx$

Note: $1 + \sqrt{x}$ is always positive so we do not need absolute value sign.

35. $y = e^x$, $y = e^{-x}$, $x = \ell n\ 2$

$$A = \int_0^{\ell n 2}(e^x - e^{-x})dx = e^x + e^{-x}\Big|_0^{\ell n 2} = (2 + \frac{1}{2}) - (1+1) = \frac{1}{2}$$

39. $y = e^x$, $y = e^{3x}$, $x = 2$

$$A = \int_0^2 (e^{3x} - e^x)dx = \frac{1}{3}e^{3x} - e^x\Big|_0^2 = (\frac{1}{3}e^6 - e^2) - (\frac{1}{3} + 1)$$

$$= \frac{1}{3}e^6 - e^2 + \frac{2}{3} \approx 127.754$$

43. Average value of $f(x) = e^x$ on $[0,1]$ is

$$\frac{1}{1-0}\int_0^1 e^x dx = e - 1$$

Exercise 5, p. 409

3. $y = (1+x^2)^{\sqrt{2}}$

$y' = 2\sqrt{2}\ x(1+x^2)^{\sqrt{2}-1}$

7. $y = \log_2 x = \dfrac{\ell n\ x}{\ell n\ 2}$

$y' = \dfrac{1}{x\ \ell n\ 2}$

11. $y = 2^{-x}\sin x$

$y' = 2^{-x}\cos x + 2^{-x}\ell n(2)(-1)\sin x$

$= 2^{-x}[\cos x - (\ell n\ 2)\sin x]$

15. $y = x^{\ell n\ x} = e^{(\ell n\ x)(\ell n\ x)}$

$y' = x^{\ell n\ x}[\ell n\ x(\frac{1}{x}) + \ell n\ x(\frac{1}{x})]$

$= x^{\ell n\ x}(\dfrac{2\ \ell n\ x}{x}) = 2(\ell n\ x)x^{\ell n\ x - 1}$

19. $y = x^{e^x} = e^{e^x \ell n\ x}$

$y' = x^{e^x}[e^x(\frac{1}{x}) + e^x \ell n\ x]$

$= e^x x^{e^x}(\dfrac{1}{x} + \ell n\ x)$

23. $y = x^{\sin x} = e^{\sin x(\ln x)}$

Use $Dx \; e^u = e^u \dfrac{du}{dx}$

$y' = x^{\sin x}[\dfrac{1}{x} \sin x + \cos x(\ln x)]$

27. $2^{xy} = x$

$2^{xy}(\ln 2)(xy'+y) = 1$, and since $2^{xy} = x$, we have

$(xy'+y) = \dfrac{1}{x \; \ln 2} \Rightarrow y' = \dfrac{1}{x^2 \; \ln 2} - \dfrac{y}{x} = \dfrac{1 - xy \; \ln 2}{x^2 \; \ln 2}$

31. $\int 2^{3x+5} dx = \dfrac{1}{3} \int 2^u du = \dfrac{2^u}{3 \; \ln 2} + C = \dfrac{2^{3x+5}}{3 \; \ln 2} + C$

Let $u = 3x + 5$; $du = 3dx$; $\dfrac{1}{3} du = dx$

35. $\displaystyle\int_3^9 \dfrac{dx}{x \; \log_3 x} = \int_3^9 \dfrac{\ln 3 \; dx}{x \; \ln x} = \ln 3 \int_3^9 \dfrac{1}{\ln x} \cdot \dfrac{dx}{x}$

$= \ln 3(\ln|\ln x|) \Big|_3^9$

$= \ln 3[\ln(\ln 9) - \ln(\ln 3)]$

$= \ln 3[\ln(\dfrac{\ln 9}{\ln 3})] = \ln 3[\ln(\dfrac{2 \; \ln 3}{\ln 3})]$

$= (\ln 3)(\ln 2)$

Use $\log_3 x = \ln x/\ln 3$ and let $u = \ln x$; $du = \dfrac{1}{x} dx$

39. $\log_b a = \dfrac{1}{\log_a b}$

<u>Proof</u>: $\log_b(a) = \dfrac{\ln a}{\ln b}$ and $\log_a(b) = \dfrac{\ln b}{\ln a}$.

Hence, $\log_b(a) = \dfrac{1}{\log_a(b)}$.

43. $\dfrac{d}{dx} f^g = \dfrac{d}{dx} e^{g \; \ln f}$

$= e^{g \; \ln f}[g(\dfrac{f'}{f}) + g' \; \ln f]$

$$= f^g(\frac{gf'}{f} + g' \ln f)$$

$$= g(x)f(x)^{g(x)-1}f'(x) + f(x)^{g(x)}g'(x)\ln f(x)$$

Exercise 6, pp. 413-414

3. $\displaystyle\lim_{n\to+\infty} (1 + \frac{1}{3n})^n = \lim_{n\to+\infty} [(1 + \frac{1}{3n})^{3n}]^{1/3} = e^{1/3}$

7. $A = Pe^r$

   (a) $1000 = Pe^{0.06} \Rightarrow P = \dfrac{1000}{e^{0.06}} = \$941.76$

   (b) If $r = 6\% = 0.06$ and the interest is compounded quarterly, then

   $$1000 = P(1 + \frac{0.06}{4})^4 = P(1.015)^4 \Rightarrow P = \frac{1000}{(1.015)^4}$$

   $$= \$942.18$$

Exercise 7, pp. 420-422

3. $\cos y \dfrac{dy}{dx} = \sin(x+y)$

   $\cos y \, dy = (\sin x \cos y + \cos x \sin y)dx$

   Not separable

7. $\ln(y^x) \dfrac{dy}{dx} = 3x^2 y$

   $\Rightarrow x \ln y \dfrac{dy}{dx} = 3x^2 \, y$

   $\Rightarrow \ln y \cdot \dfrac{1}{y} \, dy = 3x \, dx$

   $\Rightarrow \int \ln y \dfrac{1}{y} \, dy = \int 3x \, dx$

   [Substitute $u = \ln y$, $du = \dfrac{1}{y} \, dy$ on the left]

   $\Rightarrow \int u \, du = \int 3x \, dx$

$$\Rightarrow \frac{1}{2}u^2 = \frac{3}{2}x^2 + c$$

$$u^2 = 3x^2 + 2c$$

$$(\ln y)^2 = 3x^2 + k, \text{ where } k = 2c$$

11. $\frac{dy}{dx} = y\frac{(x^2-2x+1)}{y+3} \Rightarrow \frac{y+3}{y} dy = (x^2-2x+1)dx$

$\Rightarrow \int(1+\frac{3}{y})dy = \int(x^2-2x+1)dx$

$\Rightarrow y + 3\ln|y| = \frac{1}{3}x^3 - x^2 + x + c$

15. $\frac{dy}{dx} = \frac{x^2 + 2}{y}$; $y = 1$ when $x = 1$

$\Rightarrow y\ dy = (x^2+2)dx$

$\Rightarrow \int y\ dy = \int(x^2+2)dx$

$\Rightarrow \frac{1}{2}y^2 = \frac{1}{3}x^3 + 2x + c$

Substitute $y = 1$ and $x = 1$ to find c:

$\frac{1}{2} = \frac{1}{3} + 2 + c \Rightarrow c = -\frac{11}{6}$

$\Rightarrow \frac{1}{2}y^2 = \frac{1}{3}x^3 + 2x - \frac{11}{6}$

19. $\frac{dy}{dx} = \frac{e^y}{x}$; $y = 0$ when $x = 1$

$\Rightarrow e^{-y}dy = \frac{1}{x}\ dx$

$\Rightarrow \int e^{-y}dy = \int \frac{1}{x}\ dx$

$\Rightarrow -e^{-y} + c = \ln|x|$

Substitute $y = 0$ and $x = 1$ to find c:

$-e^0 + c = \ln|1| \Rightarrow c = 1$

$\Rightarrow -e^{-y} + 1 = \ln|x|$

23. We are given that the half-life of radium is 1690 years and 8 grams are present at $t = 0$. To find how much is present in 100 years, we have

$$\frac{dA}{dt} = kA \implies A(t) = Ce^{kt} \quad \text{and} \quad A(0) = 8 \implies C = 8$$

$$A(t) = 8e^{kt}$$

$$A(1690) = 4 = 8e^{1690k} \implies 1690k = \ln(\tfrac{1}{2}) \implies k = \frac{-\ln 2}{1690}$$

$$A(t) = 8e^{-\frac{\ln 2}{1690} t}$$

$$A(100) = 8e^{-\frac{100}{1690} \ln 2} = 8(\tfrac{1}{2})^{\frac{100}{1690}} = 8(\tfrac{1}{2})^{\frac{10}{169}}$$

$$\approx 8(\tfrac{1}{2})^{0.0592} \approx 7.678 \text{ g}$$

27. If $P_0 = 1500$ and $P(1 \text{ day}) = 2500$, find $P(3 \text{ days})$ if the uninhibited growth equation applies.

$$P(t) = P_0 e^{kt} = 1500\, e^{kt}$$

$$P(1) = 2500 = 1500\, e^k \implies e^k = \frac{25}{15} = \frac{5}{3} \implies k = \ln \frac{5}{3}$$

$$\implies P(t) = 1500\, e^{(\ln 5/3)t} = 1500(\tfrac{5}{3})^t$$

$$P(3) = 1500(\tfrac{5}{3})^3 = 6944.44 \approx 6944$$

31. $\dfrac{dN}{dt} = kN \qquad N(0) = 10{,}000 \qquad N(t_1) = 20{,}000$

$$N(t_1+10) = 100{,}000 \qquad N(t) = Ce^{kt} \implies N(t) = 10{,}000\, e^{kt}$$

$$20{,}000 = 10{,}000\, e^{kt_1} \implies kt_1 = \ln 2 \implies k = \frac{\ln 2}{t_1}$$

$$100{,}000 = 10{,}000\, e^{k(t_1+10)} \implies k(t_1+10) = \ln 10$$

$$\implies k = \frac{\ln 10}{t_1 + 10}$$

$$\frac{\ln 2}{t_1} = \frac{\ln 10}{t_1 + 10} \implies \ln 2(t_1+10) = t_1 \ln 10$$

$$t_1(\ln 10 - \ln 2) = 10 \ln 2$$

$$t_1 = \frac{10 \ln 2}{\ln(\frac{10}{2})} = \frac{10 \ln 2}{\ln 5}$$

Thus, $k = \frac{\ln 2}{t_1} = (\ln 2) \frac{\ln 5}{10 \ln 2} = \frac{\ln 5}{10}$

(a) $N(t) = 10,000 \; e^{(t/10) \ln 5}$ or $10,000 \; (5)^{t/10}$

(b) $N(20) = 10,000 \; (5)^{20/10} = 10,000 \; (25) = 250,000$

(c) If $N(t_1) = 20,000$, find t_1.

$$10,000 \; (5)^{t_1/10} = 20,000 \Rightarrow (5)^{t_1/10} = 2$$

$$\Rightarrow t_1 = 10 \; \frac{\ln 2}{\ln 5} \approx 4.3 \text{ min (an earlier result)}$$

35. $\frac{du}{dt} = k(u-T)$ $\qquad T = 30°C$

$$\frac{du}{u - T} k \; dt \Rightarrow \ln|u-T| = kt + C_1 \Rightarrow u - T = Ce^{kt} \Rightarrow$$

$$u(t) = T + Ce^{kt}$$

For $u(t) = 30 + Ce^{kt}$, $u(0) = 4$, and $u(2) = 10$, find $u(5)$.

$$4 = 30 + Ce^0 \Rightarrow C = -26 \Rightarrow u(t) = 30 - 26 \; e^{kt}$$

$$10 = 30 - 26 \; e^{2k} \Rightarrow -20 = -26 \; e^{2k} \Rightarrow e^{2k} = \frac{20}{26}$$

$$\Rightarrow k = \frac{1}{2} \ln \frac{10}{13}$$

$$\Rightarrow u(5) = 30 - 26e^{(5/2)\ln(10/13)} \approx 16.5°C$$

Exercise 8, pp. 429-430

3. $\frac{dy}{dx} + 2xy = e^{-x^2}$

Here $P(x) = 2x$ [as in (7.47)]

Integrating Factor $u = e^{\int P(x)dx} = e^{\int 2x \; dx} = e^{x^2}$

Multiply both sides of the D.E. by u:

$$e^{x^2}\frac{dy}{dx} + 2xe^{x^2}y = 1$$

$$\Rightarrow \frac{d}{dx}(e^{x^2}y) = 1$$

$$\Rightarrow e^{x^2}y = \int 1 \, dx$$

$$\Rightarrow e^{x^2}y = x + c \Rightarrow y = (x+c)e^{-x^2}$$

7. $y_1 = e^{2x} \Rightarrow y_1' = 2e^{2x} \Rightarrow y_1'' = 4e^{2x}$

$y_1'' + 4y_1' - 12y_1 = 4c^{2x} + 4(2e^{2x}) - 12e^{2x} = 0$

$y_2 = e^{-6x} \Rightarrow y_2' = -6e^{-6x} \Rightarrow y_2'' = 36e^{-6x}$

$y_2'' + 4y_2' - 12y = 36c^{-6x} + 4(-6e^{-6x}) - 12(e^{-6x}) = 0$

General solution: $y = c_1 e^{2x} + c_2 e^{-6x}$

11. $y_1 = \cos(\sqrt{2}x) \Rightarrow y_1' = -\sqrt{2}\sin(\sqrt{2}x) \Rightarrow y_1'' = -2\cos(\sqrt{2}x)$

$y_1'' + 2y_1 = -2\cos(\sqrt{2}x) + 2(\cos(\sqrt{2}x)) = 0$

$y_2 = \sin(\sqrt{2}x) \Rightarrow y_2' = \sqrt{2}\cos(\sqrt{2}x) \Rightarrow y_2'' = -2\sin(\sqrt{2}x)$

$y_2'' + 2y_2 = -2\sin(\sqrt{2}x) + 2(\sin(\sqrt{2}x)) = 0$

General solution: $y = c_1\cos(\sqrt{2}x) + c_2\sin(\sqrt{2}x)$

15. $y'' + 3y' - 4y = 0$

$m^2 + 3m - 4 = 0 \Rightarrow (m+4)(m-1) = 0$

$\Rightarrow m = -4$ or $m = 1$

$\Rightarrow y = c_1 e^{-4x} + c_2 e^x$

19. $y'' + 6y' = 0$

$m^2 + 6m = 0 \Rightarrow m(m+6) = 0$

$\Rightarrow m = 0$ or $m = -6$

$\Rightarrow y = c_1 e^{0x} + c_2 e^{-6x} = c_1 + c_2 e^{-6x}$

23. $y'' + 9y = 0$; $y = 2$ and $y' = 6$ when $x = \dfrac{3\pi}{2}$

$m^2 + 9 = 0 \Rightarrow (m-0)^2 + (3)^2 = 0$

$\Rightarrow \lambda - 0, \; w = 3$

$y = c_1 e^{0x} \cos 3x + c_2 e^{0x} \sin 3x$

$y = c_1 \cos 3x + c_2 \sin 3x$

$y = 2$ when $x = \dfrac{3\pi}{2} \Rightarrow 2 = c_2$

$y' = -3c_1 \sin 3x + 3c_2 \cos 3x$

$y' = 6$ when $x = \dfrac{3\pi}{2} \Rightarrow 6 = -3c_1 \Rightarrow c_1 = -2$

$\Rightarrow y = -2 \cos 3x + 2 \sin 3x$

27. $\dfrac{dq}{dt} + \dfrac{1}{RC} q = \dfrac{E}{R}$

Integrating Factor $= u = e^{\int \frac{1}{RC} dt} = e^{\frac{t}{RC}}$

Multiplying both sides of the D.E. by u:

$e^{\frac{t}{RC}} \dfrac{dq}{dt} + \dfrac{1}{RC} e^{\frac{t}{RC}} q = \dfrac{E}{R} e^{\frac{t}{RC}}$

$\Rightarrow \dfrac{d}{dt}\left(e^{\frac{t}{RC}} q\right) = \dfrac{E}{R} e^{\frac{t}{RC}}$

$\Rightarrow e^{\frac{t}{RC}} q = \int \dfrac{E}{R} e^{\frac{t}{RC}} dt$

$\Rightarrow e^{\frac{t}{RC}} q = \dfrac{E}{R} \cdot RC \, e^{\frac{t}{RC}} + K$

$\Rightarrow q = EC + Ke^{\frac{-t}{RC}}$

Since $q = 0$ when $t = 0$, we have $0 = EC + K$
or $K = -EC$

$\Rightarrow q = EC - ECe^{\frac{-t}{RC}}$

Miscellaneous Exercises, pp. 430-434

3. $y = x^{1/a} + a^{1/x}$

$y' = \frac{1}{a} x^{\frac{1}{a} - 1} + a^{\frac{1}{x}} \ln a (-\frac{1}{x^2})$

7. $g(x) = \ln(x^2 - 2x) \implies g' = \frac{1}{x^2 - 2x} \cdot (2x-2)$

or $g' = \frac{2(x-1)}{x(x-2)}$

11. $w = \ln(\sqrt{x + a} - \sqrt{x})$

$w' = \frac{1}{\sqrt{x + a} - \sqrt{x}} [\frac{1}{2\sqrt{x + a}} - \frac{1}{2\sqrt{x}}]$

$= \frac{1}{\sqrt{x + a} - \sqrt{x}} [\frac{\sqrt{x} - \sqrt{x + a}}{2\sqrt{x} \sqrt{x + a}}] = \frac{-1}{2\sqrt{x} \sqrt{x + a}}$

15. $\int \frac{e^x + 1}{e^x - 1} dx = \int \frac{e^x}{e^x - 1} dx + \int \frac{1}{e^x - 1} dx$

$= \int \frac{e^x}{e^x - 1} dx + \int \frac{e^{-x}}{1 - e^{-x}} dx$

Substitute $u = e^x - 1$, $du = e^x dx$ in the first
integral and $w = 1 - e^{-x}$, $dw = e^{-x} dx$ in the second.

$\int \frac{e^x + 1}{e^x - 1} dx = \int \frac{1}{u} du + \int \frac{1}{w} dw = \ln|u| + \ln|w| + c$

$= \ln|e^x - 1| + \ln|1 - e^{-x}| + c$

or

$\int \frac{e^x + 1}{e^x - 1} dx = \int \frac{e^{x/2} + e^{-x/2}}{e^{x/2} - e^{-x/2}} dx$

Now, let $v = e^{x/2} - e^{-x/2}$, $dv = \frac{1}{2}(e^{x/2} + e^{-x/2}) dx$

Then,

$$\int \frac{e^x + 1}{e^x - 1}dx = 2\int \frac{1}{v}dv = 2 \ln|e^{x/2} - e^{-x/2}| + K$$

19. $\ln(ab) = \ln a + \ln b$

   <u>Proof</u>: $f = \ln ax \Rightarrow f' = \frac{1}{ax} \cdot a \Rightarrow f' = \frac{1}{x}$

   and $g = \ln x \Rightarrow g' = 1/x$

   Thus, $f' = g' \Rightarrow f = g + C$ (where C is a constant)

   $\Rightarrow C = f(1) - g(1)$

   $C = \ln a - \ln 1 = \ln a$; therefore,

   $\ln ax = \ln x + \ln a$.

   Let $x = b \Rightarrow \ln ab = \ln b + \ln a$

23. $f'(x) = - f(x)$ and $f(1) = 1$

   $\frac{f'}{f} = -1 \Rightarrow \int \frac{f'}{f} dx = - \int dx \Rightarrow \ln|f| = -x + C_1$

   $\Rightarrow f = Ce^{-x}$, where $C = e^{C_1}$

   $f(1) = 1 = Ce^{-1} \Rightarrow C = e$

   Thus, $f(x) = ee^{-x} = e^{-x+1}$ or $f(x) = e^{1-x}$

27. Given that $f(x+h) = e^h f(x) + e^x f(h)$; f' exists
    for all real numbers x; $f'(0) = 2$

   (a) To show that $f(0) = 0$,

   let $x = 0 \Rightarrow f(0+h) = e^h f(0) + e^0 f(h) \Rightarrow f(h)$

   $= e^h f(0) + f(h) \Rightarrow e^h f(0) = 0 \Rightarrow f(0) = 0$

   (b) $\lim_{x \to 0} \frac{f(x) - f(0)}{x - 0} = f'(0) = 2 \Rightarrow \lim_{x \to 0} \frac{f(x)}{x} = 2$

   (c) $f'(x) = \lim_{h \to 0} \frac{f(x+h) - f(x)}{h} = \lim_{h \to 0} \frac{[e^h f(x) + e^x f(h)] - f(x)}{h}$

   $= \lim_{h \to 0} (\frac{e^h - 1}{h}) f(x) + e^x \lim_{h \to 0} \frac{f(h)}{h}$

$$f'(x) = f(x) \lim_{h \to 0} \left(\frac{e^h-1}{h}\right) + e^x \lim_{x \to 0} \frac{f(x)}{x} = f(x) + 2e^x$$

For $\lim_{h \to 0} \left(\frac{e^h-1}{h}\right) = 1$ use $F(x) = e^x$ and apply (3.15) to find $f'(0)$ and (7.20) to show that $F'(0) = e^0 = 1$.

(d) By part (c), $p = 2$.

31. $A = -xy = -x \ln x$

$A' = -x\left(\frac{1}{x}\right) - \ln x = -(1+\ln x)$

$\Longrightarrow$ CN at $x = e^{-1}$

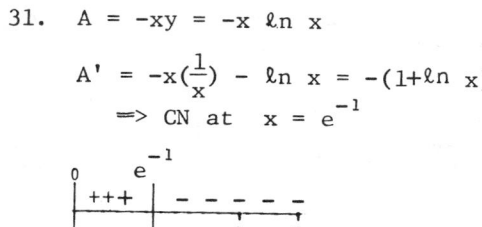

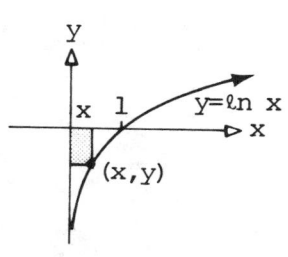

Maximum area is $A(e^{-1}) = -e^{-1}(\ln e^{-1}) = e^{-1} = \frac{1}{e}$

35. $y = \ln|x^2-1|$ for $x \in (-1,1) \Longrightarrow y = \ln(1-x^2)$

$y' = \frac{1}{1-x^2}(-2x) = \frac{-2x}{1-x^2}$

[local max at $(0,0)$]

$y'' = \frac{(1-x^2)(-2) + 2x(-2x)}{(1-x^2)^2} = \frac{-2(1+x^2)}{(1-x^2)^2}$

[concave down on $(-1,1)$] $\Longrightarrow$ (d) is true.

39. $f(x) = \frac{1}{x} + \ln x$ for $x \in [1/e, e]$

$f'(x) = \frac{-1}{x^2} + \frac{1}{x} = \frac{x-1}{x^2} \Longrightarrow$ CN at $x = 1$

(a) $f\left(\frac{1}{e}\right) = e - 1$ maximum value

$f(1) = 1 + 0 = 1$ minimum value

$f(e) = \frac{1}{e} + 1$

(b) $f'' = \dfrac{2}{x^3} - \dfrac{1}{x^2} = \dfrac{2-x}{x^3}$

$$\begin{array}{c} 1/e \qquad\quad 2 \qquad e \\ \overline{\;|\; +\;+\;+\; |\;-\;-\; |\;}\quad f'' \end{array}$$

f is concave up on $[1/e,2]$

(c)

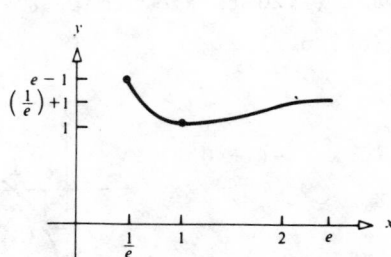

43. $\displaystyle\lim_{x\to c} \dfrac{e^x - e^c}{x-c} = \lim_{x\to c} \dfrac{f(x) - f(c)}{x-c} = f'(c),$ where

$f(x) = e^x.$ (Note that $f'(x) = e^x.$)

$\displaystyle\lim_{x\to c} \dfrac{e^x - e^c}{x-c} = e^c$

47. $y = xe^x \Rightarrow y' = e^x + xe^x = (1+x)e^x$

$\qquad\qquad x = -1$ critical number

$y' \;\dfrac{---|++++}{-1}\; \Rightarrow \;(-1, -\dfrac{1}{e})$ local minimum

$y'' = e^x + (1+x)e^x = (2+x)e^x$

$y'' \;\dfrac{----|++++}{-2}\; \Rightarrow$ y concave down on $(-\infty, -2)$
$\qquad\qquad\qquad\qquad$ concave up on $(-2, +\infty)$

$\qquad\qquad\qquad (-2, -\dfrac{2}{e^2})$ point of inflection

$\qquad\qquad\qquad \displaystyle\lim_{x\to+\infty} xe^x = +\infty$

$\qquad\qquad\qquad \displaystyle\lim_{x\to-\infty} xe^x = 0$

$\qquad\qquad\qquad$ (Use problem #44)

51. (a) To use Newton's method to solve

$10x = e^x$, let $f(x) = e^x - 10x$ and

$f'(x) = e^x - 10$. As a first guess at a root for $f(x) = 0$, we will use $c_1 = 3.5$

$$c_2 = c_1 - \frac{f(c_1)}{f'(c_1)} \approx 3.5 - \frac{-1.88455}{23.11545}$$

$c_2 \approx 3.5815_3$

$$c_3 \approx 3.58153 - \frac{f(3.58153)}{f'(3.58153)} \approx 3.58153 - \frac{.11317}{25.92847}$$

$c_3 \approx 3.57717$

$$c_4 \approx 3.57717 - \frac{f(3.57717)}{f'(3.57717)} \approx 3.57717 - \frac{.00046}{25.77216}$$

$c_4 \approx 3.57715$ (We'll stop at this point.)

Notes: $e^{3.57715} \approx 35.7714$

$10(3.57715) \approx 35.7715$

We could have started with a guess of either $c_1 = 3$ or $c_1 = 4$, and the convergence to the root simply takes more steps.

(b) There is another root of $f(x) = e^x - 10x$ near $x = .1$. Let's use Newton's method again, only now we'll start with $c_1 = .1$.

$$c_2 \approx c_1 - \frac{f(c_1)}{f'(c_1)} \approx .1 - \frac{.10517}{-8.89483}$$

$c_2 \approx .11182$

$$c_3 \approx .11182 - \frac{f(.11182)}{f'(.11182)} \approx .11182 - \frac{.00011}{-8.88169}$$

$c_3 \approx .11183$ (We'll stop at this point.)

Note: $e^{.11183} \approx 1.1183$

$10(.11183) \approx 1.1183$

55. (a) $W = \int_{V_1}^{V_2} pdV = \int_{V_1}^{V_2} \frac{nRT}{V} dV = nRT \ \ell n |V| \ \Big|_{V_1}^{V_2}$

$$= nRT[\ell n |V_2| - \ell n |V_1|] = nRT \ \ell n (\frac{V_2}{V_1}),$$

since $\left|\frac{V_2}{V_1}\right| = \frac{V_2}{V_1}$.

(b) $W = \int_{V_1}^{V_2} pdV = \int_{V_1}^{V_2} \frac{K}{V^\gamma} dV = K \int_{V_1}^{V_2} V^{-\gamma} dV = K \ \frac{V^{1-\gamma}}{1-\gamma} \ \Big|_{V_1}^{V_2}$

$$= \frac{K}{1-\gamma} [V_2^{1-\gamma} - V_1^{1-\gamma}], \quad \gamma \neq 1$$

59. $2y'' + 3y' - 2y = 0$

$\Rightarrow 2m^2 + 3m - 2 = 0$

$\Rightarrow (2m-1)(m+2) = 0$

$\Rightarrow m = \frac{1}{2}$ or -2

$\Rightarrow$ General solution: $y = c_1 e^{\frac{1}{2}x} + c_2 e^{-2x}$

63. $2y'' - 4y' - y = 0$

$\Rightarrow 2m^2 - 4m - 1 = 0$

$\Rightarrow m = \frac{4 \pm \sqrt{16 - 4(2)(-1)}}{2(2)} = \frac{4 \pm \sqrt{24}}{4}$

$\Rightarrow m = 1 + \frac{1}{2}\sqrt{6}$ or $1 - \frac{1}{2}\sqrt{6}$

$\Rightarrow$ General solution: $y = c_1 e^{(1 + \frac{1}{2}\sqrt{6})x} + c_2 e^{(1 - \frac{1}{2}\sqrt{6})x}$

Chapter Eight

Exercise 1, pp. 440-442

3. $y = \tan 5x$

$y' = 5 \sec^2 5x$

7. $y = e^{-4x} \sin 3x$

$y' = e^{-4x}(\cos 3x)(3) + (\sin 3x)(e^{-4x})(-4)$

$\quad = e^{-4x}(3 \cos 3x - 4 \sin 3x)$

11. $y = \sin(x^2)$

$y' = 2x \cos(x^2)$

15. $y = \sec(x^2+2x-1)$

$y' = \sec(x^2+2x-1) \tan(x^2+2x-1)(2x+2)$

19. $y = x^2 \sin 4x$

$y' = x^2(\cos 4x)(4) + (\sin 4x)(2x)$

$\quad = 2x(2x \cos 4x + \sin 4x)$

23. $y = x \tan x$

$y' = x \sec^2 x + \tan x$

27. $y = \sec x \tan x$

$y' = \sec x \sec^2 x + \tan x(\sec x \tan x)$

$\quad = \sec x(\sec^2 x + \tan^2 x)$

31. $x \sec y + y \tan x = x$

$x(\sec y \tan y)y' + \sec y + y \sec^2 x + y'\tan x = 1$

$y' = \dfrac{1 - \sec y - y \sec^2 x}{x \sec y \tan y + \tan x}$

35. $y = \sec^2(x^3+x)$

$y' = 2 \sec(x^3+x) \sec(x^3+x) \tan(x^3+x)(3x^2+1)$

$\quad = 2 \sec^2(x^3+x) \tan(x^3+x)(3x^2+1)$

39. Note: $\sin x - \cos x = \sqrt{2}[(\sin x)\frac{\sqrt{2}}{2} - (\cos x)\frac{\sqrt{2}}{2}]$

$= \sqrt{2}[\sin x \cos \frac{\pi}{4} - \cos x \sin \frac{\pi}{4}] = \sqrt{2} \sin(x - \frac{\pi}{4})$

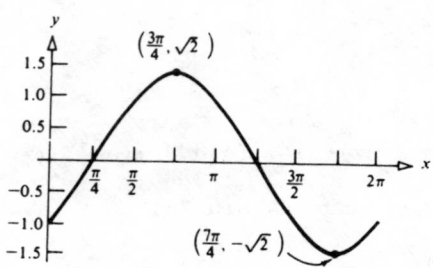

43. $f(x) = \sin x - \tan x \Rightarrow f'(x) = \cos x - \sec^2 x$. Slope at the x-intercepts alternates between 0 and -2.

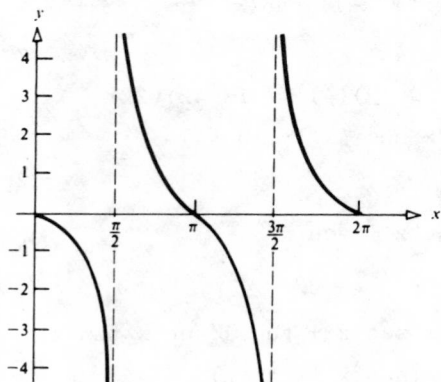

47.

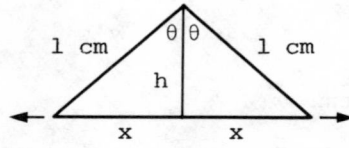

$\frac{dx}{dt} = 1$ cm/min.

$\sin \theta = \frac{x}{1}$, $\cos \theta = \frac{h}{1}$

Area $= A = 2(\frac{1}{2}hx) = hx = \cos \theta \sin \theta$

$$\frac{dA}{dt} = [-\sin^2\Theta + \cos^2\Theta]\frac{d\Theta}{dt}$$

Since $\sin\Theta = x$, $\cos\Theta\frac{d\Theta}{dt} = \frac{dx}{dt}$

or, $\frac{d\Theta}{dt} = \sec\Theta\frac{dx}{dt}$

Thus,

$$\frac{dA}{dt} = [-\sin^2\Theta + \cos^2\Theta]\sec\Theta\frac{dx}{dt}$$

When $\Theta = 30° = \frac{\pi}{6}$, we have

$$\frac{dA}{dt} = [-(\frac{1}{2})^2 + (\frac{\sqrt{3}}{2})^2]\frac{2}{\sqrt{3}}(1)\frac{cm^2}{min}$$

$$= \frac{1}{\sqrt{3}}\frac{cm^2}{min}$$

51. $\frac{d}{dx}(\csc x) = \frac{d}{dx}(\frac{1}{\sin x}) = \frac{\sin x \frac{d}{dx}(1) - 1\frac{d}{dx}(\sin x)}{\sin^2 x}$

$= \frac{-\cos x}{\sin^2 x} = -(\frac{1}{\sin x})(\frac{\cos x}{\sin x}) = -\csc x \cot x$

Exercise 2, p. 444-445

3. $\int \sec 5x\, dx = \frac{1}{5}\int \sec u\, du$ (use $u = 5x$)

$= \frac{1}{5}\ell n|\sec u + \tan u| + C$

$= \frac{1}{5}\ell n|\sec 5x + \tan 5x| + C$

7. $\int \sec 4x \tan 4x\, dx = \frac{1}{4}\int \sec u \tan u\, du$ (use $u = 4x$)

$= \frac{1}{4}\sec u + C = \frac{\sec(4x)}{4} + C$

11. $\int (\tan x)^{1/2}\sec^2 x\, dx = \int u^{1/2}du$ (use $u = \tan x$)

$= \frac{2}{3}u^{3/2} + C = \frac{2}{3}(\tan x)^{3/2} + C$

15. $\int \sec(3x-1)\tan(3x-1)du = \dfrac{\sec(3x-1)}{3} + C$ (use u = 3x-1)

19. $\int_0^{\pi/4} (1+\sec^2 x)dx = (x+\tan x)\big|_0^{\pi/4} = \dfrac{\pi}{4} + 1$

23. $\int_{\pi/2}^{3\pi/4} \pi\csc^2 x \, dx = -\pi\cot x\big|_{\pi/2}^{3\pi/4} = -\pi\cot(\dfrac{3\pi}{4}) = \pi$

27. $\int \csc x \, dx = \int \csc x[\dfrac{\csc x - \cot x}{\csc x - \cot x}]dx = \int \dfrac{1}{u} du$

$\qquad\qquad = \ln|u| + C = \ln|\csc x - \cot x| + C$

$\qquad\qquad (u = \csc x - \cot x,$

$\qquad\qquad du = (-\csc x \cot x + \csc^2 x)dx)$

31. $\int \sec x \, dx = \int \dfrac{dx}{\sin(\dfrac{\pi}{2} - x)} = \int \csc(\dfrac{\pi}{2} - x)dx$

$\quad = -\ln|\tan(\dfrac{\pi}{4} - \dfrac{x}{2})| + C$, by Problem 30. The latter

$\quad$ equals $\ln|\cot(\dfrac{\pi}{4} - \dfrac{x}{2})| + C = \ln|\tan[\dfrac{\pi}{2} - (\dfrac{\pi}{4} - \dfrac{x}{2})]| + C$

$\quad = \ln|\tan(\dfrac{x}{2} + \dfrac{\pi}{4})| + C$

Exercise 3, p. 450-451

3. $y = \sin^{-1}(-1)$

$\quad \sin y = -1$ and $-\dfrac{\pi}{2} \leqslant y \leqslant \dfrac{\pi}{2}$

$\quad y = -\pi/2$

7. $\sin^{-1}(\dfrac{\sqrt{2}}{2}) + \sin^{-1}(-\dfrac{\sqrt{2}}{2}) = \dfrac{\pi}{4} + (-\dfrac{\pi}{4}) = 0$

11. $\tan^{-1}(\sin 0) = \tan^{-1}(0) = 0$

15. $\sec^{-1}(5)$ = angle θ as shown.

$\quad$ So $\tan[\sec^{-1}(5)] = \sqrt{24} = 2\sqrt{6}$,

$\quad$ by the Pythagorean theorem.

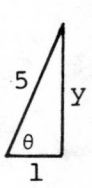

19. Apply $\sin(A+B) = \sin A \cos B + \cos A \sin B$

$\sin[\cos^{-1}(\frac{1}{3}) + \sin^{-1}(\frac{2}{3})]$

$= \sin[\cos^{-1}(\frac{1}{3})]\cos[\sin^{-1}(\frac{2}{3})]$

$\quad + \cos[\cos^{-1}(\frac{1}{3})]\sin[\sin^{-1}(\frac{2}{3})]$

$\cos^{-1}(\frac{1}{3}) = \theta \Rightarrow \frac{1}{3} = \cos \theta$

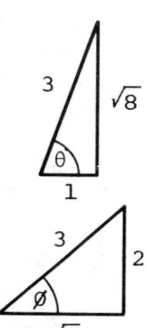

$\Rightarrow \sin[\cos^{-1}(\frac{1}{3})] = \frac{\sqrt{8}}{3}$

$\sin^{-1}(\frac{2}{3}) = \emptyset \Rightarrow \frac{2}{3} = \sin \emptyset$

$\Rightarrow \cos[\sin^{-1}(\frac{2}{3})] = \frac{\sqrt{5}}{3}$

Therefore,

$\sin[\cos^{-1}(\frac{1}{3}) + \sin^{-1}(\frac{2}{3})] = \frac{\sqrt{8}}{3} \cdot \frac{\sqrt{5}}{3} + \frac{1}{3} \cdot \frac{2}{3}$

$= \frac{\sqrt{40} + 2}{9}$

23. For any angle θ, $\sin 2\theta = 2 \sin \theta \cos \theta$. Thus,
$\sin(2 \cos^{-1}x) = 2 \sin(\cos^{-1}x)\cos(\cos^{-1}x)$. Now,
with a proof very similar to Problem 19, one may
show $\sin(\cos^{-1}x) = \sqrt{1-x^2}$. Therefore, $\sin(2 \cos^{-1}x)$
$= 2x\sqrt{1-x}$

27. Set $\cos^{-1}(-x) = y$. Then $0 \le y \le \pi$ and $\cos(y) = -x$.
Since $\cos(\pi-\theta) = -\cos \theta$ in general, the latter
gives $\cos(\pi-y) = x$, with $0 \le \pi-y \le \pi$. But this
says $\cos^{-1}x = \pi-y$, or $y = \pi-\cos^{-1}x$, i.e. $\cos^{-1}(-x)$
$= \pi - \cos^{-1}x$.

Exercise 4, pp. 457-458

3. Set $y = \cos^{-1}u$ and $u = x^2$. Then $y' = \frac{dy}{du} \cdot \frac{du}{dx}$

$= (\frac{-1}{\sqrt{1 - u^2}})(2x) = \frac{-2x}{\sqrt{1 - x^4}}$, $|x| < 1$.

7. $y = \sin^{-1}x + \cos^{-1}x$

$$y' = \frac{1}{\sqrt{1-x^2}} - \frac{1}{\sqrt{1-x^2}} = 0, \quad |x| < 1.$$

11. Set $y = \tan^{-1}u$ and $u = \frac{2x-1}{2x}$. Then $y' = \frac{dy}{du} \cdot \frac{du}{dx}$

$$= (\frac{1}{1+u^2})[\frac{4x - (4x-2)}{4x^2}] = [\frac{1}{1 + (\frac{2x-1}{2x})^2}][\frac{2}{4x^2}]$$

$$= (\frac{4x^2}{8x^2 - 4x + 1})(\frac{2}{4x^2}) = \frac{2}{8x^2 - 4x + 1}, \quad x \neq 0$$

15. Set $y = \sin^{-1}u$ and $u = 1 - x^2$. Then $y' = \frac{dy}{du} \cdot \frac{du}{dx}$

$$= (\frac{1}{\sqrt{1-u^2}})(-2x) = (\frac{1}{\sqrt{1 - (1-x^2)^2}})(-2x) = \frac{-2x}{\sqrt{2x^2 - x^4}}$$

$$= \begin{cases} \dfrac{-2}{\sqrt{2-x^2}} & \text{if } 0 < x < \sqrt{2} \\ \text{undefined} & \text{if } x = 0 \\ \dfrac{2}{\sqrt{2-x^2}} & \text{if } -\sqrt{2} < x < 0 \end{cases}$$

19. Set $y = \sin^{-1}u$ and $u = (1-x^2)^{1/2}$. Using the chain rule twice,

$$y' = \frac{dy}{du} \cdot \frac{du}{dx} = (\frac{1}{\sqrt{1-u^2}})(\frac{1}{2}(1-x^2)^{-1/2})(-2x)$$

$$= (\frac{1}{\sqrt{1 - (1-x^2)}})(\frac{-x}{\sqrt{1-x^2}}) = \frac{-x}{\sqrt{x^2 - x^4}}$$

$$= \begin{cases} \dfrac{-1}{\sqrt{1-x^2}} & \text{if } 0 < x < 1 \\ \text{undefined} & \text{if } x = 0 \\ \dfrac{1}{\sqrt{1-x^2}} & \text{if } -1 < x < 0 \end{cases}$$

23. Set $y = \sin^{-1}u$ and $u = \frac{x-1}{x+1}$. Then $y' = \frac{dy}{du} \cdot \frac{du}{dx}$

$$= (\frac{1}{\sqrt{1-u^2}})[\frac{(x+1) - (x-1)}{(x+1)^2}] = [\frac{1}{\sqrt{1 - (\frac{x-1}{x+1})^2}}][\frac{2}{(x+1)^2}]$$

$$= [\sqrt{\frac{(x+1)^2}{4x}}][\frac{2}{(x+1)^2}] = \frac{1}{\sqrt{x}(x+1)} \; , \; x > 0$$

27. Set $y = \tan^{-1}u$ and $u = \ln x$. Then $y' = \frac{dy}{du} \cdot \frac{du}{dx}$

$$= (\frac{1}{1 + u^2})(\frac{1}{x}) = [\frac{1}{1 + (\ln x)^2}](\frac{1}{x}), \; x > 0$$

31. Set $y = \tan^{-1}u$ and $u = \sin x$. Then $y' = \frac{dy}{du} \cdot \frac{du}{dx}$

$$= (\frac{1}{1 + u^2})(\cos x) = \frac{\cos x}{1 + \sin^2 x}$$

35. $\frac{d}{dx}(\sin^{-1}y + \cos^{-1}x) = \frac{d}{dx}(y)$

$$\frac{1}{\sqrt{1 - y^2}} \, y' - \frac{1}{\sqrt{1 - x^2}} = y'$$

$$y'[\frac{1}{\sqrt{1 - y^2}} - 1] = \frac{1}{\sqrt{1 - x^2}}$$

$$y' = \frac{\sqrt{1 - y^2}}{\sqrt{1 - x^2}(1 - \sqrt{1 - y^2})}$$

39. $\int \frac{dx}{x^2 + 25} = \frac{1}{5} \int \frac{\frac{1}{5} dx}{(\frac{x}{5})^2 + 1} = \frac{1}{5} \tan^{-1}(\frac{x}{5}) + C$

43. $\int \frac{dx}{\sqrt{16 - 9x^2}} = \frac{1}{3} \int \frac{\frac{3}{4} dx}{\sqrt{1 - (\frac{3x}{4})^2}} = \frac{1}{3} \sin^{-1}(\frac{3x}{4}) + C$

47. $\int \frac{\sin x \, dx}{\sqrt{4 - \cos^2 x}} = - \sin^{-1}[\frac{\cos x}{2}] + C$, using (8.29) and substituting $u = \cos x$, $du = -\sin x \, dx$

51. $\int \frac{2dx}{2x^2 + 2x + 3} = \int \frac{dx}{x^2 + x + \frac{3}{2}} = \int \frac{dx}{(x^2 + x + \frac{1}{4}) + \frac{5}{4}}$

$$= \int \frac{dx}{(x + \frac{1}{2})^2 + \frac{5}{4}} = \frac{2\sqrt{5}}{5} \tan^{-1}[\frac{2\sqrt{5}}{5}(x + \frac{1}{2})] + C, \quad \text{using}$$

(8.28) with $a = \frac{\sqrt{5}}{2}$ and substituting $u = x + \frac{1}{2}$

55. $\int \dfrac{dx}{\sqrt{2x-x^2}} = \int \dfrac{dx}{\sqrt{1-(x^2-2x+1)}}$

$\qquad = \int \dfrac{dx}{\sqrt{1-(x-1)^2}} \qquad \left[\text{use} \begin{array}{l} u = x - 1 \\ du = dx \end{array} \right]$

$\qquad = \int \dfrac{du}{\sqrt{1-u^2}} = \sin^{-1}u + c$

$\qquad = \sin^{-1}(x-1) + c$

59. $\int_0^1 \dfrac{dx}{5x^2+1} = \dfrac{1}{5} \int_0^1 \dfrac{dx}{x^2+1/5} = \dfrac{\sqrt{5}}{5} \tan^{-1}(\sqrt{5}x)\Big|_0^1 \text{ [using 8.28]}$

$\qquad = \dfrac{\sqrt{5}}{5} \tan^{-1}(\sqrt{5})$

63. By the chain rule, $\dfrac{d}{dx}[\tan^{-1}(\cot x)] = \dfrac{1}{1+\cot^2 x}(-\csc^2 x)$

$\qquad\qquad\qquad\qquad\qquad\qquad = -\dfrac{\csc^2 x}{\csc^2 x} = -1$

67. $\dfrac{d}{dx}(\sin^{-1}x + \cos^{-1}x) = \dfrac{1}{\sqrt{1-x^2}} - \dfrac{1}{\sqrt{1-x^2}} = 0.$ Thus,

$\sin^{-1}x + \cos^{-1}x$ is constant on $(-1,1)$. Setting $x = 0$,
we see this constant is $\pi/2$. Since also
$\sin^{-1}(-1) + \cos^{-1}(-1) = \pi/2 = \sin^{-1}(1) + \cos^{-1}(1)$,
we have $\sin^{-1}x + \cos^{-1}x = \pi/2$ on $[-1, 1]$

71. $\int \dfrac{dx}{\sqrt{a^2 - x^2}} = \int \dfrac{\frac{1}{a}\,dx}{\sqrt{1 - (\frac{x}{a})^2}} = \sin^{-1}(\dfrac{x}{a}) + C,$ using 8.25

and substituting $u = \dfrac{x}{a}$, $du = \dfrac{1}{a}dx$

3. Given that $m = 1$ gram and $k = 4$ dynes/cm, then $\omega = \sqrt{\dfrac{k}{m}} = 2$ rad/sec, and $x(t) = a \cos(2t) + b \sin(2t)$. Now if $x(0) = -1$ and $x'(0) = 0$, we have $x(t) = -\cos(2t)$. Thus, the velocity is $x'(t) = 2 \sin(2t)$.

7. The equation $\dfrac{1}{2} mR^2 \dfrac{d^2\theta}{dt^2} = -k\theta$ becomes $\dfrac{d^2\theta}{dt^2} + (\dfrac{2k}{mR^2})\theta = 0$

Since this has the same form as (8.31), we obtain the general solution

$\theta(t) = a \cos(\omega t) + b \sin(\omega t)$, where $\omega = \dfrac{1}{R} \sqrt{\dfrac{2k}{m}}$

With initial conditions $\theta(0) = \theta_0$, $\theta'(0) = 0$, we obtain

$$\theta(t) = \theta_0 \cos[\dfrac{1}{R} \sqrt{\dfrac{2k}{m}} \, t]$$

$$\text{Period} = \dfrac{2\pi}{\omega} = 2\pi R \sqrt{\dfrac{m}{2k}}$$

Exercise 6, pp. 468-469

3. $\sinh(-A) = \dfrac{e^{-A} - e^{A}}{2} = -\sinh(A)$

7. We use the results in Problems 5 and 6 to find

$$
\begin{aligned}
\cosh(3x) &= \cosh(2x) \cosh x + \sinh(2x) \sinh x \\
&= (\cosh^2 x + \sinh^2 x)\cosh x + 2 \sinh^2 x \cosh x \\
&= \cosh^3 x + 3 \sinh^2 x \cosh x \\
&= \cosh^3 x + 3(\cosh^2 x - 1)\cosh x \\
&= 4 \cosh^3 x - 3 \cosh x
\end{aligned}
$$

11. $y = \cosh(x^2+1)$. By the chain rule,

$y' = 2x \sinh(x^2+1)$

15. $y = \tanh^2 x$. By the chain rule,

$y' = 2 \tanh x \, \mathrm{sech}^2 x$

19. $y = \tanh(x^2)$. By the chain rule,

$y' = 2x \, \mathrm{sech}^2(x^2)$

23. Method 1:

$$\int \sinh^2 x \, dx = \int \left(\frac{e^x - e^{-x}}{2}\right)^2 dx = \frac{1}{4}\int [e^{2x} - 2 + e^{-2x}]dx$$

$$= \frac{1}{4}[\frac{1}{2}e^{2x} - 2x - \frac{1}{2}e^{-2x}] + c$$

$$= \frac{1}{4}[\frac{e^{2x} - e^{-2x}}{2}] - \frac{1}{2}x + c$$

$$= \frac{1}{4} \sinh(2x) - \frac{1}{2}x + c$$

Method 2: We can develop an analogous formula to $\sin^2 x = \frac{1 - \cos 2x}{2}$ (see Chapter 8, Section 2) only

for hyperbolic functions. From problem 6,
$\cosh(2x) = \cosh(x+x) = \cosh^2 x + \sinh^2 x$. From (8.43) we have

$\cosh^2 x - \sinh^2 x = 1$, thus $\cosh(2x) = (1+\sinh^2 x) + \sinh^2 x$

$= 1 + 2\sinh^2 x$. Hence, $\sinh^2 x = \frac{\cosh(2x) - 1}{2}$

$$\int \sinh^2 x \, dx = \int \frac{\cosh(2x) - 1}{2}dx = \frac{1}{2}\int [\cosh(2x) - 1]dx$$

$$= \frac{1}{2}[\frac{1}{2} \sinh(2x) - x] + c = \frac{1}{4} \sinh(2x) - \frac{1}{2}x + c$$

27. $\int \sqrt{\cosh x + 1} \, dx = \int \sqrt{\frac{e^x + e^{-x} + 2}{2}} \, dx$

$$= \frac{\sqrt{2}}{2} \int \sqrt{(e^{x/2} + e^{-x/2})^2} \, dx = \frac{\sqrt{2}}{2} \int (e^{x/2} + e^{-x/2})dx$$

$$= \sqrt{2}(e^{x/2} - e^{-x/2}) + C = 2\sqrt{2} \sinh(\frac{x}{2}) + C$$

31.

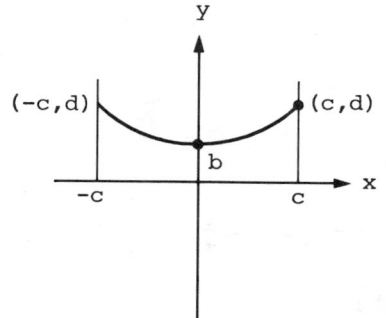

Length of rope = L

$y = a \cosh(\frac{x}{a}) + (b-a)$

Arc Length $= \int_{-c}^{c} \sqrt{1 + (\frac{dy}{dx})^2} \, dx$

$= \int_{-c}^{c} \sqrt{1 + \sinh^2(\frac{x}{a})} \, dx$

$= \int_{-c}^{c} \sqrt{\cosh^2(\frac{x}{a})} \, dx$ [Use (8.43)]

$= \int_{-c}^{c} \cosh(\frac{x}{a}) \, dx$ [since $\cosh(\frac{x}{a}) \geq 1$ for all x]

$= a \sinh(\frac{x}{a}) \Big|_{-c}^{c} = a \sinh(\frac{c}{a}) - a \sinh(\frac{-c}{a})$

$= 2a \sinh(\frac{c}{a})$ [Use (8.48)]

We are given that the arc length is L, that is,

$L = 2a \sinh(\frac{c}{a})$

Now, since (c,d) is on the curve, we have

$d = a \cosh(\frac{c}{a}) + (b-a)$

Hence, $\frac{d - b + a}{a} = \cosh(\frac{c}{a})$

$\Rightarrow \frac{(d-b+a)^2}{a^2} = \cosh^2(\frac{c}{a})$

$$\Rightarrow \frac{(d-b+a)^2}{a^2} = 1 + \sinh^2\left(\frac{c}{a}\right) \quad \text{[use (8.43)]}$$

From our arc length result:

$$\Rightarrow \frac{(d-b+a)^2}{a^2} = 1 + \frac{L^2}{4a^2}$$

$$\Rightarrow (d-b+a)^2 = a^2 + \frac{1}{4}L^2$$

$$\Rightarrow (d-b)^2 + 2a(d-b) + a^2 - a^2 = \frac{1}{4}L^2$$

$$\Rightarrow 2a(d-b) = \frac{1}{4}L^2 - (d-b)^2$$

$$\Rightarrow a = \frac{L^2}{8(d-b)} - \frac{(d-b)}{2}$$

35. $\int_0^1 \pi\left[a\cosh\left(\frac{x}{a}\right) + (b-a)\right]^2 dx$

$$= \pi\int_0^1 \left[a^2\cosh^2\left(\frac{x}{a}\right) + \frac{2(b-a)}{a}\cosh\left(\frac{x}{a}\right) + (b-a)^2\right]dx$$

[Apply the formula developed in problem 24, method 2.]

$$= \pi\int_0^1 \left[a^2\left(\frac{\cosh\left(\frac{2x}{a}\right)+1}{2}\right) + \frac{2(b-a)}{a}\cosh\left(\frac{x}{a}\right) + (b-a)^2\right]dx$$

$$= \pi\left[\frac{a^2}{2}\cdot\frac{a}{2}\sinh\left(\frac{2x}{a}\right) + \frac{a^2}{2}x + \frac{2(b-a)}{a}\cdot a\sinh\left(\frac{x}{a}\right) + (b-a)^2 x\right]\Bigg|_0^1$$

$$= \pi\left[\frac{a^3}{4}\sinh\left(\frac{2}{a}\right) + \frac{a^2}{2} + 2(b-a)\sinh\left(\frac{1}{a}\right) + (b-a)^2\right]$$

39. $y = a\cosh\left(\frac{x}{a}\right) + (b-a)$

$$\frac{dy}{dx} = \sinh\left(\frac{x}{a}\right)$$

$$\frac{d^2y}{dx^2} = \frac{1}{a}\cosh\left(\frac{x}{a}\right)$$

$$\frac{1}{a}\sqrt{1 + (\frac{dy}{dx})^2} = \frac{1}{a}\sqrt{1 + [\sinh(\frac{x}{a})]^2} = \frac{1}{a}\sqrt{1 + \sinh^2(\frac{x}{a})}$$

$$= \frac{1}{a}\sqrt{\cosh^2(\frac{x}{a})} \quad \text{[Use (8.43)]}$$

$$= \frac{1}{a}\cosh(\frac{x}{a}) \quad \text{[since } \cosh(\frac{x}{a}) \geqslant 1 \text{ for all x]}$$

$$= \frac{d^2y}{dx^2}$$

Exercise 7, pp. 470-471

3. Let $y = \tanh^{-1}u$ and $u = x^2 - 1$. Then $y' = \frac{dy}{du}\frac{du}{dx}$

$$= (\frac{1}{1 - u^2})(2x) = \frac{2x}{1 - (x^2-1)^2} = \frac{2x}{2x^2 - x^4} = \frac{2}{2x - x^3}$$

7. Let $y = \tanh^{-1}u$ and $u = \cos 2x$. Then $y' = \frac{dy}{du}\frac{du}{dx}$

$$= (\frac{1}{1 - u^2})(-2 \sin 2x) = (\frac{1}{1 - \cos^2(2x)})(-2 \sin 2x)$$

$$= -2 \csc(2x), \quad x \neq \frac{n\pi}{2}$$

11. Let $y = \cosh^{-1}x$ for $x \geq 1$. Then $\cosh y$

$$= \frac{1}{2}(e^y + e^{-y}) = x. \text{ Thus, } (e^y)^2 - 2x(e^y) + 1 = 0$$

$$\Longrightarrow e^y = \frac{2x \pm \sqrt{4x^2 - 4}}{2} = x \pm \sqrt{x^2 - 1} \quad \text{or}$$

$y = \ln(x \pm \sqrt{x^2 - 1})$. Since $y \geq 0$, the minus sign

does not apply, so $y = \ln(x + \sqrt{x^2 - 1})$ for $x \geq 1$

15. $\tanh(\tanh^{-1}x) = x$ for $|x| < 1$

Differentiating both sides with respect to x,

$$\text{sech}^2(\tanh^{-1}x) \frac{d}{dx}(\tanh^{-1}x) = 1$$

Since $\text{sech}^2u = 1 - \tanh^2u$, $\text{sech}^2[\tanh^{-1}x] = 1 - x^2$

Thus, $\frac{d}{dx}[\tanh^{-1}x] = \frac{1}{1 - x^2}$ for $|x| < 1$

19. $\int_2^3 \dfrac{1}{\sqrt{x^2-1}}\,dx = \cosh^{-1}x\Big|_2^3 = \ell n(x+\sqrt{x^2-1})\Big|_2^3$

[Use the results of problems 14 and 11.]

Thus,

$\int_2^3 \dfrac{1}{\sqrt{x^2-1}}\,dx = \ell n(3+\sqrt{8}) - \ell n(2+\sqrt{3}) = \ell n(\dfrac{3+\sqrt{8}}{2+\sqrt{3}})$

Miscellaneous Exercises, pp. 471-473

3. $y = \tanh(\dfrac{1}{x})$. By the chain rule,

$y' = -\dfrac{1}{x^2}\,\text{sech}^2(\dfrac{1}{x})$, $x \neq 0$

7. $y = \sin^{-1}(x-1) + (x-1)\sqrt{2x-x^2}$

We use the product and chain rules on the second term:

$y' = \dfrac{1}{\sqrt{1-(x-1)^2}} + (x-1)\dfrac{d}{dx}\sqrt{2x-x^2}$

$\qquad + (\sqrt{2x-x^2})\dfrac{d}{dx}(x-1)$

$\quad = \dfrac{1}{\sqrt{2x-x^2}} + \dfrac{(x-1)(2-2x)}{2\sqrt{2x-x^2}} + \sqrt{2x-x^2} = 2\sqrt{2x-x^2}$

(after expanding the middle term and noting cancellations), $0 < x < 2$

11. $y = \dfrac{8x}{x^2+4} - 4\tan^{-1}(\dfrac{1}{2}x) + x$

We use the quotient rule for the first term and the chain rule for the second:

$y' = \dfrac{8(x^2+4) - (8x)(2x)}{(x^2+4)^2} - (\dfrac{4}{1+\frac{x^2}{4}})(\dfrac{1}{2}) + 1$

$\quad = \dfrac{32-8x^2}{(x^2+4)^2} - \dfrac{8(x^2+4)}{(x^2+4)^2} + \dfrac{(x^2+4)^2}{(x^2+4)^2} = \dfrac{(x^2-4)^2}{(x^2+4)^2}$

15. $y = \sin^{-1}(\dfrac{x}{a}) + \dfrac{\sqrt{a^2-x^2}}{x}$. Using the chain rule for

the first term and the quotient rule for the second,

$$y' = \frac{1}{\sqrt{1 - (x/a)^2}} \frac{d}{dx}\left(\frac{x}{a}\right) + \frac{x \frac{d}{dx}(\sqrt{a^2-x^2}) - \sqrt{a^2 - x^2}}{x^2}$$

$$y' = \frac{1}{a\sqrt{1 - \frac{x^2}{a^2}}} + \left(\frac{\frac{-x^2}{\sqrt{a^2 - x^2}} - \sqrt{a^2 - x^2}}{x^2}\right)$$

$$= \frac{1}{\sqrt{a^2-x^2}} + \frac{-a^2}{x^2\sqrt{a^2 \cdot x^2}} = -\frac{\sqrt{a^2 \cdot x^2}}{x^2}, \quad 0 < |x| < |a|$$

19. $y = x \tan^{-1}x - \ell n\sqrt{1 + x^2} = x \tan^{-1}x - \frac{1}{2} \ell n(1+x^2)$

We use the product rule for the first term and the chain rule for the second:

$$y' = x \frac{d}{dx}(\tan^{-1}x) + \tan^{-1}x \frac{d}{dx}(x) - \frac{1}{2(1+x^2)} \frac{d}{dx}(1+x^2)$$

$$= \frac{x}{1 + x^2} + \tan^{-1}x - \frac{x}{1 + x^2} = \tan^{-1}x$$

23. $y = \sin^{-1}(e^x)$. By the chain rule,

$$y' = \frac{e^x}{\sqrt{1 - e^{2x}}}, \quad x < 0$$

27. $$\lim_{x \to 0} \left(\frac{\sinh x}{x}\right) = \lim_{x \to 0} \left(\frac{\sinh x - \sinh 0}{x - 0}\right)$$

$$= \frac{d}{dx}(\sinh x)\Big|_{x=0} = \cosh 0 = 1$$

31. $y = x - \cosh^{-1}x$, domain: $x \geq 1$

$$y' = 1 - \frac{1}{\sqrt{x^2 - 1}} = 0 \quad \text{only for} \quad x = \sqrt{2}$$

Since $y' < 0$ for $x < \sqrt{2}$ and $y' > 0$ for $x > \sqrt{2}$, $x = \sqrt{2}$ gives a minimum. Thus, the minimum value is $\sqrt{2} - \cosh^{-1}\sqrt{2}$.

35. Area shown $= \int_{-a}^{a}\sqrt{a^2 - x^2}\, dx$

$$= 2\int_{0}^{a}\sqrt{a^2 - x^2}\, dx$$

$$= x\sqrt{a^2 - x^2} + a^2\sin^{-1}\left(\frac{x}{a}\right)\Big|_{0}^{a}$$

$$= a^2\sin^{-1}(1) = \frac{\pi a^2}{2}.$$

This is correct, since from elementary geometry, a semicircle of radius a has area $\frac{\pi a^2}{2}$.

39. $\displaystyle\int_{2/\sqrt{3}}^{2} \frac{dx}{x\sqrt{x^2 - 1}} = \sec^{-1}(x)\Big|_{2/\sqrt{3}}^{2} = \sec^{-1}(2) - \sec^{-1}(2/\sqrt{3})$

$$= \pi/3 - \pi/6 = \pi/6$$

43. Area $\displaystyle= \int_{-r}^{r} \frac{dx}{1 + x^2} = 2\int_{0}^{r} \frac{dx}{1 + x^2} = 2\tan^{-1}x\Big|_{0}^{r} = 2\tan^{-1}r$

As $r \to +\infty$, this area approaches $\displaystyle\lim_{r \to +\infty}[2\tan^{-1}r] = \pi$.

47. $\displaystyle\int_{a}^{b}\sqrt{1 + (\frac{dy}{dx})^2}\ dx = \int_{a}^{b}\sqrt{1 + \sec^4 x}\ dx$

51. We have $\displaystyle\int_{-\pi/2}^{k}\cos x\ dx = 3\int_{k}^{\pi/2}\cos x\ dx,$ or

$$\sin k - \sin(-\pi/2) = 3(\sin \pi/2 - \sin k)$$
$$\sin k + 1 = 3 - 3\sin k$$
$$4\sin k = 2$$
$$\sin k = 1/2 \quad \text{and} \quad k = \pi/6$$

55. (a) $y = gd(x) = \tan^{-1}(\sinh x)$

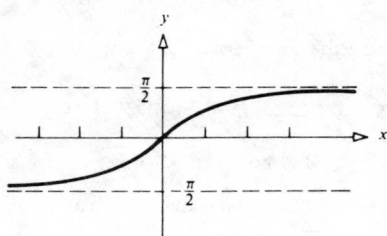

Note that $y' = \dfrac{\cosh x}{1 + \sinh^2 x} = \mathrm{sech}\ x$

(b) Let $y = \tan^{-1}(\sinh x)$ for $-\dfrac{\pi}{2} < y < \dfrac{\pi}{2}$.

Then for $-\dfrac{\pi}{2} < y < \dfrac{\pi}{2}$, we have

$$\tan y = \sinh x$$
$$1 + \tan^2 y = 1 + \sinh^2 x$$

$$\sec^2 y = \cosh^2 x$$

This gives sec y = cosh x, since both sides are positive if $-\pi/2 < y < \pi/2$. Hence, cos y = sech x.

Finally, sin y = tan y cos y = sinh x sech x = tanh x.

59. $x(t) = A \sin(\omega t + \phi_o)$. We are given that amplitude $A = 24$, period $= 2\pi/\omega = 4$, so $\omega = \pi/2$, and thus

$$x(t) = 24 \sin[\frac{\pi}{2}t + \phi_o].$$

Since $x(0) = 24$, $\phi_o = \frac{\pi}{2}$ and

$$x(t) = 24 \sin[\frac{\pi}{2}t + \frac{\pi}{2}] = 24 \cos[\frac{\pi}{2}t]$$

(a) $x(0.5) = 24 \cos(\frac{\pi}{4})$ cm $= +12\sqrt{2}$ cm

(b) acceleration $a(t) = x''(t) = -6\pi^2 \cos[\frac{\pi}{2}t]$ cm/sec^2

When $t = 0.5$, $a = -3\pi^2\sqrt{2}$ cm/sec^2

Since mass = 10 g, Force = Mass × Acceleration $= -30\pi^2\sqrt{2}$ dynes.

Direction is downward and magnitude is $30\pi^2\sqrt{2}$ dynes.

(c) Set $24 \cos[\frac{\pi}{2}T] = -12$. Then $\cos[\frac{\pi}{2}T] = -\frac{1}{2}$ or $\frac{\pi}{2}T = \frac{2\pi}{3}$ and $T = \frac{4}{3}$ sec.

(d) $v(t) = x'(t) = -12\pi\sin[\frac{\pi}{2}t]$. At $t = \frac{4}{3}$ sec, velocity is thus $-6\pi\sqrt{3}$ cm/sec.

Chapter Nine

Exercise 1, p. 470

3. $\int x(3x^2-5)^5 dx \qquad u = 3x^2 - 5; \quad du = 6x\ dx$

$= \frac{1}{6} \int u^5 du = (\frac{1}{6})\frac{1}{6} u^6 + C = \frac{1}{36}(3x^2-5)^6 + C$

7. $\int \frac{x}{(1-9x^2)^{1/2}} dx = u = 1 - 9x^2; \quad du = -18x\ dx$

$= \frac{-1}{18} \int u^{-1/2} du = (-\frac{1}{18})2\ u^{1/2} + C = -\frac{1}{9}(1-9x^2)^{1/2} + C$

11. $\int \frac{e^x}{1 + 2e^x} dx \qquad u = 1 + 2e^x; \quad du = 2e^x dx$

$= \frac{1}{2} \int \frac{du}{u} = \frac{1}{2}\ \ell n|u| + C = \frac{1}{2}\ \ell n(1+2e^x) + C$

15. $\int x\ e^{x^2} dx \qquad u = x^2; \quad du = 2x\ dx; \quad \frac{1}{2} du = x\ dx$

$= \frac{1}{2} \int e^u du = \frac{1}{2} e^u + C = \frac{1}{2} e^{x^2} + C$

19. $\int \sin x \cos x\ dx \qquad u = \sin x; \quad du = \cos x\ dx$

$= \int u\ du = \frac{1}{2} u^2 + C = \frac{1}{2} \sin^2 x + C$

23. $\int \frac{1 + \cos(2x)}{\sin^2(2x)} dx = \int [\frac{1}{\sin^2(2x)} + \frac{\cos 2x}{\sin^2 2x}]\ dx$

$= \frac{1}{2} \int \csc^2(u)\ du + \frac{1}{2} \int \csc(u) \cot(u)\ du$

$= -\frac{1}{2} \cot u - \frac{1}{2} \csc u + C = -\frac{1}{2}(\cot 2x + \csc 2x) + C,$
where $u = 2x, du = 2dx$

27. $\int \frac{\tan t}{\cos t} dt = \int \sec t \tan t\ dt = \sec t + C$

31. $\int \frac{\sin^2(3x)}{1 + \cos(3x)} dx = \int [1 - \cos(3x)]dx$

$$= x - \frac{1}{3} \sin(3x) + C$$

35. $\int \frac{\sinh x}{1 - \cosh x} \, dx$ $u = 1 - \cosh x; \quad du = -\sinh x \, dx$

$= -\int \frac{du}{u} = -\ell n|u| + C = -\ell n|1 - \cosh x| + C$

Exercise 2, pp. 484-485

3. $\int x \cos x \, dx = x \sin x - \int \sin x \, dx$

$= x \sin x + \cos x + C$

$\quad u = x \qquad dv = \cos x \, dx$
$\quad du = dx \qquad v = \sin x$

7. $\int \cot^{-1} x \, dx = x \cot^{-1} x + \int \frac{x}{x^2 + 1} \, dx$

$= x \cot^{-1} x + \frac{1}{2} \ell n (x^2 + 1) + C$

$\quad u = \cot^{-1} x \qquad dv = dx$
$\quad du = \frac{-1}{x^2 + 1} \qquad v = x$

11. $\int e^x \sin x \, dx$

$= e^x \sin x - \int e^x \cos x \, dx$ $\qquad u = \sin x \qquad dv = e^x dx$

$= e^x \sin x - e^x \cos x - \int e^x \sin x \, dx$ $\qquad du = \cos x \, dx \quad v = e^x$

$2 \int e^x \sin x \, dx = e^x (\sin x - \cos x) + C$ $\qquad u = \cos x \qquad dv = e^x dx$

$\int e^x \sin x \, dx = \frac{1}{2} e^x (\sin x - \cos x) + C$ $\qquad du = -\sin x \, dx \quad v = e^x$

15. $\int x^2 \sin x \, dx$

$\qquad\qquad\qquad\qquad\qquad\qquad u = x^2 \qquad dv = \sin x \, dx$

$= -x^2 \cos x + 2 \int x \cos x \, dx$ $\qquad du = 2x \, dx \quad v = -\cos x$

$= -x^2 \cos x + 2x \sin x - 2 \int \sin x \, dx$

$\qquad\qquad\qquad\qquad\qquad\qquad u = x \qquad dv = \cos x \, dx$

$= -x^2 \cos x + 2x \sin x + 2 \cos x + C$ $\qquad du = dx \qquad v = \sin x$

19. $\int_0^2 x^2 e^{-3x} dx$

$= -\frac{x^2}{3} e^{-3x} \Big|_0^2 + \frac{2}{3} \int_0^2 x e^{-3x} dx$ $\qquad u = x^2 \qquad dv = e^{-3x} dx$

$\qquad\qquad\qquad\qquad\qquad\qquad du = 2x \, dx \quad v = -\frac{1}{3} e^{-3x}$

$= -\frac{4}{3} e^{-6} + \frac{2}{3} [-\frac{1}{3} x e^{-3x} \Big|_0^2 + \frac{1}{3} \int_0^2 e^{-3x} dx]$

$$u = x \qquad dv = e^{-3x}dx$$
$$du = dx \qquad v = -\frac{1}{3}e^{-3x}$$

$$= -\frac{4}{3}e^{-6} - \frac{2}{9}(2e^{-6}) + \frac{2}{9}(-\frac{1}{3})e^{-3x}\Big|_0^2$$

$$= -\frac{4}{3}e^{-6} - \frac{4}{9}e^{-6} - \frac{2}{27}(e^{-6}-1) = e^{-6}(-\frac{4}{3} - \frac{4}{9} - \frac{2}{27}) + \frac{2}{27}$$

$$= \frac{2}{27} - e^{-6}(\frac{36}{27} + \frac{12}{27} + \frac{2}{27}) = \frac{2}{27} - \frac{50}{27}e^{-6}$$

23. $\int x^n \ln x \, dx$

$$u = \ln x \qquad dv = x^n dx$$

$$= \frac{x^{n+1}}{n+1}\ln x - \frac{1}{n+1}\int x^n dx \qquad du = \frac{1}{x}dx \qquad v = \frac{x^{n+1}}{n+1}$$

$$= \frac{x^{n+1}\ln x}{n+1} - \frac{x^{n+1}}{(n+1)^2} + C$$

27. $\int xe^x \cos x \, dx$

$$u = \cos x \qquad dv = xe^x dx$$
$$du = -\sin x \, dx \qquad v = xe^x - e^x$$

$= xe^x \cos x - e^x \cos x + \int xe^x \sin x \, dx - \int e^x \sin x \, dx$

Now evaluate $\int xe^x \sin x \, dx$ letting

$$u = \sin x \qquad dv = xe^x dx$$
$$du = \cos x \, dx \qquad v = xe^x - e^x$$

$\int xe^x \sin x \, dx = xe^x \sin x - e^x \sin x - \int xe^x \cos x \, dx$
$+ \int e^x \cos x \, dx$

From formulas 125 and 126 in the Table of Integrals, we have $\int e^x \sin x \, dx = \frac{e^x}{2}(\sin x - \cos x)$

$$\int e^x \cos x \, dx = \frac{e^x}{2}(\sin x + \cos x)$$

Substituting and collecting terms, we have
$\int xe^x \cos x \, dx = xe^x \cos x - e^x \cos x + xe^x \sin x$

$- e^x \sin x - \int xe^x \cos x \, dx + \dfrac{e^x}{2} (\sin x + \cos x)$

$- \dfrac{e^x}{2} (\sin x - \cos x) + C$

$2 \int xe^x \cos x \, dx = xe^x \cos x + xe^x \sin x - e^x \sin x + C$

$\qquad\qquad\qquad = e^x (x \cos x + x \sin x - \sin x) + C$

$\int xe^x \cos x \, dx = \dfrac{e^x}{2} [x(\cos x + \sin x) - \sin x] + C$

31. $\int x^2 \tan^{-1} x \, dx$ $\qquad\qquad u = \tan^{-1} x$ $\qquad\qquad dv = x^2 dx$

$\qquad\qquad\qquad\qquad\qquad\qquad du = \dfrac{1}{1 + x^2} \, dx$ $\qquad v = \dfrac{1}{3} x^3$

$= \dfrac{1}{3} x^3 \tan^{-1} x - \dfrac{1}{3} \int \dfrac{x^3}{1 + x^2} \, dx$ $\quad u = x^2$ $\qquad\qquad dv = \dfrac{x}{1 + x^2} \, dx$

$\qquad\qquad\qquad\qquad\qquad\qquad\qquad du = 2x \, dx$ $\qquad v = \dfrac{1}{2} \ln(1+x^2)$

Note: $\dfrac{x^3}{1+x^2} = x - \dfrac{x}{1+x^2}$, by long division.

Thus $\int x^2 \tan^{-1} x \, dx = \dfrac{1}{3} x^3 \tan^{-1} x - \dfrac{1}{3} \int x \, dx + \dfrac{1}{3} \int \dfrac{x}{1+x^2} \, dx$

$\qquad = \dfrac{1}{3} x^3 \tan^{-1} x - \dfrac{1}{6} x^2 + \dfrac{1}{6} \ln(1+x^2) + C$

35. $\int x^n \sin^{-1} x \, dx$ $\qquad u = \sin^{-1} x$ $\qquad\qquad dv = x^n dx$

$\qquad\qquad\qquad\qquad\qquad\qquad du = \dfrac{1}{\sqrt{1 - x^2}} \, dx$ $\qquad v = \dfrac{1}{n + 1} x^{n+1}$

$= \dfrac{x^{n+1}}{n + 1} \sin^{-1} x - \dfrac{1}{n + 1} \int \dfrac{x^{n+1}}{\sqrt{1 - x^2}} \, dx$

39. $\int \sin^n x \, dx \qquad u = \sin^{n-1} x \qquad\qquad\qquad dv = \sin x \, dx$

$\qquad\qquad\qquad du = (n-1)\sin^{(n-2)} x \cos x \, dx \qquad v = -\cos x$

$= -\sin^{n-1} x \cos x + (n-1) \int \sin^{n-2} x \cos^2 x \, dx$

$= -\sin^{n-1} x \cos x + (n-1) \int \sin^{n-2} x (1-\sin^2 x) dx$

$= -\sin^{n-1} x \cos x + (n-1) \int \sin^{(n-2)} x \, dx$

$\qquad\qquad\qquad\qquad - (n-1) \int \sin^n x \, dx$

Thus, $n \int \sin^n x \, dx = -\sin^{n-1} x \cos x$

$+ (n-1) \int \sin^{(n-2)} x \, dx$ so $\int \sin^n x \, dx$

$= -\dfrac{\sin^{n-1} x \cos x}{n} + \dfrac{n-1}{n} \int \sin^{n-2} x \, dx$

43. $y = \cos x$, $x \in [0, \pi/2]$. $\qquad\qquad$ Revolve about y-axis.

$V = 2\pi \int_0^{\pi/2} x \cos x \, dx \qquad\qquad u=x,\ dv = \cos x \, dx$
$\qquad\qquad\qquad\qquad\qquad\qquad\qquad du = dx,\ v = \sin x$

$= 2\pi[(x \sin x)\big|_0^{\pi/2} - \int_0^{\pi/2} \sin x \, dx] = 2\pi[\pi/2 + (\cos x)\big|_0^{\pi/2}]$

$= (\dfrac{\pi}{2}+0-1)2\pi = \pi^2 - 2\pi = \pi(\pi-2)$

47. (a) $\int_0^{\pi/2} \sin^6 x \, dx = \dfrac{(5)(3)(1)}{(6)(4)(2)}(\dfrac{\pi}{2}) = \dfrac{15}{96}\pi = \dfrac{5\pi}{32}$

(b) $\int_0^{\pi/2} \sin x^5 dx = \dfrac{(4)(2)}{(5)(3)(1)} = \dfrac{8}{15}$

(c) $\int_0^{\pi/2} \cos^8 x \, dx = \dfrac{(7)(5)(3)(1)}{(8)(6)(4)(2)}(\dfrac{\pi}{2}) = \dfrac{105\pi}{768} = \dfrac{35\pi}{256}$

(d) $\int_0^{\pi/2} \cos^6 x \, dx = \dfrac{15\pi}{32} = \dfrac{5\pi}{32}$

Exercise 3, pp. 490-491

3. $\int \dfrac{\sin x}{\cos^2 x} \, dx \qquad u = \cos x; \quad du = -\sin x \, dx$

$= -\int u^{-2} du = \dfrac{1}{u} + C = \dfrac{1}{\cos x} + C = \sec x + C$

7. $\int \sin^2 3x \, dx \qquad \sin^2 A = \dfrac{1}{2} - \dfrac{1}{2} \cos 2A$

$= \dfrac{1}{2} \int (1-\cos 6x) dx = \dfrac{1}{2} x - \dfrac{1}{12} \sin 6x + C$

11. $\int \sin^5 3x \, dx = \int (1-\cos^2 3x)^2 \sin 3x \, dx$

$= \int (1-2\cos^2 3x + \cos^4 3x) \sin 3x \, dx$

$= -\dfrac{1}{3} \int (1-2u^2+u^4) du$ (using $u = \cos 3x$,

$\qquad\qquad\qquad\qquad\qquad du = -3 \sin 3x \, dx$)

$= -\dfrac{1}{3} u + \dfrac{2}{9} u^3 - \dfrac{1}{15} u^5 + C$

$= -\dfrac{1}{3} \cos 3x + \dfrac{2}{9} \cos^3 3x - \dfrac{1}{15} \cos^5 3x + C$

15. $\int \sin^3x \cos^5x \, dx = \int \cos^5x(1-\cos^2x) \sin x \, dx$

$= \int \cos^5x \sin x \, dx - \int \cos^7x \sin x \, dx$

$= -\frac{1}{6} \cos^6x + \frac{1}{8} \cos^8x + C$

19. $\int \sin^{1/2}x \cos^3x \, dx = \int \sin^{1/2}x \cos^2x \cos x \, dx$

$= \int \sin^{1/2}x(1-\sin^2x) \cos x \, dx$

$= \int \sin^{1/2}x \cos x \, dx - \int \sin^{5/2}x \cos x \, dx$

$= \frac{2}{3} \sin^{3/2}x - \frac{2}{7} \sin^{7/2}x + C$

23. $\int \tan^3x \, dx = \int (\sec^2x-1) \tan x \, dx$

$= \int \tan x \sec^2x \, dx - \int \tan x \, dx$

$= \frac{1}{2} \tan^2x + \ln|\cos x| + C$

27. $\int \csc^3x \cot^5x \, dx = \int \csc^2x(\csc^2x-1)^2 \csc x \cot x \, dx$

$= \int (\csc^6x - 2 \csc^4x + \csc^2x) \csc x \cot x \, dx$

$u = \csc x; \quad du = -\csc x \cot x$

$= -\int (u^6-2u^4 + u^2)du = -\frac{1}{7} u^7 + \frac{2}{5} u^5 - \frac{1}{3} u^3 + C$

$= -\frac{1}{7} \csc^7x + \frac{2}{5} \csc^5x - \frac{1}{3} \csc^3x + C$

31. (Use $2 \sin A \cos B = \sin(A+B) + \sin(A-B)$)

$\int \sin 3x \cos x \, dx = \frac{1}{2} \int \sin 4x \, dx + \frac{1}{2} \int \sin 2x \, dx$

$= -\frac{1}{8} \cos 4x - \frac{1}{4} \cos 2x + C$

35. (Use $2 \cos A \cos B = \cos(A-B) + \cos(A+B)$)

$\int \cos 2x \cos x \, dx = \frac{1}{2} \int \cos x \, dx + \frac{1}{2} \int \cos 3x \, dx$

$= \frac{1}{2} \sin x + \frac{1}{6} \sin 3x + C$

39. $\int \sec^n x\, dx$, n even   Let n = 2k, and write

$\sec^n x = \sec^{2k} x = (\sec^2 x)^{k-1} \sec^2 x = (1+\tan^2 x)^{k-1} \sec^2 x$

$\quad = [1 + (k-1)\tan^2 x + \dfrac{(k-1)(k-2)}{2}\tan^4 x + \cdots$

$\quad\quad + (k-1)\tan^{2k-4} x + \tan^{2k-2} x]\sec^2 x$

Thus, $\int \sec^n x\, dx = \tan x + \dfrac{k-1}{3}\tan^3 x$

$\quad + \dfrac{(k-1)(k-2)}{2\cdot 5}\tan^5 x + \cdots + \dfrac{k-1}{2k-3}\tan^{2k-3} x$

$\quad + \dfrac{1}{2k-1}\tan^{2k-1} x + C$

43. $a(t) = \cos^2 t \sin t$ m/sec^2;   s(0) = 0; v(0) = 5 m/sec

$v(t) = \int \cos^2 t \sin t\, dt$ u = cos t; du = - sin t dt

$= -\dfrac{1}{3}\cos^3 t + C$ and v(0) = 5 =>

$5 = -\dfrac{1}{3}(1) + C$ => $C = 5 + \dfrac{1}{3} = \dfrac{16}{3}$

$v(t) = -\dfrac{1}{3}\cos^3 t + \dfrac{16}{3}$ => $s(t) = -\dfrac{1}{3}\int \cos^3 t\, dt + \dfrac{16}{3}\int dt$

$= -\dfrac{1}{3}\int (1-\sin^2 t)\cos t\, dt + \dfrac{16}{3}t + k$ u = sin t;
$\quad\quad\quad\quad\quad\quad\quad\quad\quad\quad\quad\quad\quad\quad\quad\quad\quad$ du = cos t dt

$s(t) = -\dfrac{1}{3}\sin t + (\dfrac{1}{3})(\dfrac{1}{3})\sin^3 t + \dfrac{16}{3}t + k$

$s(t) = \dfrac{1}{9}\sin^3 t - \dfrac{1}{3}\sin t + \dfrac{16}{3}t + k$ but s(0) = 0 =>

$0 = 0-0+0+k$ => k = 0 => $s(t) = \dfrac{1}{9}\sin^3 t - \dfrac{1}{3}\sin t + \dfrac{16}{3}t$

Exercise 4, pp. 495-496

3. $\int \dfrac{\sqrt{9-x^2}}{x^2}\, dx$

$x = 3\sin\theta$, $9 - x^2 = 9(1-\sin^2\theta) = 9\cos^2\theta$

$dx = 3\cos\theta\, d\theta$, $\sqrt{9-x^2} = 3\cos\theta$

$$= \int \frac{3 \cos \theta}{9 \sin^2\theta} \, 3 \cos d\theta = \int \frac{\cos^2\theta}{\sin^2\theta} \, d\theta = \int \cot^2\theta \, d\theta$$

$$= \int (\csc^2\theta - 1) d\theta = -\cot\theta - \theta + C = \frac{-\cos \theta}{\sin \theta} - \theta + C$$

$$= -\frac{\sqrt{9 - x^2}}{x} - \sin^{-1} \frac{x}{3} + C$$

7. $\int \dfrac{dx}{(x^2+4)^{3/2}}$

$x = 2 \tan \theta, \qquad x^2 + 4 = 4(\tan^2\theta + 1) = 4 \sec^2\theta$

$dx = 2 \sec^2\theta \, d\theta, \ \sqrt{x^2 + 4} = 2 \sec \theta$

$$= \int \frac{2 \sec^2\theta \, d\theta}{8 \sec^3\theta}$$

$$= \frac{1}{4} \int \cos \theta \, d\theta$$

$$= \frac{1}{4} \sin \theta + C$$

$$= \frac{x}{4\sqrt{x^2 + 4}} + C$$

$$\sin \theta = \frac{x}{\sqrt{x^2 + 4}}$$

11. $\int \sqrt{16 - x^2} \, dx \qquad x = 4 \sin \theta, \qquad 16 - x^2 = 16 \cos^2\theta$

$\qquad\qquad\qquad\qquad\qquad dx = 4 \cos \theta \, d\theta, \ \sqrt{16 - x^2} = 4 \cos \theta$

$$= \int 16 \cos^2\theta \, d\theta = \frac{16}{2} \int (1 + \cos 2\theta) d\theta = 8\theta + 4 \sin 2\theta +$$

$$= 8 \sin^{-1}(\tfrac{x}{4}) + 8(\tfrac{x}{4})(\frac{\sqrt{16 - x^2}}{4}) + C \quad \text{(using } \sin 2\theta$$
$$\qquad\qquad\qquad\qquad\qquad\qquad\qquad\qquad\qquad = 2 \sin \theta \cos \theta)$$

$$= 8 \sin^{-1}(\tfrac{x}{4}) + \frac{1}{2} x\sqrt{16 - x^2} + C$$

15. $\int \dfrac{x^2}{\sqrt{16 - x^2}} \, dx \qquad x = 4 \sin \theta, \qquad 16 - x^2 = 16 \cos^2\theta$

$\qquad\qquad\qquad\qquad\qquad dx = 4 \cos \theta \, d\theta, \ \sqrt{16 - x^2} = 4 \cos \theta$

$$= \int \frac{16 \sin^2\theta}{4 \cos \theta} \, 4 \cos \theta \, d\theta = 16 \int \sin^2\theta \, d\theta$$

$$= 8 \int (1-\cos 2\theta)d\theta = 8\theta - 4 \sin 2\theta + C \quad \textbf{(use } \sin2\theta =2 \ \sin\theta \ \cos\theta)$$

$$= 8\theta - 8 \sin \theta \cos \theta + C = 8 \sin^{-1}(\tfrac{x}{4}) - \tfrac{1}{2} x\sqrt{16 - x^2} + C$$

19. $\int \dfrac{dx}{(x^2-9)^{3/2}}$ $\qquad$ $x = 3 \sec \theta,$ $\qquad\qquad$ $x^2 - 9 = 9 \tan^2\theta$

$\qquad\qquad\qquad\qquad\qquad\quad dx = 3 \sec \theta \tan \theta \, d\theta, \qquad (x^2-9)^{1/2} = 3 \tan \theta$

$$= \int \frac{3 \sec \theta \tan \theta \, d\theta}{27 \tan^3\theta}$$

$$= \frac{1}{9} \int \frac{\sec \theta}{\tan^2\theta} \, d\theta$$

$$= \frac{1}{9} \int (\frac{1}{\cos \theta})(\frac{\cos^2\theta}{\sin^2\theta}) d\theta$$

$$= \frac{1}{9} \int \frac{\cos \theta}{\sin^2\theta} \, d\theta$$

$$= \frac{1}{9}(- \frac{1}{\sin \theta}) + C$$

$$= \frac{-x}{9\sqrt{x^2 - 9}} + C \qquad\qquad \sin \theta = \sqrt{x^2 - 9}/x$$

23. $\int \dfrac{dx}{\sqrt{(x-1)^2 - 4}}$ $\qquad$ $x - 1 = 2 \sec \theta$

$\qquad\qquad\qquad\qquad\quad (x-1)^2 - 4 = 4 \tan^2\theta$

$\qquad\qquad\qquad\qquad\quad dx = 2 \sec \theta \tan \theta \, d\theta$

$\qquad\qquad\qquad\qquad\quad \sqrt{(x-1)^2 - 4} = 2 \tan \theta$

$$= \int \frac{2 \sec \theta \tan \theta \, d\theta}{2 \tan \theta} = \int \sec \theta \, d\theta$$

$$= \ln|\sec \theta + \tan \theta| + K = \ln\left|\frac{x - 1 + \sqrt{(x-1)^2 - 4}}{2}\right| + K$$

$$= \ln|x - 1 + \sqrt{(x-1)^2 - 4}| - \ln 2 + K$$

$$= \ln|x - 1 + \sqrt{(x-1)^2 - 4}| + C, \text{ where } C = -\ln 2 + K$$

27. $\dfrac{dx}{\sqrt{(x+1)^2 - 4}}$ $\left[\begin{array}{ll} x + 1 = 2 \sec \theta & (x+1)^2 - 4 = 4 \tan^2\theta \\ dx = 2 \sec \theta \tan \theta \, d\theta & \sqrt{(x+1)^2-4} = 2 \tan \theta \end{array}\right]$

$$= \int \frac{2 \sec \theta \tan \theta \, d\theta}{2 \tan \theta} = \int \sec \theta \, d\theta$$

$$= \ln|\sec \theta + \tan \theta| + K = \ln\left|\frac{(x+1) + \sqrt{(x+1)^2 - 4}}{2}\right| + K$$

$$= \ln|x + 1 + \sqrt{(x+1)^2 - 4}| + C, \text{ where } C = -\ln 2 + K$$

31. First let's look at the indefinite integral:

$$\begin{cases} 2x = \tan \theta, \ 2 \, dx = \sec^2\theta d\theta \\ \sqrt{1 + 4x^2} = \sqrt{1 + \tan^2\theta} = \sec \theta \end{cases}$$

$$\int \frac{x^3}{\sqrt{1 + 4x^2}} \, dx = \int \frac{\frac{1}{8}\tan^3\theta}{\sec \theta}\left(\frac{1}{2} \sec^2\theta \, d\theta\right)$$

$$= \frac{1}{16}\int \tan^3\theta \sec \theta \, d\theta$$

$$= \frac{1}{16}\int \tan^2\theta(\tan \theta \sec \theta \, d\theta)$$

$$= \frac{1}{16}\int (\sec^2\theta-1)(\tan \theta \sec \theta) \, d\theta$$

$(u = \sec \theta, \ du = \sec \theta \tan \theta \, d\theta)$

$$= \frac{1}{16}\int (u^2-1)du = \frac{1}{16}\left(\frac{1}{3}u^3-u\right) + C$$

$$= \frac{1}{48} \sec^3\theta - \frac{1}{16} \sec \theta + C$$

$$= \frac{1}{48}(\sqrt{1 + 4x^2})^3 - \frac{1}{16} \sqrt{1 + 4x^2} + C$$

Now, the definite integral:

$$\int_1^2 \frac{x^3}{\sqrt{1 + 4x^2}} \, dx = \left[\frac{1}{48}(\sqrt{1 + 4x^2})^3 - \frac{1}{16} \sqrt{1 + 4x^2}\right]\Big|_1^2$$

$$= \left[\frac{17\sqrt{17}}{48} - \frac{\sqrt{17}}{16}\right] - \left[\frac{5\sqrt{5}}{48} - \frac{\sqrt{5}}{16}\right] = \frac{14\sqrt{17} - 2\sqrt{5}}{48} = \frac{7\sqrt{17} - \sqrt{5}}{24}$$

35. $\int \dfrac{dx}{\sqrt{x^2 - a^2}}$

$$x = a \sec \theta$$
$$dx = a \sec \theta \tan \theta \, d\theta$$
$$x^2 - a^2 = a^2 \tan^2 \theta$$
$$\sqrt{x^2 - a^2} = a \tan \theta$$

$$\int \dfrac{a \sec \theta \tan \theta \, d\theta}{a \tan \theta} = \int \sec \theta \, d\theta$$

$$= \ln|\sec \theta + \tan \theta| + C$$

$$= \ln \left| \dfrac{x}{a} + \dfrac{\sqrt{x^2 - a^2}}{a} \right| + C$$

39. If $y = 5x - x^2$ then $y' = 5 - 2x$ and $\sqrt{1 + y'^2}$ $= \sqrt{1 + (5-2x)^2}$. Note that $y = 0$ if $x = 0$ or $x = 5$, so that

$$L = \int_0^5 \sqrt{1 + (5-2x)^2} \, dx$$

$$u = 5 - 2x \qquad x = 0 \implies u = 5$$
$$du = -2dx \qquad x = 5 \implies u = -5$$

$$= \dfrac{-1}{2} \int_5^{-5} \sqrt{1 + u^2} \, du = \dfrac{1}{2} \int_{-5}^5 \sqrt{u^2 + 1} \, du$$

In Problem 36 , we showed that

$$\int \sqrt{x^2 + a^2} \, dx = \dfrac{1}{2} x\sqrt{x^2 + a^2} + \dfrac{a^2}{2} \ln \left| \dfrac{\sqrt{x^2 + a^2} + x}{a} \right| + C$$

Thus,

$$L = \dfrac{1}{2} \int_{-5}^5 \sqrt{u^2+1} \, du = \dfrac{1}{2} \left(\dfrac{1}{2} u\sqrt{u^2 + 1} + \dfrac{1}{2} \ln|u + \sqrt{u^2 + 1}| \right) \Big|_{-5}^5$$

$$= \dfrac{1}{4}(5\sqrt{26}) + \dfrac{1}{4} \ln(5+\sqrt{26}) - \dfrac{1}{4}(-5\sqrt{26}) - \dfrac{1}{4} \ln(-5+\sqrt{26})$$

$$= \dfrac{5}{2} \sqrt{26} + \dfrac{1}{4} \ln \left[\dfrac{\sqrt{26} + 5}{\sqrt{26} - 5} \right] \approx 13.9$$

3. $\int \dfrac{dx}{\sqrt{8 + 2x - x^2}} = \int \dfrac{dx}{\sqrt{8 - (x^2-2x)}}$

$\qquad\qquad = \int \dfrac{dx}{\sqrt{9 - (x^2-2x+1)}} \qquad \begin{array}{l} u = x - 1 \\ du = dx \end{array}$

$\qquad\qquad = \int \dfrac{du}{\sqrt{3^2 - u^2}} \qquad = \sin^{-1}(\dfrac{u}{3}) + C$

$\qquad\qquad = \sin^{-1}(\dfrac{x-1}{3}) + C$

7. $\int \dfrac{dx}{\sqrt{24 - 2x - x^2}} = \int \dfrac{dx}{\sqrt{24 - (x^2+2x)}}$

$\qquad\qquad = \int \dfrac{dx}{\sqrt{25 - (x^2+2x+1)}} \qquad \left\{\begin{array}{l} u = x + 1 \\ du = dx \end{array}\right\}$

$\qquad\qquad = \int \dfrac{du}{\sqrt{5^2 - u^2}}$

$\qquad\qquad = \sin^{-1}(\dfrac{u}{5}) + C = \sin^{-1}(\dfrac{x+1}{5}) + C$

11. $\displaystyle\int_1^3 \dfrac{dx}{\sqrt{x^2 - 2x + 5}} = \int_1^3 \dfrac{dx}{\sqrt{(x^2-2x+1) + 4}}$

$\qquad\qquad = \displaystyle\int_1^3 \dfrac{dx}{\sqrt{(x-1)^2 + 2^2}} \qquad \left\{\begin{array}{l} x - 1 = 2\tan\theta \\ dx = 2\sec^2\theta\ d\theta \\ \sqrt{\tan^2\theta + 1} = \sec\theta \\ x = 1 \Rightarrow \theta = 0 \\ x = 3 \Rightarrow \theta = \dfrac{\pi}{4} \end{array}\right.$

$\qquad\qquad = \displaystyle\int_1^{\pi/4} \dfrac{2\sec^2\theta\ d\theta}{2\sqrt{\tan^2\theta + 1}}$

$\qquad\qquad = \displaystyle\int_0^{\pi/4} \dfrac{\sec^2\theta\ d\theta}{\sec\theta} = \int_0^{\pi/4} \sec\theta\ d\theta$

$\qquad\qquad = \ln|\sec\theta + \tan\theta| \Big|_0^{\pi/4}$

$\qquad\qquad = \ln|\sqrt{2}+1| - \ln|1+0| = \ln(\sqrt{2}+1)$

15. $\int \dfrac{dx}{\sqrt{(x+h)^2 + k}} = \int \dfrac{\sqrt{k}\ \sec^2\theta\ d\theta}{\sqrt{k\ \tan^2\theta + k}}$

$\left\{\begin{array}{l} \text{For } k > 0, \\ x + h = \sqrt{k}\ \tan\theta \\ dx = \sqrt{k}\ \sec^2\theta\ d\theta \\ \sqrt{\tan^2\theta + 1} = \sec\theta \end{array}\right.$

$= \int \dfrac{\sqrt{k}\ \sec^2\theta\ d\theta}{\sqrt{k(\tan^2\theta + 1)}} = \int \dfrac{\sec^2\theta}{\sec\theta}\ d\theta$

$= \int \sec\theta\ d\theta = \ln|\sec\theta + \tan\theta| + C$

$= \ln\left|\dfrac{\sqrt{(x+h)^2 + k} + x + h}{\sqrt{k}}\right| + C$

$\tan\theta = \dfrac{x + h}{\sqrt{k}} \Rightarrow$

$\sec\theta = \dfrac{\sqrt{(x+h)^2 + k}}{\sqrt{k}}$

$= \ln|\sqrt{(x+h)^2 + k} + x + h| - \ln(\sqrt{k}) + C$

$= \ln|\sqrt{(x+h)^2 + k} + x + h| - M$

where $M = C - \ln(\sqrt{k})$
is an arbitrary constant

Note: The absolute values are no longer necessary
since $k > 0$ guarantees that $\sqrt{(x+h)^2 + k} > |x + h|$
and hence $\sqrt{(x+h)^2 + k} + x + h > 0$

Exercise 6, pp. 504-505

3. $\int \dfrac{x^3 + 3x - 4}{x - 2}$

Using long division,

$= \int \left(x^2 + 2x + 7 + \dfrac{10}{x-2}\right)dx$

$= \dfrac{1}{3}x^3 + x^2 + 7x + 10\ \ln|x - 2| + C$

$\begin{array}{r} x^2\ +\ 2x\ +\ 7 \\ x-2\ \overline{\smash{)}\ x^3\qquad +\ 3x\ -\ 4} \\ \underline{x^3\ -\ 2x^2} \\ 2x^2\ +\ 3x \\ \underline{2x^2\ -\ 4x} \\ 7x\ -\ 4 \\ \underline{7x\ -14} \\ 10 \end{array}$

7. Let $I = \int \dfrac{x\ dx}{(x-1)(x-2)} = A \int \dfrac{dx}{x - 1} + B \int \dfrac{dx}{x - 2}$

$\dfrac{x}{(x-1)(x-2)} = \dfrac{A}{x - 1} + \dfrac{B}{x - 2} \Rightarrow x = (A+B)x - 2A - B$

$$A + B = 1 \qquad -A = 1 \qquad A = -1$$
$$-2A - B = 0 \qquad\qquad\qquad B = 2$$

$$I = -\ln|x - 1| + 2\ln|x - 2| + C = \ln\left|\frac{(x-2)^2}{(x-1)}\right| + C$$

11. Let $I = \int \dfrac{(x-3)}{(x+2)(x+1)^2}\, dx$

$$= A \int \frac{dx}{x+2} + B \int \frac{dx}{x+1} + C \int \frac{dx}{(x+1)^2}$$

$$\frac{x-3}{(x+2)(x+1)^2} = \frac{A}{x+2} + \frac{B}{x+1} + \frac{C}{(x+1)^2} \implies$$

$$x - 3 = A(x^2+2x+1) + B(x^2+3x+2) + C(x+2)$$

$$A + B \qquad\quad = 0 \qquad\qquad\qquad A = -5$$
$$2A + 3B + C = 1 \qquad\qquad\qquad B = 5$$
$$A + 2B + 2C = -3 \qquad\qquad\qquad C = -4$$

$$I = -5\ln|x + 2| + 5\ln|x + 1| + \frac{4}{x+1} + C$$

$$= 5\ln\left|\frac{x+1}{x+2}\right| + \frac{4}{x+1} + C$$

15. Let $I = \int \dfrac{x\, dx}{(x+3)(x-1)} = A \int \dfrac{dx}{(x+3)} + B \int \dfrac{dx}{(x-1)}$

$$\frac{x}{(x+3)(x-1)} = \frac{A}{(x+3)} + \frac{B}{(x-1)} \implies$$

$$x = A(x-1) + B(x+3) = (A+B)x - A + 3B$$

$$A + B = 1 \qquad\qquad A = 3/4$$
$$-A + 3B = 0 \qquad\qquad B = 1/4$$

$$I = \frac{3}{4}\ln|x + 3| + \frac{1}{4}\ln|x - 1| + C$$

$$= \frac{1}{4}\ln\left|(x+3)^3(x-1)\right| + C$$

19. Let $I = \int \dfrac{x^2\, dx}{x^3 - 4x^2 + 5x - 2} = \int \dfrac{x^2\, dx}{(x-2)(x-1)^2}$

$$= A \int \frac{dx}{x-2} + B \int \frac{dx}{x-1} + C \int \frac{dx}{(x-1)^2}$$

$$\frac{x^2}{(x-2)(x-1)^2} = \frac{A}{x-2} + \frac{B}{x-1} + \frac{C}{(x-1)^2}$$

$$\Rightarrow x^2 = A(x^2-2x+1) + B(x^2-3x+2) + C(x-2)$$

$$= (A+B)x^2 + (-2A-3B+C)x + A + 2B - 2C$$

$$\begin{array}{ll} A + B = 1 & A = 4 \\ -2A - 3B + C = 0 & B = -3 \\ A + 2B - 2C = 0 & C = -1 \end{array}$$

$$I = 4 \ln|x - 2| - 3 \ln|x - 1| + \frac{1}{x-1} + C$$

$$= \ln\left|\frac{(x-2)^4}{(x-1)^3}\right| + \frac{1}{x-1} + C$$

23. Let $I = \int_0^1 \frac{dx}{x^2 - 9} = \frac{1}{6} \int_0^1 \frac{dx}{x-3} - \frac{1}{6} \int_0^1 \frac{dx}{x+3}$

From (9.17), $\frac{A}{x-a} + \frac{B}{x+a} = \frac{A}{x-3} + \frac{B}{x+3}$ and

$A = 1/(2a) = 1/6 \qquad B = -1/(2a) = -1/6$

$$I = \left(\frac{1}{6} \ln|x-3| - \frac{1}{6} \ln|x+3|\right)\Big|_0^1 = \frac{1}{6} \ln\left|\frac{x-3}{x+3}\right|\Big|_0^1$$

$$= \frac{1}{6}\left(\ln\frac{1}{2} - \ln 1\right) = \frac{1}{6} \ln\frac{1}{2} = -\frac{\ln 2}{6}$$

27. $y = \frac{4}{x^2 - 4} \quad x \in [3,5] \quad$ From (9.17)

$A = 1/4, \; B = -1/4.$

$$\text{Area} = \int_3^5 \frac{4 \, dx}{x^2 - 4} = 4 \int_3^5 \frac{dx}{x^2 - 4}$$

$$= \frac{4}{4} \int_3^5 \frac{dx}{x-2} - \frac{4}{4} \int_3^5 \frac{dx}{x+2}$$

$$\text{Area} = \left(\ln|x-2| - \ln|x+2|\right)\Big|_3^5$$

$$= \left(\ln\left|\frac{x-2}{x+2}\right|\right)\Big|_3^5 = \ln\frac{3}{7} - \ln\frac{1}{5} = \ln\frac{15}{7}$$

3. Let $I = \int \dfrac{2x + 1}{x^3 - 1}\, dx = \int \dfrac{2x + 1}{(x-1)(x^2+x+1)}\, dx$

$\qquad = A \int \dfrac{dx}{x - 1} + \int \dfrac{Bx + C}{x^2 + x + 1}\, dx$

$\dfrac{2x + 1}{(x-1)(x^2+x+1)} = \dfrac{A(x^2+x+1) + (Bx+C)(x-1)}{(x-1)(x^2+x+1)}$

$\Rightarrow 2x + 1 = Ax^2 + Ax + A + Bx^2 - Bx + Cx - C$

$\begin{array}{lll} A + B = 0 & A = 1 & Bx + C = -x = \frac{-1}{2}(2x+1-1) \\ A - B + C = 2 & B = -1 & \\ A \quad - C = 1 & C = 0 & \end{array}$

$I = \int \dfrac{dx}{x - 1} - \dfrac{1}{2} \int \dfrac{2x + 1 - 1}{x^2 + x + 1}\, dx$

$\qquad = \ell n|x - 1| - \dfrac{1}{2} \int \dfrac{2x + 1}{x^2 + x + 1}\, dx \qquad\qquad w = x^2 + x + 1$

$\qquad\qquad\qquad\qquad\qquad\qquad\qquad\qquad\qquad\qquad dw = 2x + 1$

$\qquad\qquad + \dfrac{1}{2} \int \dfrac{dx}{(x + \frac{1}{2})^2 + (\frac{\sqrt{3}}{2})^2}$

$\qquad\qquad\qquad\qquad\qquad\qquad\qquad\qquad u = x + \dfrac{1}{2} \qquad\qquad a =$

$\qquad\qquad\qquad\qquad\qquad\qquad\qquad\qquad du = dx$

$\qquad = \ell n|x - 1| - \dfrac{1}{2} \ell n|x^2 + x + 1| + \dfrac{1}{\sqrt{3}} \tan^{-1} \dfrac{2x + 1}{\sqrt{3}} + C$

$\qquad = \ell n \dfrac{|x - 1|}{\sqrt{x^2 + x + 1}} + \dfrac{1}{\sqrt{3}} \tan^{-1} \dfrac{2x + 1}{\sqrt{3}} + C$

7. Let $I = \int \dfrac{x^2 + 2x + 3}{(x+1)(x^2+2x+4)}\, dx$

$\qquad = A \int \dfrac{dx}{x + 1} + \int \dfrac{Bx + C}{x^2 + 2x + 4}\, dx$

$x^2 + 2x + 3 = A(x^2+2x+4) + (Bx+C)(x+1)$

$\begin{array}{llll} A + B = 1 & A = 2/3 & Bx + C = \frac{1}{3}x + \frac{1}{3} \\ 2A + B + C = 2 & B = 1/3 & \\ 4A \quad + C = 3 & C = 1/3 & = \frac{1}{6}(2x+2) \end{array}$

$$I = \frac{2}{3} \int \frac{dx}{x + 1} + \frac{1}{6} \int \frac{2x + 2}{x^2 + 2x + 4} \, dx$$

$$= \frac{2}{3} \ell n |x + 1| + \frac{1}{6} \ell n |x^2 + 2x + 4| + C$$

$$= \frac{1}{3} \ell n \left| (x+1)^2 \sqrt{x^2 + 2x + 4} \right| + C$$

11. Let $I = \int \frac{x^3}{(x^2+16)^3} \, dx = \int \frac{Ax + B}{x^2 + 16} \, dx + \int \frac{Cx + D}{(x^2+16)^2} \, dx$

$$+ \int \frac{Ex + F}{(x^2+16)^3} \, dx$$

$x^3 = (Ax+B)(x^4+32x^2+256) + (Cx + D)(x^2+16) + Ex + F$

$\quad = Ax^5 + 32Ax^3 + 256Ax + Bx^4 + 32Bx^2 + 256B + Cx^3$

$\quad + 16 Cx + Dx^2 + 16D + Ex + F$

A	= 0	A = 0	Ax + B = 0
B	= 0	B = 0	Cx + D = x
32A + C	= 1	C = 1	Ex + F = -16x
32B + D	= 0	D = 0	
256A + 16C + E	= 0	E = -16	
256B + 16D + F	= 0	F = 0	

$$I = \int \frac{x}{(x^2+16)^2} \, dx - 16 \int \frac{x \, dx}{(x^2+16)^3}$$

Let $u = x^2 + 16$, $du = 2x \, dx$, $\frac{1}{2} du = x \, dx$.

$$I = \frac{1}{2} \int \frac{du}{u^2} - \frac{16}{2} \int \frac{du}{u^3} = \frac{1}{2}(-\frac{1}{u}) - 8(-\frac{1}{2})(\frac{1}{u^2}) + C$$

$$= [-\frac{1}{2}(\frac{1}{x^2+16})] + 4[\frac{1}{(x^2+16)^2}] + C$$

$$= -\frac{1}{2}[\frac{x^2 + 16}{(x^2+16)^2}] + 4[\frac{1}{(x^2+16)^2}] + C$$

$$= \frac{1}{(x^2+16^2)^2} (-\frac{x^2}{2} - 8 + 4) + C$$

$$= \frac{1}{(x^2+16^2)^2} (\frac{-x^2 - 8}{2}) + C = \frac{-(x^2+8)}{2(x^2+16^2)^2} + C$$

15. $y = \dfrac{8}{x^3 + 1}$ $x \in [0,2]$

Area $= 8 \displaystyle\int_0^2 \dfrac{dx}{x^3 + 1} = 8 \displaystyle\int_0^2 \dfrac{dx}{(x+1)(x^2-x+1)}$

$\qquad = 8A \displaystyle\int_0^2 \dfrac{dx}{x + 1} + 8 \displaystyle\int_0^2 \dfrac{Bx + C}{x^2 - x + 1}\, dx$

$\dfrac{1}{(x+1)(x^2-x+1)} = \dfrac{A}{x + 1} + \dfrac{Bx + C}{x^2 - x + 1}$

$\Longrightarrow 1 = Ax^2 - Ax + A + Bx^2 + Bx + Cx + C$

$A + B = 0$	$A = 1/3$	$Bx + C = -\dfrac{1}{3}x + \dfrac{2}{3}$
$-A + B + C = 0$	$B = -1/3$	
$A + C = 1$	$C = 2/3$	$= -\dfrac{1}{3}(x-2)$

Area $= \dfrac{8}{3} \displaystyle\int_0^2 \dfrac{dx}{x + 1} - \dfrac{8}{3} \displaystyle\int_0^2 \dfrac{x - 2}{x^2 - x + 1}\, dx$

To evaluate the second integral, write $x - 2$

as $\dfrac{1}{2}(2x-4) = \dfrac{1}{2}[(2x-1) - 3]$, and $x^2 - x + 1$ as

$(x - \dfrac{1}{2})^2 + (\dfrac{\sqrt{3}}{2})^2$ so that the integral becomes

$-\dfrac{4}{3} \displaystyle\int_0^2 \dfrac{(2x-1)}{x^2 - x + 1}\, dx + 4 \displaystyle\int \dfrac{dx}{(x - \dfrac{1}{2})^2 + (\dfrac{\sqrt{3}}{2})^2}$

Area $= \dfrac{8}{3}(\ell n|x+1|) \Big|_0^2 - \dfrac{4}{3}(\ell n|x^2-x+1|) \Big|_0^2$

$\qquad + 4(\dfrac{2}{\sqrt{3}}) \tan^{-1}(\dfrac{2x-1}{\sqrt{3}}) \Big|_0^2$

$\qquad = \dfrac{8}{3} \ell n\, 3 - \dfrac{4}{3} \ell n\, 3 + \dfrac{8}{\sqrt{3}} \tan^{-1}\sqrt{3} - \dfrac{8}{\sqrt{3}} \tan^{-1}(-\dfrac{1}{\sqrt{3}})$

$\qquad = \dfrac{4}{3} \ell n\, 3 + \dfrac{8}{\sqrt{3}}(\dfrac{\pi}{3}) - \dfrac{8}{\sqrt{3}}(\dfrac{-\pi}{6}) = \dfrac{4}{3} \ell n\, 3 + \dfrac{8\pi}{3\sqrt{3}} + \dfrac{4\pi}{3\sqrt{3}}$

$\qquad = \dfrac{4}{3} \ell n\, 3 + \dfrac{4\pi}{\sqrt{3}} \approx 8.72$

Note: In each case we use the substitutions $\sin x = 2z/(1+z^2)$, $\cos x = (1-z^2)/(1+z^2)$, $dx = 2dz/(1+z^2)$, and $z = \tan(x/2)$.

3. $\displaystyle \int \frac{dx}{1 - \cos x} = \int \frac{\dfrac{2dz}{1 + z^2}}{1 - \dfrac{1 - z^2}{1 + z^2}} = \int \frac{2dz}{1 + z^2 - (1-z^2)}$

$\displaystyle = \int \frac{2dz}{2z^2} = -\frac{1}{z} + C = \frac{-1}{\tan(x/2)} + C$

7. Let $\displaystyle I = \int \frac{\sin x \, dx}{3 + \cos x}$

$\displaystyle \frac{\sin x}{3 + \cos x} = [\frac{2z}{1 + z^2}][\frac{1 + z^2}{3(1+z^2) + (1-z^2)}]$

$\displaystyle = \frac{2z}{2z^2 + 4} = \frac{z}{z^2 + 2}$

$\displaystyle I = \int \frac{2z \, dz}{(1+z^2)(2+z^2)} = \int \frac{Az + B}{z^2 + 1} \, dz + \int \frac{Cz + D}{z^2 + 2} \, dz$

$2z = (Az+B)(z^2+2) + (Cz+D)(z^2+1)$

$= Az^3 + 2Az + Bz^2 + 2B + Cz^3 + Cz + Dz^2 + D$

$A + C = 0$	$A = 2$	$Az + B = 2z$
$B + D = 0$	$B = 0$	$Cz + D = -2z$
$2A + C = 2$	$C = -2$	
$2B + D = 0$	$D = 0$	

$\displaystyle I = \int \frac{2z \, dz}{z^2 + 1} - \int \frac{2z \, dz}{z^2 + 2} = \ln|z^2 + 1| - \ln|z^2 + 2| + C$

$\displaystyle = \ln(\frac{1 + z^2}{2 + z^2}) + C = \ln[\frac{1 + \tan^2(x/2)}{2 + \tan^2(x/2)}] + C$

Using $\tan^2(x/2) = \dfrac{1 - \cos x}{1 + \cos x}$, we can express the

answer as $\ln \left| \dfrac{1 + \dfrac{1 - \cos x}{1 + \cos x}}{2 + \dfrac{1 - \cos x}{1 + \cos x}} \right| + C = \ln \left| \dfrac{2}{3 + \cos x} \right| + C$

$= \ln 2 - \ln|3 + \cos x| + C = -\ln|3 + \cos x| + K$

[Note: Substituting $u = 3 + \cos x$ in the original integral would produce the same result more easily.]

11. Let $I = \int \dfrac{\cot x \, dx}{1 + \sin x}$

$\cot x = \dfrac{1 - z^2}{2z}$; $1 + \sin x = 1 + \dfrac{2z}{1 + z^2}$

$$= \dfrac{1 + z^2 + 2z}{1 + z^2} = \dfrac{(z+1)^2}{1 + z^2}$$

$I = \int \dfrac{1 - z^2}{2z} \left[\dfrac{1 + z^2}{(1+z)^2}\right] \dfrac{2dz}{1 + z^2} = \int \dfrac{1 - z^2}{z(1+z)^2} \, dz$

$\quad = \int \dfrac{(1-z)(1+z)}{z(1+z)^2} \, dz = \int \dfrac{1 - z}{z(1+z)} \, dz$

$\quad = A \int \dfrac{dz}{z} + B \int \dfrac{dz}{1 + z} \qquad A = 1 \qquad\qquad A = 1$

$\qquad\qquad\qquad\qquad\qquad\qquad\qquad A + B = -1 \qquad B = -2$

$1 - z = A(1+z) + B(z)$

$I = \int \dfrac{dz}{z} - 2 \int \dfrac{dz}{1 + z} = \ln|z| - 2 \ln|1 + z| + C$

$\quad = \ln\left|\tan(x/2)\right| - 2 \ln\left|1 + \tan(x/2)\right| + C$

$\quad = \ln\left|\dfrac{\tan(x/2)}{[1 + \tan(x/2)]^2}\right| + C$

15. Let $I = \int \dfrac{\cos x}{2 - \cos x} \, dx$

$2 - \cos x = 2 - \left(\dfrac{1 - z^2}{1 + z^2}\right) = \dfrac{2 + 2z^2 - 1 + z^2}{1 + z^2}$

$\qquad\qquad = \dfrac{3z^2 + 1}{1 + z^2}$

$I = \int \left(\dfrac{1 - z^2}{1 + z^2}\right)\left(\dfrac{1 + z^2}{3z^2 + 1}\right)\left(\dfrac{2dz}{1 + z^2}\right) = \int \dfrac{2 - 2z^2}{(z^2+1)(3z^2+1)} \, dz$

$\quad = \int \dfrac{Az + B}{z^2 + 1} \, dz + \int \dfrac{Cz + D}{3z^2 + 1} \, dz$

$2 - 2z^2 = (Az + B)(3z^2+1) + (Cz + D)(z^2+1)$

$\qquad\quad = 3Az^3 + Az + 3Bz^2 + B + Cz^3 + Cz + Dz^2 + D$

$3A + C = 0$	$A = 0$	$Az + B = -2$
$3B + D = -2$	$B = -2$	$Cz + D = 4$
$A + C = 0$	$C = 0$	
$B + D = 2$	$D = 4$	

$$I = -\int \frac{2dz}{z^2 + 1} + \int \frac{4dz}{3z^2 + 1} = -2 \tan^{-1}z + \frac{4}{\sqrt{3}} \tan^{-1}(\sqrt{3}\, z)$$

For the definite integral, we have $x = \pi/2 \Rightarrow$
$z = \tan \pi/4 = 1$ and $x = 0 \Rightarrow z = \tan(0) = 0$

$$\int_0^{\pi/2} \frac{\cos x}{2 - \cos x}\, dx = (\frac{4}{\sqrt{3}} \tan^{-1}(\sqrt{3}\, z) - 2 \tan^{-1}z) \Big|_0^1$$

$$= (\frac{4}{\sqrt{3}})(\frac{\pi}{3}) - 2(\frac{\pi}{4}) = (\frac{4\sqrt{3}}{9} - \frac{1}{2})\pi$$

19. We know the formula $\int \csc x\, dx = \ell n |\csc x - \cot x| + C$,

and $\csc x - \cot x = \dfrac{1}{\sin x} - \dfrac{\cos x}{\sin x} = \dfrac{1 - \cos x}{\sin x}$. Also,

$$(\frac{1 - \cos x}{\sin x})^2 = \frac{(1-\cos x)^2}{\sin^2 x} = \frac{(1-\cos x)^2}{1 - \cos^2 x}$$

$$= \frac{(1-\cos x)^2}{(1-\cos x)(1+\cos x)} = \frac{1 - \cos x}{1 + \cos x} ,$$

when $\cos x \neq 1$,
so that $\csc x - \cot x = \sqrt{\dfrac{1 - \cos x}{1 + \cos x}}$ is valid for

all x for which the separate terms are defined.

Exercise 9, p. 512

3. Let $I = \int \dfrac{dx}{x - \sqrt[3]{x}}$ $u = \sqrt[3]{x}$; $u^3 = x$; $dx = 3u^2 du$

$$I = \int \frac{3u^2 du}{u^3 - u} = \int \frac{3u\, du}{u^2 - 1} = \frac{3}{2} \ell n |u^2 - 1| + C$$

$$= \frac{3}{2} \ell n |x^{2/3} - 1| + C$$

7. Let $I = \int \dfrac{dx}{\sqrt[3]{2 + 3x}}$ $u = \sqrt[3]{2 + 3x}$; $u^3 = 2 + 3x$;
$x = \dfrac{1}{3}u^3 - \dfrac{2}{3}$; $dx = u^2 du$

$$I = \int \frac{u^2 du}{u} = \int u\, du = \frac{1}{2}u^2 + C = \frac{1}{2}(2+3x)^{2/3} + C$$

[Note: The substitution $w = 2 + 3x$ will also work.]

11. Let $I = \int \dfrac{dx}{\sqrt{x} + 1}$ $u = \sqrt{x} + 1$; $u - 1 = \sqrt{x}$;

$$x = (u-1)^2; \quad dx = 2(u-1)du$$

$I = \int \dfrac{2(u-1)du}{u} = \int (2 - \dfrac{2}{u})du = 2u - 2\,\ell n|u|$

$\quad = 2\sqrt{x} + 2 - 2\,\ell n|\sqrt{x} + 1| + K = 2[\sqrt{x} - \ell n|\sqrt{x} + 1|] + C$

15. $y = \sqrt{x + 1} + x$; $x \in [0,3]$ Revolve about x-axis.

$V = \pi \int_0^3 f^2(x)dx = \pi \int_0^3 [(x+1) + 2x\sqrt{x + 1} + x^2]\,dx$

$\quad = \pi \int_0^3 (x^2+x+1)dx + 2\pi \int_0^3 x\sqrt{x + 1}\,dx$

To evaluate the second integral, let

$u = \sqrt{x + 1}$ $x = u^2 - 1$ [Note: the substitution

$u^2 = x + 1$ $dx = 2u\,du$ $w = x + 1$ and $x = w - 1$

                                               will also work.]

Then, $\int x\sqrt{x + 1}\,dx = \int (u^2-1)u\,2u\,du = 2\int (u^4-u^2)du$

$$= \dfrac{2}{5}u^5 - \dfrac{2}{3}u^3 + C$$

$$= \dfrac{2}{5}(x+1)^{5/2} - \dfrac{2}{3}(x+1)^{3/2} + C$$

$V = \pi(\dfrac{x^3}{3} + \dfrac{x^2}{2} + x)\Big|_0^3 + 2\pi[\dfrac{2}{5}(x+1)^{5/2} - \dfrac{2}{3}(x+1)^{3/2}]\Big|_0^3$

$\quad = \pi(9 + \dfrac{9}{2} + 3) + 2\pi[\dfrac{2}{5}(2)^5 - \dfrac{2}{3}(2)^3 - \dfrac{2}{5} + \dfrac{2}{3}]$

$\quad = \dfrac{495 + 464}{30}\,\pi = \dfrac{959}{30}\,\pi$

Exercise il, pp. 520-523

3. Let $I = \int_1^2 \dfrac{dx}{x}$; $n = 4$

x	1	5/4	3/2	7/4	2
f(x)	1	4/5	2/5	4/7	1/2
2f(x)		8/5	4/3	8/7	

$I \approx \dfrac{1}{8}(1 + \dfrac{8}{5} + \dfrac{4}{3} + \dfrac{8}{7} + \dfrac{1}{2})$

$\quad \approx \dfrac{1}{8}(1.5 + 1.6 + 1.33333 + 1.14286) \approx 0.69702$

7. Let $I = \int_{\pi/2}^{\pi} \frac{\sin x}{x}\,dx$; $n = 3$

x	$\frac{\pi}{2}$	$\frac{2\pi}{3}$	$\frac{5\pi}{6}$	π
f(x)	$\frac{2}{\pi}$	$\frac{3\sqrt{3}}{4\pi}$	$\frac{3}{5\pi}$	0
2f(x)		$\frac{3\sqrt{3}}{2\pi}$	$\frac{6}{5\pi}$	

$$I \approx \frac{\pi}{12}\left(\frac{2}{\pi} + \frac{3\sqrt{3}}{2\pi} + \frac{6}{5\pi} + 0\right) = \frac{1}{6} + \frac{\sqrt{3}}{8} + \frac{1}{10} \approx 0.48317$$

11. Let $I = \int_{0}^{4} x^2\,dx$; $n = 8$

x	f(x)	c	cf(x)
0	0	1	0
1/2	1/4	4	1
1	1	2	2
3/2	9/4	4	9
2	4	2	8
5/2	25/4	4	25
3	9	2	18
7/2	49/4	4	49
4	16	1	16

$$\Sigma = 128$$

$$I \approx \frac{4 - 0}{(3)8}\,(128)$$

$$= \frac{1}{6}(128)$$

$$\approx 21.333$$

15. Let $I = \int_{0}^{2} \frac{1}{\sqrt{1 + x}}\,dx$; $n = 6$

x	f(x)	c	cf(x)
0	1	1	1
1/3	$\sqrt{3}/2$	4	3.46410
2/3	$\sqrt{3/5}$	2	1.54919
1	$\sqrt{2}/2$	4	2.82843
4/3	$\sqrt{3/7}$	2	1.30931
5/3	$\sqrt{3/8}$	4	2.44949
2	$1/\sqrt{3}$	1	0.57735

$$I \approx \frac{2}{18}[\Sigma\ cf(x)]$$

$$\approx \frac{1}{9}(13.17787)$$

$$\approx 1.46421$$

19. Let $I = \int_0^1 e^{-x^2} dx$; $n = 4$

x	f(x)	c	cf(x)
0	1	1	1
1/4	$e^{-1/16}$	4	3.75765
1/2	$e^{-1/4}$	2	1.55760
3/4	$e^{-9/16}$	4	2.27913
1	e^{-1}	1	0.36788

$I \approx \dfrac{1}{12}[\Sigma\ cf(x)]$

$\approx \dfrac{1}{12}(8.96226)$

≈ 0.74686

23. Let $I = \int_4^7 \sqrt{9 + x^2}\ dx$; $n = 6$

x	f(x)	c_1	$c_1 f(x)$	c_2	$c_2 f(x)$
4	$\sqrt{25}$	1	5.0	1	5.0
4.5	$\sqrt{29.25}$	2	10.81665	4	21.63331
5	$\sqrt{34}$	2	11.66190	2	11.66190
5.5	$\sqrt{39.25}$	2	12.52996	4	25.05993
6	$\sqrt{45}$	2	13.41641	2	13.41641
6.5	$\sqrt{51.25}$	2	14.31782	4	28.63564
7	$\sqrt{58}$	1	7.61577	1	7.61577

Trapezoidal rule: $I \approx \dfrac{7 - 4}{12}\ [\Sigma\ c_1 f(x)]$

$= \dfrac{1}{4}(75.35851) \approx 18.83963$

Simpson's rule: $I \approx \dfrac{7 - 4}{18}\ [\Sigma\ c_2 f(x)]$

$\approx \dfrac{1}{6}(113.02296) \approx 18.83716$

27. $V = \dfrac{150 - 0}{12}\ (105 + 236 + 284 + 240 + 220 + 180 + 78)$

$= \dfrac{150(1343)}{12} = 16,787.5\ m^3$

31. Trapezoidal rule: $V \approx \dfrac{25 - 0}{20}(0 + 5020 + 7720 +$

+ 9740 + 10,320 + 11,180 + 11,620 + 12,420 + 13,780
+ 15,360 + 8270) $= \dfrac{5}{4}(105,430) = 131,787.5 \text{ m}^3$

Simpson's rule: $V \approx \dfrac{25 - 0}{30}(0 + 10,040 + 7720$

+ 19,480 + 10,320 + 22,360 + 11,620 + 24,840 + 13,780
+ 30,720 + 8270) $= \dfrac{5}{6}[159,150] = 132,625 \text{ m}^3$

35. For $\int_0^4 x^2 dx$, we have $b - a = 4$ and $n = 8$.

$f(x) = x^2$; $f'(x) = 2x$; $f''(x) = 2 \Longrightarrow |f''(x)|$

$= 2$ for $x \in [0,4] \Longrightarrow M = 2$. Hence,

$E \leqslant \dfrac{(b-a)^3 M}{12\ n^2} = \dfrac{4^3(2)}{12(64)} = \dfrac{1}{6} = 0.1667$

39. The arc length of $y = \sin x$ from $x = 0$ to
$x = \pi/2$ is given by

$L = \int_0^{\pi/2} \sqrt{1 + \cos^2 x}\ dx$. Using the trapezoidal rule

with $n = 3$ and $f(x) = \sqrt{1 + \cos^2 x}$ gives

$L \approx \dfrac{\pi}{12}[\Sigma\ cf(x)] = \dfrac{\pi}{12}(7.29603) = 1.9101$

x	f(x)	c	cf(x)
0	$\sqrt{2}$	1	1.41421
$\pi/6$	$\sqrt{7/4}$	2	2.64575
$\pi/3$	$\sqrt{5/4}$	2	2.23607
$\pi/2$	$\sqrt{1}$	1	1.000

43. Let $I = \int_0^\pi f(x)dx$ for $f(x) = \begin{cases} (\sin x)/x & \text{if } x \neq 0 \\ 1 & \text{if } x = 0 \end{cases}$;

$$n = 6$$

x	$f(x)$	c_1	$c_1 f(x)$
0	1	1	1
$\pi/6$	$3/\pi$	2	1.90986
$\pi/3$	$3\sqrt{3}/2\pi$	2	1.65399
$\pi/2$	$2/\pi$	2	1.27324
$2\pi/3$	$3\sqrt{3}/4\pi$	2	0.82699
$5\pi/6$	$3/5\pi$	2	0.38197
π	0	1	0

Using the trapezoidal rule, we have

$$I \approx \frac{\pi}{12} [\Sigma\ c_1 f(x)] = \frac{\pi}{12}(7.04605) = 1.84465$$

Miscellaneous, pp. 528-532

3. $\int \dfrac{dx}{x^2 + 4x + 20} = \int \dfrac{dx}{(x+2)^2 + 4^2} = \dfrac{1}{4}\ \tan^{-1}(\dfrac{x+2}{4}) + C$

7. Let $I = \int \dfrac{\sqrt{x}}{1 + x}\ dx$ and $u = \sqrt{x}$ so that $x = u^2$,

   $dx = 2u\ du$, and $1 + x = u^2 + 1$. Then division

   gives $\dfrac{2u^2}{u^2 + 1} = 2 - \dfrac{2}{u^2 + 1}$. Thus,

   $I = \int \dfrac{2u^2}{u^2 + 1}\ du = \int (2 - \dfrac{2}{u^2 + 1})du = 2u - 2\ \tan^{-1}u + C$

   $= 2\sqrt{x} - 2\ \tan^{-1}\sqrt{x} + C$

11. $\int \cot^2\theta\ d\theta = \int (\csc^2\theta - 1)d\theta = -\cot\theta - \theta + C$

15. $\int x \sin 2x\ dx \qquad u = x \qquad dv = \sin 2x\ dx$

    $\qquad\qquad\qquad\qquad du = dx \qquad v = -\dfrac{1}{2}\cos 2x$

    $= -\dfrac{x}{2}\cos 2x + \dfrac{1}{2} \int \cos 2x\ dx = \dfrac{-x}{2}\cos 2x + \dfrac{1}{4}\sin 2x + C$

19. $\int t^3\sqrt{2-t}\ dt$ $u = \sqrt{2-t}$, $u^2 = 2 - t$

   $= \int(2-u^2)^3 u(-2u\,du)$ $t = 2 - u^2$ $dt = -2u\ du$

   $= \int (8-12u^2+6u^4-u^6)(u)(-2u)du$

   $= \int (2u^8-12u^6+24u^4-16u^2)du$

   $= \frac{2}{9}u^9 - \frac{12}{7}u^7 + \frac{24}{5}u^5 - \frac{16}{3}u^3 + C$

   $= \frac{2}{9}(2-t)^{9/2} - \frac{12}{7}(2-t)^{7/2} + \frac{24}{5}(2-t)^{5/2} - \frac{16}{3}(2-t)^{3/2} + C$

23. $\int \frac{dy}{\sqrt{2y+1}}$ $u = 2y + 1;$ $du = 2dy$

   $= \frac{1}{2}\int \frac{du}{\sqrt{u}} = \frac{1}{2}\int u^{-1/2}du = u^{1/2} + C = \sqrt{2y+1} + C$

27. $\int \tanh 2v\ dv$ $u = \cosh 2v;$ $du = 2\sinh 2v\ dv$

   $= \int \frac{\sinh 2v}{\cosh 2v}\ dv = \frac{1}{2}\int \frac{du}{u} = \frac{1}{2}\ln|u| + C$

   $= \frac{1}{2}\ln|\cosh 2v| + C$

31. We write $\dfrac{x}{x^4-4} = \dfrac{x}{(x^2-2)(x^2+2)} = \dfrac{Ax+B}{x^2-2} + \dfrac{Cx+D}{x^2+2}$.

   This gives $x = (Ax+B)(x^2+2) + (Cx+D)(x^2-2)$,   or

   $x = (A+C)x^3 + (B+D)x^2 + (2A-2C)x + (2B-2D)$.

   Hence, $\left\{\begin{matrix}A + C = 0\\ 2A - 2C = 1\end{matrix}\right\}$ gives $A = \frac{1}{4}$ and $C = -\frac{1}{4}$

   while $\left\{\begin{matrix}B + D = 0\\ 2B - 2D = 0\end{matrix}\right\}$ gives $B = D = 0$.

   We then have

   $\int \frac{x}{x^4-4}\ dx = \frac{1}{4}\int \frac{x}{x^2-2}\ dx - \frac{1}{4}\int \frac{x}{x^2+2}\ dx$

$$= \frac{1}{8} \ell n |x^2 - 2| - \frac{1}{8} \ell n |x^2 + 2| + C$$

$$= \frac{1}{8} \ell n \left| \frac{x^2 - 2}{x^2 + 2} \right| + C$$

35. If $u = \frac{\pi}{4} - \theta$, then $du = -d\theta$. Hence,

$$\int \tan(\frac{\pi}{4} - \theta) d\theta = -\int \tan u \, du = \ell n |\cos u| + C$$

$$= \ell n \left| \cos(\frac{\pi}{4} - \theta) \right| + C$$

39. Since $\sqrt{\dfrac{a + x}{a - x}} = \sqrt{\dfrac{a + x}{a - x}} \sqrt{\dfrac{a + x}{a + x}} = \sqrt{\dfrac{(a+x)^2}{a^2 - x^2}}$

$$= \frac{a + x}{\sqrt{a^2 - x^2}} = \frac{a}{\sqrt{a^2 - x^2}} + \frac{x}{\sqrt{a^2 - x^2}} \, ,$$

we have

$$\int \sqrt{\frac{a + x}{a - x}} \, dx = a \int \frac{dx}{\sqrt{a^2 - x^2}} + \frac{1}{2} \int \frac{2x \, dx}{\sqrt{a^2 - x^2}}$$

$$= a \sin^{-1} \frac{x}{a} - \sqrt{a^2 - x^2} + C$$

43. $\int (3y^2 - 6y)^3 (y-1) dy \qquad u = 3y^2 - 6y; \ du = 6(y-1)dy$

$$= \frac{1}{6} \int u^3 du = \frac{1}{24} u^4 + C = \frac{1}{24}(3y^2 - 6y)^4 + C$$

47. Using $\cos^3 3x = \cos^2 3x \cos 3x = (1 - \sin^2 3x)\cos 3x$
and $u = \sin 3x$, $du = 3 \cos 3x \, dx$

$$\int \cos^3 3x \, dx = \int (1 - u^2)\frac{1}{3} du = \frac{1}{3} u - \frac{1}{9} u^3 + C$$

$$= \frac{1}{3} \sin 3x - \frac{1}{9} \sin^3 3x + C$$

51. Integration by parts and then substitution gives:

$$\int x^2 \sin^{-1} x \, dx = \frac{1}{3} x^3 \sin^{-1} x - \frac{1}{3} \int \frac{x^3}{\sqrt{1 - x^2}} \, dx$$

$$\left[\begin{array}{ll} u = \sin^{-1} x & dv = x^2 dx \\ du = \dfrac{1}{\sqrt{1 - x^2}} \, dx & v = \dfrac{1}{3} x^3 \end{array} \right] \left[\begin{array}{l} w = \sqrt{1 - x^2} \, , \ w^2 = 1 - x^2 \\ x^2 = 1 - w^2, \ 2x dx = -2w dw \end{array} \right]$$

$$= \frac{1}{3} x^3 \sin^{-1} x - \frac{1}{3} \int \frac{(1-w^2)(-w\,dw)}{w} = \frac{1}{3} x^3 \sin^{-1} x +$$

$$\frac{1}{3} \int (1-w^2)\,dw$$

$$= \frac{1}{3} x^3 \sin^{-1} x + \frac{1}{3} \sqrt{1-x^2} - \frac{1}{9}(\sqrt{1-x^2})^3 + C$$

55. Since $\cos^2 x = \frac{1}{2} + \frac{1}{2} \cos 2x$, we have

$$\int x \cos^2 x \, dx = \frac{1}{2} \int x(1+\cos 2x)\,dx$$

$$= \frac{1}{2} \int x \, dx + \frac{1}{2} \int x \cos 2x \, dx$$

Using integration by parts gives

$$\int x \cos 2x \, dx = \frac{1}{2} x \sin 2x - \frac{1}{2} \int \sin 2x \, dx$$

$$= \frac{1}{2} x \sin 2x + \frac{1}{4} \cos 2x + C$$

$$u = x \qquad\qquad dv = \cos 2x \, dx$$
$$du = dx \qquad\qquad v = \frac{1}{2} \sin 2x$$

Thus, $\int x \cos^2 x \, dx = \frac{1}{4} x^2 + \frac{1}{4} x \sin 2x + \frac{1}{8} \cos 3x + C$

59. We have $\dfrac{w^3}{1-w^2} = \dfrac{-w^3}{w^2-1} = -w - \dfrac{w}{w^2-1}$, so that

$$\int \frac{w^3 \, dw}{1-w^2} = -\int (w + \frac{w}{w^2-1})\,dw = -\frac{1}{2} w^2 - \frac{1}{2} \ln |w^2 - 1| + C$$

63. Letting $x = a \sec \theta$ gives $x^2 - a^2 = a^2(\sec^2\theta - 1)$
$= a^2 \tan^2\theta$, so that $\sqrt{x^2 - a^2} = a \tan \theta$ and
$dx = a \sec \theta \tan \theta \, d\theta$. Hence,

$$\int \frac{dx}{x\sqrt{x^2-a^2}} = \int \frac{a \sec \theta \tan \theta \, d\theta}{(a \sec \theta)(a \tan \theta)} = \frac{1}{a} \int d\theta$$

$$= \frac{1}{a} \theta + C = \frac{1}{a} \sec^{-1}(\frac{x}{a}) + C$$

67. Letting $x = a \cosh t$ gives $x^2 - a^2 = a^2(\cosh^2 t - 1)$
$= a^2 \sinh^2 t$, so that $\sqrt{x^2 - a^2} = a \sinh t$ and
$dx = a \sinh t \, dt$. Hence,

$$\int \frac{dx}{\sqrt{x^2 - a^2}} = \int \frac{a \sinh t \; dt}{a \sinh t} = \int dt = t + C$$

$$= \cosh^{-1}\left(\frac{x}{a}\right) + C$$

Also,

$$t = \cosh^{-1}\left(\frac{x}{a}\right) \implies \cosh t = \frac{x}{a} \implies \frac{e^t + e^{-t}}{2} = \frac{x}{a}$$

$$\implies ae^{2t} + a = 2xe^t \implies ae^{2t} - 2xe^t + a = 0$$

$$\implies e^t = \frac{2x \pm \sqrt{4x^2 - 4a^2}}{2a} = \frac{x \pm \sqrt{x^2 - a^2}}{a}$$

$$\implies t = \ell n \left| \frac{x + \sqrt{x^2 - a^2}}{a} \right| = \cosh^{-1}\left(\frac{x}{a}\right). \quad \text{We discard the}$$

negative inside the log since $\cosh^{-1}\left(\frac{x}{a}\right) \geq 0$.

This gives the formula

$$\int \frac{dx}{\sqrt{x^2 - a^2}} = \cosh^{-1}\left(\frac{x}{a}\right) + C_1 = \ell n \left(\frac{x + \sqrt{x^2 - a^2}}{a}\right) + C$$

71. If $f''(x) = \frac{1}{3}(x+5)^{-1/2}$, then

$$f'(x) = \frac{1}{3} \int (x+5)^{-1/2} dx = \frac{2}{3}(x+5)^{1/2} + C_1,$$

and $f(x) = \int \left[\frac{2}{3}(x+5)^{1/2} + C_1\right] dx = \frac{4}{9}(x+5)^{3/2} + C_1 x + C_2.$

Property (ii) states that $f(4) = 2$ and $f'(4)$
$= \tan 45° = 1$. Using $f'(4) = 1$ gives

$$1 = \frac{2}{3}(9)^{1/2} + C_1, \quad \text{or} \quad C_1 = 1 - 2 = -1. \quad \text{Hence,}$$

$f(x) = \frac{4}{9}(x+5)^{3/2} - x + C_2.$ Using $f(4) = 2$, we

have $2 = \frac{4}{9}(9)^{3/2} - 4 + C_2$, or $C_2 = 6 - 12 = -6.$

Hence, $f(x) = \frac{4}{9}(x+5)^{3/2} - x - 6.$

75. (a) $A = \int_0^{\pi/4} (\sec x - 2 \sin x) dx$

$A = \ln|\sec x + \tan x| + 2 \cos x \Big|_0^{\pi/4}$

$A = \ln|\sqrt{2} + 1| + \sqrt{2} - \ln|1 + 0| - 2$

$A = \ln(\sqrt{2}+1) + \sqrt{2} - 2$

$2 \sin x = \sec x \implies 2 \sin x \cos x = 1$

or $\sin 2x = 1 \implies 2x = \frac{\pi}{2} \implies x = \frac{\pi}{4}$

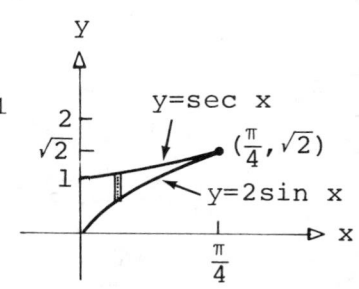

(b) $V = \pi \int_0^{\pi/4} [f^2(x) - g^2(x)] dx$

$= \pi \int_0^{\pi/4} (\sec^2 x - 4 \sin^2 x) dx$

$= \pi \int_0^{\pi/4} (\sec^2 x - 2 + 2 \cos 2x) dx$

$= \pi(\tan x - 2x + \sin 2x) \Big|_0^{\pi/4}$

$= \pi[(1 - \frac{\pi}{2} + 1) - (0)] = \pi(2 - \frac{\pi}{2})$

79. Using integration by parts merely results in an identity where the integral cancels and no information is gained.

$\int e^x \cosh x\, dx = e^x \cosh x - \int e^x \sinh x\, dx$

$= e^x(\cosh x - \sinh x) + \int e^x \cosh x\, dx$

$\left[\begin{array}{ll} u = \cosh x & dv = e^x dx \\ du = \sinh x\, dx & v = e^x \end{array}\right] \left[\begin{array}{ll} u = \sinh x & dv = e^x dx \\ du = \cosh x\, dx & v = e^x \end{array}\right]$

Using $\cosh x = \frac{1}{2}(e^x + e^{-x})$, we have

$\int e^x \cosh x\, dx = \frac{1}{2} \int (e^{2x}+1) dx = \frac{1}{4}e^{2x} + \frac{1}{2}x + C.$

83. Problem 83 is to be solved using a computer.

Chapter Ten

Exercise 1, pp. 536-537

Note: An asterisk above the equal sign indicates application of L'Hospital's rule.

3. $\lim\limits_{x \to 1} \dfrac{2x^3 + 5x^2 - 4x - 3}{x^3 + x^2 - 10x + 8} \overset{*}{=} \lim\limits_{x \to 1} \dfrac{6x^2 + 10x - 4}{3x^2 + 2x - 10} = -12/5$

7. $\lim\limits_{x \to 1} \dfrac{\ln x}{x^2 - 1} \overset{*}{=} \lim\limits_{x \to 1} \dfrac{\frac{1}{x}}{2x} = 1/2$

11. $\lim\limits_{x \to \pi} \dfrac{1 + \cos x}{\sin 2x} \overset{*}{=} \lim\limits_{x \to \pi} \dfrac{-\sin x}{2 \cos 2x} = 0$

15. $\lim\limits_{x \to 0^+} \dfrac{\cot x}{\cot 2x} \overset{*}{=} \lim\limits_{x \to 0^+} \dfrac{-\csc^2 x}{-2 \csc^2 2x} = \lim\limits_{x \to 0^+} \dfrac{\frac{-1}{\sin^2 x}}{\frac{-2}{\sin^2 2x}}$

$= \lim\limits_{x \to 0^+} \dfrac{\sin^2 2x}{2 \sin^2 x} = \lim\limits_{x \to 0^+} \dfrac{4 \sin^2 x \cos^2 x}{2 \sin^2 x}$

$= \lim\limits_{x \to 0^+} \dfrac{2 \cos^2 x}{1} = 2$

19. $\lim\limits_{x \to 0} \dfrac{\sin^2 x}{x} \overset{*}{=} \lim\limits_{x \to 0} \dfrac{2 \sin x \cos x}{1} = 0$

23. $\lim\limits_{x \to 0} \dfrac{\cos x - 1}{\cos 2x - 1} \overset{*}{=} \lim\limits_{x \to 0} \dfrac{-\sin x}{-2 \sin 2x} \overset{*}{=} \lim\limits_{x \to 0} \dfrac{-\cos x}{-4 \cos 2x} = \dfrac{1}{4}$

27. $\lim\limits_{x \to 0} \dfrac{e^x - e^{-x} - 2 \sin x}{3x^3} \overset{*}{=} \lim\limits_{x \to 0} \dfrac{e^x + e^{-x} - 2 \cos x}{9x^2}$

$\overset{*}{=} \lim\limits_{x \to 0} \dfrac{e^x - e^{-x} + 2 \sin x}{18x}$

$\overset{*}{=} \lim\limits_{x \to 0} \dfrac{e^x + e^{-x} + 2 \cos x}{18} = \dfrac{2}{9}$

31. $\displaystyle\lim_{x\to1^-}\frac{\ln(1-x)}{\cot\pi x}\overset{*}{=}\lim_{x\to1^-}\frac{\dfrac{-1}{1-x}}{-\pi\csc^2\pi x}=\lim_{x\to1^-}\frac{\dfrac{-1}{1-x}}{\dfrac{-\pi}{\sin^2\pi x}}$

$\displaystyle=\lim_{x\to1^-}\frac{\sin^2\pi x}{\pi(1-x)}\overset{*}{=}\lim_{x\to1^-}\frac{2\pi\,\sin\pi x\,\cos\pi x}{-\pi}=0$

35. $\displaystyle\lim_{x\to+\infty}\frac{x^\alpha}{e^x}\overset{*}{=}\lim_{x\to+\infty}\frac{\alpha x^{\alpha-1}}{e^x}\overset{*}{=}\lim_{x\to+\infty}\frac{\alpha(\alpha-1)x^{\alpha-2}}{e^x}\overset{*}{=}\cdots=0$

That is, apply L'Hospital's rule k times, where $\alpha-k=0$ or $\alpha-k<0$. Then the numerator is constant and the denominator approaches infinity (for $\alpha>0$).

39. $\displaystyle\lim_{x\to\pi/4}\frac{1+\cos4x}{\sec^2x-2\tan x}\overset{*}{=}\lim_{x\to\pi/4}\frac{-4\sin4x}{2\sec^2x\tan x-2\sec^2x}$

$\displaystyle\overset{*}{=}\lim_{x\to\pi/4}\frac{-16\cos4x}{4\sec^2x\tan^2x+2\sec^4x-4\sec^2x\tan x}$

$\displaystyle=\frac{16}{8}=2$

43. $\displaystyle\lim_{x\to\pi/4}\frac{\cos^2 2x}{1-\tan x}\overset{*}{=}\lim_{x\to\pi/4}\frac{-4\cos2x\sin2x}{-\sec^2x}=0$

47. $\displaystyle\lim_{x\to0}\frac{x-\tan^{-1}x}{x^3}\overset{*}{=}\lim_{x\to0}\frac{1-\dfrac{1}{1+x^2}}{3x^2}\overset{*}{=}\lim_{x\to0}\frac{\dfrac{2x}{(1+x^2)^2}}{6x}$

$\displaystyle=\lim_{x\to0}\frac{1}{3(1+x^2)^2}=\frac{1}{3}$

51. $\displaystyle\lim_{x\to0}\frac{\tan^{-1}x}{\sin^{-1}x}\overset{*}{=}\lim_{x\to0}\frac{\dfrac{1}{1+x^2}}{\dfrac{1}{\sqrt{1-x^2}}}=1$

55. $\displaystyle\lim_{x\to0}\frac{\ln(\cosh x)}{x}\overset{*}{=}\lim_{x\to0}\frac{\dfrac{1}{\cosh x}(\sinh x)}{1}=0$

59. $\displaystyle\lim_{x \to +\infty} \frac{(\ln x)^{\beta}}{x^{\alpha}} \overset{*}{=} \lim_{x \to +\infty} \frac{\beta(\ln x)^{\beta-1}(1/x)}{\alpha(x^{\alpha-1})} \overset{*}{=} \lim_{x \to +\infty} \frac{\beta(\ln x)^{\beta-}}{\alpha\, x^{\alpha}}$

$$\overset{*}{=} \lim_{x \to +\infty} \frac{\beta(\beta-1)(\ln x)^{\beta-2}(\frac{1}{x})}{(\alpha)(\alpha)x^{\alpha-1}}$$

$$= \lim_{x \to +\infty} \frac{\beta(\beta-1)(\ln x)^{\beta-2}}{\alpha^2\, x^{\alpha}} \overset{*}{=} \cdots = 0$$

That is, apply L'Hospital's rule k times, where $\beta - k = 0$ or $\beta - k < 0$. Then the numerator is constant and the denominator approaches positive infinity (for $\alpha, \beta > 0$).

63. $\displaystyle\lim_{x \to c} \frac{x^{\alpha} - c^{\alpha}}{x^{\beta} - c^{\beta}} \overset{*}{=} \lim_{x \to c} \frac{\alpha\, x^{\alpha-1}}{\beta\, x^{\beta-1}} = \lim_{x \to c} \frac{\alpha\, x^{\alpha-\beta}}{\beta}$

$$= \frac{\alpha}{\beta} c^{\alpha-\beta} \qquad (\text{for } \alpha \neq 0,\ \beta \neq 0,\ \text{and } c > 0)$$

67. Let $x = 1/u$. Then, as $x \to -\infty$, $u \to 0^-$, and

$$\lim_{x \to -\infty} \frac{f(x)}{g(x)} = \lim_{u \to 0^-} \frac{f(1/u)}{g(1/u)} \overset{*}{=} \lim_{u \to 0^-} \frac{(-1/u^2)f'(1/u)}{(-1/u^2)g'(1/u)}$$

$$= \lim_{x \to -\infty} \frac{f'(x)}{g'(x)}$$

Exercise 2, pp. 541-542

Note: An asterisk above the equal sign indicates application of L'Hospital's rule.

3. $\displaystyle\lim_{x \to +\infty} [x(e^{1/x}-1)] = \lim_{x \to +\infty} \frac{e^{1/x} - 1}{1/x}$

$$\overset{*}{=} \lim_{x \to +\infty} \frac{e^{1/x}(-1/x^2)}{-1/x^2}$$

$$= \lim_{x \to +\infty} \frac{e^{1/x}}{1} = 1$$

7. $\displaystyle\lim_{x\to\pi/2}[\tan x \ \ell n(\sin x)] = \lim_{x\to\pi/2} \frac{\ell n(\sin x)}{\cot x}$

$\displaystyle\overset{*}{=} \lim_{x\to\pi/2} \frac{(1/\sin x)\cos x}{-\csc^2 x}$

$\displaystyle= \lim_{x\to\pi/2} \frac{\cos x/\sin x}{-1/\sin^2 x}$

$\displaystyle= \lim_{x\to\pi/2} (-\cos x \sin x) = 0$

11. $\displaystyle\lim_{x\to-\infty}(x^2 e^x) = \lim_{x\to-\infty} \frac{x^2}{e^{-x}} \overset{*}{=} \lim_{x\to-\infty} \frac{2x}{-e^{-x}} \overset{*}{=} \lim_{x\to-\infty} \frac{2}{e^{-x}} = 0$

15. $\displaystyle\lim_{x\to1^+}(1-x)\tan(\frac{\pi}{2}x) = \lim_{x\to1^+} \frac{1-x}{\cot(\frac{\pi}{2}x)}$

$\displaystyle\overset{*}{=} \lim_{x\to1^+} \frac{-1}{-\frac{\pi}{2}\csc^2(\frac{\pi}{2}x)} = \frac{-1}{\frac{-\pi}{2}} = \frac{2}{\pi}$

19. $\displaystyle\lim_{x\to0}(\cot x - \frac{1}{x}) = \lim_{x\to0}(\frac{\cos x}{\sin x} - \frac{1}{x})$

$\displaystyle= \lim_{x\to0} \frac{x\cos x - \sin x}{x\sin x}$

$\displaystyle\overset{*}{=} \lim_{x\to0} \frac{\cos x - x\sin x - \cos x}{\sin x + x\cos x}$

$\displaystyle= \lim_{x\to0} \frac{-x\sin x}{\sin x + x\cos x}$

$\displaystyle\overset{*}{=} \lim_{x\to0} \frac{-\sin x - x\cos x}{\cos x + \cos x - x\sin x} = \frac{0}{2} = 0$

23. $\lim\limits_{x \to 1} (\dfrac{1}{\ln x} - \dfrac{1}{x - 1}) = \lim\limits_{x \to 1} \dfrac{x - 1 - \ln x}{x \ln x - \ln x}$

$$\overset{*}{=} \lim\limits_{x \to 1} \dfrac{1 - \dfrac{1}{x}}{\ln x + 1 - \dfrac{1}{x}}$$

$$\overset{*}{=} \lim\limits_{x \to 1} \dfrac{\dfrac{1}{x^2}}{\dfrac{1}{x} + \dfrac{1}{x^2}} = \dfrac{1}{2}$$

27. $y = (\sin x)^{\tan x} \Rightarrow \ln y = \tan x \, \ln(\sin x)$

$\lim\limits_{x \to \pi/2} \ln y = \lim\limits_{x \to \pi/2} [\tan x \, \ln(\sin x)] = 0$ by Problem 7

Thus, $\lim\limits_{x \to \pi/2} y = e^0 = 1$

31. $y = (1+x^2)^{1/x} \Rightarrow \ln y = \dfrac{1}{x} \ln(1+x^2)$

$\lim\limits_{x \to +\infty} \ln y = \lim\limits_{x \to +\infty} \dfrac{\ln(1+x^2)}{x} \overset{*}{=} \lim\limits_{x \to +\infty} \dfrac{\dfrac{2x}{1 + x^2}}{1}$

$$= \lim\limits_{x \to +\infty} \dfrac{2x}{1 + x^2} \overset{*}{=} \lim\limits_{x \to +\infty} \dfrac{2}{2x} = 0$$

Thus, $\lim\limits_{x \to +\infty} y = e^0 = 1$

35. $y = (\dfrac{\sin x}{x})^{1/x} \Rightarrow \ln y = \dfrac{1}{x} \ln(\dfrac{\sin x}{x})$

$\lim\limits_{x \to 0} \ln y = \lim\limits_{x \to 0} \dfrac{\ln(\dfrac{\sin x}{x})}{x}$

$$\overset{*}{=} \lim\limits_{x \to 0} \dfrac{\dfrac{1}{(\sin x)/x} (\dfrac{x \cos x - \sin x}{x^2})}{1}$$

$$= \lim\limits_{x \to 0} \dfrac{x \cos x - \sin x}{x \sin x}$$

$$\overset{*}{=} \lim_{x \to 0} \frac{\cos x - x \sin x - \cos x}{\sin x + x \cos x}$$

$$= \lim_{x \to 0} \frac{-x \sin x}{\sin x + x \cos x}$$

$$\overset{*}{=} \lim_{x \to 0} \frac{-\sin x - x \cos x}{\cos x + \cos x - x \sin x} = 0$$

Thus, $\lim_{x \to 0} y = e^0 = 1$

39. $P(x)$ polynomial function of degree $k \Rightarrow \dfrac{d^k}{dx^k} P(x) = M,$

a constant

$$\lim_{x \to +\infty} \frac{P(x)}{e^x} \overset{*}{=} \lim_{x \to +\infty} \frac{M}{e^x} \quad \text{after} \quad k \quad \text{applications of}$$

L'Hospital's rule. Therefore, $\lim_{x \to +\infty} \dfrac{P(x)}{e^x} = 0$

43. $f'(0) = \lim_{x \to 0} \dfrac{f(x) - f(0)}{x - 0} = \lim_{x \to 0} \dfrac{e^{-1/x^2}}{x} = 0 \quad$ using

Problem 41 for $x \to 0^+$ and a similar argument for $x \to 0^-$.

Exercise 3, pp. 549-551

3. Not improper

7. Improper; $x/(x^2-1)$ is not defined at 1

11. $\displaystyle\int_0^{+\infty} e^{2x}dx = \lim_{b \to +\infty} \int_0^b e^{2x}dx = \lim_{b \to +\infty} \left. (\tfrac{1}{2} e^{2x}) \right|_0^b$

$$= \lim_{b \to +\infty} (\tfrac{1}{2} e^{2b} - \tfrac{1}{2}) = +\infty; \text{ integral diverges}$$

15. $\int_0^1 \frac{dx}{\sqrt{x}} = \lim_{a \to 0^+} \int_a^1 \frac{dx}{\sqrt{x}} = \lim_{a \to 0^+} (2\sqrt{x})\Big|_a^1 = \lim_{a \to 0^+} (2-2\sqrt{a}) = 2;$

integral converges

19. $\int_0^a \frac{dx}{\sqrt{a-x}} = \lim_{b \to a^-} \int_0^b \frac{dx}{\sqrt{a-x}} = \lim_{b \to a^-} (-2\sqrt{a-x})\Big|_0^b$

$\qquad\qquad = \lim_{b \to a^-} (-2\sqrt{a-b} + 2\sqrt{a}) = 2\sqrt{a};$

integral converges

23. $\int_0^1 \frac{x \, dx}{(1-x^2)^2} = \lim_{b \to 1^-} \int_0^b \frac{x \, dx}{(1-x^2)^2} = \lim_{b \to 1^-} (\frac{1}{2(1-x^2)})\Big|_0^b$

$\qquad\qquad = \lim_{b \to 1^-} [\frac{1}{2(1-b^2)} - \frac{1}{2}] = +\infty ;$

integral diverges

27. $\int_0^{\pi/4} \tan 2x \, dx = \lim_{b \to \pi/4^-} \int_0^b \tan 2x \, dx$

$\qquad\qquad = \lim_{b \to \pi/4^-} (\frac{1}{2} \ln|\sec 2x|)\Big|_0^b$

$\qquad\qquad = \lim_{b \to \pi/4^-} (\frac{1}{2} \ln|\sec 2b| - \frac{1}{2} \ln|\sec 0|)$

$\qquad\qquad = +\infty ;$ integral diverges

31. $\int_0^{2a} \frac{dx}{(x-a)^2} = \int_0^a \frac{dx}{(x-a)^2} + \int_a^{2a} \frac{dx}{(x-a)^2} ;$ integral

diverges since $\int_0^a \frac{dx}{(x-a)^2} = \lim_{b \to a^-} \int_0^b \frac{dx}{(x-a)^2}$

$= \lim_{b \to a^-} (-\frac{1}{x-a})\Big|_0^b = \lim_{b \to a^-} (-\frac{1}{b-a} - \frac{1}{a}) = +\infty .$ Part

of the original integral diverges, so the whole
integral diverges.

35. $\int_{-\infty}^{1} \dfrac{x \, dx}{\sqrt{2 - x}} = \lim\limits_{a \to -\infty} \int_{a}^{1} \dfrac{x \, dx}{\sqrt{2 - x}}$ (substitute $u = 2 - x$)

$$= \lim\limits_{a \to -\infty} \left[\dfrac{2}{3}(\sqrt{2-x})^3 - 4\sqrt{2-x}\right]\Big|_{a}^{1}$$

$$= \lim\limits_{a \to -\infty} \left[\dfrac{2}{3} - 4 - \dfrac{2}{3}(\sqrt{2-a})^3 + 4\sqrt{2-a}\right]$$

$$= \dfrac{-10}{3} + \lim\limits_{a \to -\infty} \left[\sqrt{2-a}\left(-\dfrac{2}{3}(2-a) + 4\right)\right]$$

$$= \dfrac{-10}{3} + \lim\limits_{a \to -\infty} \left[\sqrt{2-a}\left(\dfrac{8}{3} + \dfrac{2}{3}a\right)\right]$$

$$= -\infty \; ; \text{ integral diverges}$$

39. $\int_{0}^{1} \dfrac{dx}{1 - x^2} = \lim\limits_{b \to 1^{-}} \int_{0}^{b} \dfrac{dx}{1 - x^2}$ (Use partial fractions.)

$$= \lim\limits_{b \to 1^{-}} \left[\dfrac{1}{2}\,\ln\left|\dfrac{1 + x}{1 - x}\right|\right]\Big|_{0}^{b}$$

$$= \lim\limits_{b \to 1^{-}} \left[\dfrac{1}{2}\ln\left|\dfrac{1 + b}{1 - b}\right| - \dfrac{1}{2}\ln|1|\right]$$

$$= +\infty; \text{ integral diverges}$$

43. $\int_{0}^{+\infty} \sin \pi x \, dx = \lim\limits_{b \to +\infty} \int_{0}^{b} \sin \pi x \, dx$

$$= \lim\limits_{b \to +\infty} \left(-\dfrac{1}{\pi} \cos \pi x\right)\Big|_{0}^{b}$$

$$= \lim\limits_{b \to +\infty} \left(-\dfrac{1}{\pi} \cos \pi b + \dfrac{1}{\pi}\right)$$

does not exist, since $\cos \pi b$ oscillates between -1 and $+1$; integral diverges.

47. $\displaystyle\int_{2a}^{+\infty} \frac{dx}{x^2 - a^2} = \lim_{b \to +\infty} \int_{2a}^{b} \frac{dx}{x^2 - a^2}$ (Use partial fractions.)

$\displaystyle = \lim_{b \to +\infty} \left[\frac{1}{2a} \ln\left| \frac{x - a}{x + a} \right| \right] \Big|_{2a}^{b}$

$\displaystyle = \lim_{b \to +\infty} \left[\frac{1}{2a} \ln\left| \frac{b - a}{b + a} \right| - \frac{1}{2a} \ln\left(\frac{1}{3}\right) \right]$

$\displaystyle = \frac{1}{2a} \ln(1) - \frac{1}{2a}(\ln 1 - \ln 3)$

$\displaystyle = \frac{1}{2a} \ln 3$; integral converges

51. $\displaystyle\int_{0}^{a} \frac{dx}{\sqrt{a^2 - x^2}} = \lim_{b \to a^-} \int_{0}^{b} \frac{dx}{\sqrt{a^2 - x^2}} = \lim_{b \to a^-} \left(\sin^{-1} \frac{x}{a} \right) \Big|_{0}^{b}$

$\displaystyle = \lim_{b \to a^-} \left[\sin^{-1} \frac{b}{a} - \sin^{-1}(0) \right]$

$\displaystyle = \sin^{-1} 1 - 0 = \frac{\pi}{2}$; integral converges

55. $\displaystyle\int_{0}^{\pi} \frac{1}{1 - \cos x} \, dx = \lim_{a \to 0^+} \int_{a}^{\pi} \frac{1}{1 - \cos x} \, dx$ (Use $z = \tan \frac{x}{2}$; see Section 9.8

$\displaystyle = \lim_{a \to 0^+} \left[-\cot\left(\frac{x}{2}\right) \right] \Big|_{a}^{\pi}$

$\displaystyle = \lim_{a \to 0^+} \left[-\cot\left(\frac{\pi}{2}\right) + \cot\left(\frac{a}{2}\right) \right]$

$\displaystyle = +\infty$; integral diverges

59. $\displaystyle\int_{0}^{2} \frac{1}{(x-1)^{1/3}} \, dx = \lim_{a \to 1^-} \int_{0}^{a} \frac{1}{(x-1)^{1/3}} \, dx + \lim_{b \to 1^+} \int_{b}^{2} \frac{1}{(x-1)^{1/3}} \, dx$

$\displaystyle = \lim_{a \to 1^-} \left[\frac{3}{2}(x-1)^{2/3} \Big|_{0}^{a} \right] + \lim_{b \to 1^+} \left[\frac{3}{2}(x-1)^{2/3} \Big|_{b}^{2} \right]$

$\displaystyle = \lim_{a \to 1^-} \left[\frac{3}{2}(a-1)^{2/3} - \frac{3}{2} \right] + \lim_{b \to 1^+} \left[\frac{3}{2} - \frac{3}{2}(b-1)^{2/3} \right]$

$\displaystyle = \left[0 - \frac{3}{2} \right] + \left[\frac{3}{2} - 0 \right] = 0$; integral converges

63. Volume $= \pi \int_{1}^{+\infty} (\frac{1}{\sqrt{x}})^2 dx = \lim_{b \to +\infty} \pi \int_{1}^{b} \frac{1}{x} dx$

$= \lim_{b \to +\infty} (\pi \ln|x|) \Big|_{1}^{b} = \lim_{b \to +\infty} [\pi \ln b - \pi \ln 1]$

$= +\infty.$ Volume does not exist (i.e., infinite)

67. $\int xe^{-x} dx = -xe^{-x} + \int e^{-x} dx = -xe^{-x} - e^{-x} + C$

(Use $u = x$, $dv = e^{-x} dx$ and integration by parts.)

$\int_{0}^{+\infty} xe^{-x} dx = \lim_{b \to +\infty} (-xe^{-x} - e^{-x}) \Big|_{0}^{b}$

$= \lim_{b \to +\infty} [\frac{-b}{e^b} - \frac{1}{e^b} + 0 + 1] = 1;$

integral converges (Use $\lim_{b \to +\infty} \frac{-b}{e^b} \stackrel{*}{=} \lim_{b \to +\infty} \frac{-1}{e^b} = 0.$)

71. $\int_{0}^{+\infty} te^{-t^2} dt = \lim_{b \to +\infty} \int_{0}^{b} te^{-t^2} dt = \lim_{b \to +\infty} (-\frac{1}{2} e^{-t^2}) \Big|_{0}^{b}$

$= \lim_{b \to +\infty} (-\frac{1}{2e^{b^2}} + \frac{1}{2}) = \frac{1}{2}$

75. $F = \frac{rIm}{10} \int_{-\infty}^{+\infty} \frac{dy}{(r^2+y^2)^{3/2}}$ (substitute $y = r \tan\theta$)

$= \lim_{a \to -\infty} [(\frac{rIm}{10}) \frac{y}{r^2\sqrt{r^2 + y^2}}] \Big|_{a}^{0} + \lim_{b \to +\infty} [(\frac{rIm}{10}) \frac{y}{r^2\sqrt{r^2 + y^2}}] \Big|_{0}^{b}$

$= \frac{Im}{10r} \lim_{a \to -\infty} (\frac{-a}{\sqrt{r^2 + a^2}}) + \frac{Im}{10r} \lim_{b \to +\infty} (\frac{b}{\sqrt{r^2 + b^2}})$

$= \frac{IM}{10r} + \frac{IM}{10r} = \frac{IM}{5r}$

(For $a < 0$, $\frac{-a}{\sqrt{r^2 + a^2}} = \frac{\sqrt{a^2}}{\sqrt{r^2 + a^2}} = \frac{1}{\sqrt{\frac{r^2}{a^2} + 1}}$ and for $b > 0$, $b/\sqrt{r^2+b^2} = \sqrt{b^2}/\sqrt{r^2+b^2} = 1/\sqrt{(r^2/b^2)+1}$

Exercise 4, pp. 560-561

3. $f(x) = x^3 + x^2 - 8$ $f(1) = -6$

    $f'(x) = 3x^2 + 2x$ $f'(1) = 5$

    $f''(x) = 6x + 2$ $f''(1) = 8$

  $f^{(3)}(x) = 6$ $f^{(3)}(1) = 6$

   $f(x) = -6 + \dfrac{5}{1!}(x-1) + \dfrac{8}{2!}(x-1)^2 + \dfrac{6}{3!}(x-1)^3$

               $= -6 + 5(x-1) + 4(x-1)^2 + (x-1)^3$

7. $f(x) = 2x^4 - 6x^3 + x$ $f(-1) = 7$

    $f'(x) = 8x^3 - 18x^2 + 1$ $f'(-1) = -25$

    $f''(x) = 24x^2 - 36x$ $f''(-1) = 60$

  $f^{(3)}(x) = 48x - 36$ $f^{(3)}(-1) = -84$

  $f^{(4)}(x) = 48$ $f^{(4)}(-1) = 48$

  $f^{(5)}(x) = 0$

    $P_4(x) = 7 + \dfrac{-25}{1!}(x+1) + \dfrac{60}{2!}(x+1)^2 + \dfrac{-84}{3!}(x+1) + \dfrac{48}{4!}(x+1)$

            $= 7 - 25(x+1) + 30(x+1)^2 - 14(x+1)^3 + 2(x+1)^4$

    $R_5(x) = \dfrac{0}{5!}(x+1)^5 = 0$

11. $f(x) = \ln x$ $f(1) = 0$

    $f'(x) = \dfrac{1}{x}$ $f'(1) = 1$

    $f''(x) = \dfrac{-1}{x^2}$ $f''(1) = -1$

  $f^{(3)}(x) = \dfrac{2}{x^3}$ $f^{(3)}(1) = 2$

  $f^{(4)}(x) = \dfrac{-6}{x^4}$ $f^{(4)}(1) = -6$

$$f^{(5)}(x) = \frac{24}{x^5} \qquad\qquad f^{(5)}(1) = 24$$

$$f^{(6)}(x) = \frac{-120}{x^6}$$

$$P_5(x) = 0 + \frac{1}{1!}(x-1) + \frac{-1}{2!}(x-1)^2 + \frac{2}{3!}(x-1)^3 + \frac{-6}{4!}(x-1)^4$$
$$+ \frac{24}{5!}(x-1)^5$$

$$= (x-1) - \frac{1}{2}(x-1)^2 + \frac{1}{3}(x-1)^3 - \frac{1}{4}(x-1)^4 + \frac{1}{5}(x-1)^5$$

$$R_6(x) = \frac{-120/u^6}{6!}(x-1)^6 = -\frac{1}{6u^6}(x-1)^6$$

15. $f(x) = \cos x \qquad\qquad f(0) = 1$

$\quad f'(x) = -\sin x \qquad\qquad f'(0) = 0$

$\quad f''(x) = -\cos x \qquad\qquad f''(0) = -1$

$\quad f^{(3)}(x) = \sin x \qquad\qquad f^{(3)}(0) = 0$

$\quad f^{(4)}(x) = \cos x \qquad\qquad f^{(4)}(0) = 1$

$\quad f^{(5)}(x) = -\sin x \qquad\qquad f^{(5)}(0) = 0$

$\quad f^{(6)}(x) = -\cos x \qquad\qquad f^{(6)}(0) = -1$

$\quad f^{(7)}(x) = \sin x$

$$P_6(x) = 1 + \frac{0}{1!}(x-0) + \frac{-1}{2!}(x-0)^2 + \frac{0}{3!}(x-0)^3$$
$$+ \frac{1}{4!}(x-0)^4 + \frac{0}{5!}(x-0)^5$$
$$+ \frac{-1}{6!}(x-0)^6$$

$$= 1 - \frac{1}{2!}x^2 + \frac{1}{4!}x^4 - \frac{1}{6!}x^6$$

$$R_7(x) = \frac{\sin u}{7!}x^7$$

19. $f(x) = \dfrac{1}{1-x}$ $f(0) = 1$

$f'(x) = \dfrac{1}{(1-x)^2}$ $f'(0) = 1$

$f''(x) = \dfrac{2}{(1-x)^3}$ $f''(0) = 2$

$f^{(3)}(x) = \dfrac{6}{(1-x)^4}$ $f^{(3)}(0) = 6$

$f^{(4)}(x) = \dfrac{24}{(1-x)^5}$ $f^{(4)}(0) = 24$

$f^{(5)}(x) = \dfrac{120}{(1-x)^6}$

$P_4(x) = 1 + \dfrac{1}{1!}(x-0) + \dfrac{2}{2!}(x-0)^2 + \dfrac{6}{3!}(x-0)^3 + \dfrac{24}{4!}(x-0$

$= 1 + x + x^2 + x^3 + x^4$

$R_5(x) = \dfrac{120/(1-u)^6}{5!}(x-0)^5 = \dfrac{x^5}{(1-u)^6}$

23. In Problem 12, we found that for $f(x) = \ell n(1+x)$, $P_4(x) = x - \dfrac{x^2}{2} + \dfrac{x^3}{3} - \dfrac{x^4}{4}$. Letting $x = 0.1$ and $n = 4$, we have

$\ell n(1.1) \approx 0.1 - \dfrac{1}{2}(0.1)^2 + \dfrac{1}{3}(0.1)^3 - \dfrac{1}{4}(0.1)^4 \approx 0.0953$

$\left| R_5(0.1) \right| = \left| \dfrac{24}{(1+u)^5}(\dfrac{1}{5!})(0.1)^5 \right| = \left| (\dfrac{0.00001}{5}) \dfrac{1}{(1+u)^5} \right|$

for u between 0 and 0.1.

So $\left| R_5(0.1) \right| < \left| (\dfrac{0.00001}{5}) \dfrac{1}{(1+0)^5} \right| = 0.000002$

< 0.00001 as desired.

27. Using the results of Problem 15 with $x = 1° = \pi/180$ radians and $n = 2$, we have

$\cos(1°) = \cos(\dfrac{\pi}{180}) \approx 1 - \dfrac{1}{2!}(\dfrac{\pi}{180})^2 \approx 0.999847$

$$\left| R_3\left(\frac{\pi}{180}\right) \right| = \left| \frac{\sin u}{3!}\left(\frac{\pi}{180}\right)^3 \right| \quad \text{for} \quad u \quad \text{between} \quad 0$$

and $\frac{\pi}{180}$.

So $\left| R_3\left(\frac{\pi}{180}\right) \right| < \left| \frac{1}{3!}\left(\frac{\pi}{180}\right)^3 \right| = 0.0000009 < 0.0001$,
as desired.

31. Using Example 3, with $x = \frac{1}{2}$ and $n = 3$,

we have

$$e^{1/2} \approx 1 + \frac{1}{2} + \frac{1}{2!}\left(\frac{1}{2}\right)^2 + \frac{1}{3!}\left(\frac{1}{2}\right)^3 \approx 1.6458$$

$$\left| R_4\left(\frac{1}{2}\right) \right| = \left| \frac{e^u}{4!}\left(\frac{1}{2}\right)^4 \right| \quad \text{for} \quad u \quad \text{between} \quad 0 \quad \text{and} \quad \frac{1}{2}$$

(use $e^u < e^1 < 3$).

So $\left| R_4\left(\frac{1}{2}\right) \right| < \left| \frac{3}{4!}\left(\frac{1}{2}\right)^4 \right| = 0.0078 < 0.01$, as desired.

35.

$$f(x) = e^{-x^2} \qquad\qquad f(0) = 1$$

$$f'(x) = -2xe^{-x^2} \qquad\qquad f'(0) = 0$$

$$f''(x) = (-2+4x^2)e^{-x^2} \qquad\qquad f''(0) = -2$$

$$P_2(x) = 1 + \frac{0}{1!}(x-0) + \frac{-2}{2!}(x-0)^2 = 1 - x^2$$

39.

$$f(x) = \sqrt{1 + x} \qquad\qquad f(0) = 1$$

$$f'(x) = \frac{1}{2}(1+x)^{-1/2} \qquad\qquad f'(0) = \frac{1}{2}$$

$$f''(x) = -\frac{1}{4}(1+x)^{-3/2} \qquad\qquad f''(0) = -\frac{1}{4}$$

$$f^{(3)}(x) = \frac{3}{8}(1+x)^{-5/2} \qquad\qquad f^{(3)}(0) = \frac{3}{8}$$

$$P_3(x) = 1 + \frac{\frac{1}{2}}{1!}(x-0) + \frac{-\frac{1}{4}}{2!}(x-0)^2 + \frac{\frac{3}{8}}{3!}(x-0)^3$$

$$= 1 + \frac{1}{2}x - \frac{1}{8}x^2 + \frac{1}{16}x^3$$

3. $\displaystyle\lim_{x\to\pi/2} \frac{\sec^2 x}{\sec^2 3x} = \lim_{x\to\pi/2} \frac{\cos^2 3x}{\cos^2 x} \stackrel{*}{=} \lim_{x\to\pi/2} \frac{-6 \cos 3x \sin 3x}{-2 \cos x \sin x}$

$$\stackrel{*}{=} \lim_{x\to\pi/2} \frac{18 \sin^2 3x - 18 \cos^2 3x}{2 \sin^2 x - 2 \cos^2 x}$$

$$= \frac{18(-1)^2}{2(1)^2} = 9$$

7. $\displaystyle\int_0^1 \frac{\sin\sqrt{x}}{\sqrt{x}}\,dx = \lim_{a\to 0^+} \int_a^1 \frac{\sin\sqrt{x}}{\sqrt{x}}\,dx = \lim_{a\to 0^+} (-2 \cos\sqrt{x})\Big|_a^1$

$$= \lim_{a\to 0^+} [-2 \cos(1) + 2 \cos\sqrt{a}]$$

$$= 2 - 2 \cos(1); \text{ integral converges}$$

11. $\displaystyle\int_0^1 \frac{x\,dx}{\sqrt{1 - x^2}} = \lim_{b\to 1^-} \int_0^b \frac{x\,dx}{\sqrt{1 - x^2}} = \lim_{b\to 1^-} (- \sqrt{1 - x^2})\Big|_0^b$

$$= \lim_{b\to 1^-} [- \sqrt{1 - b^2} + 1] = 1; \text{ integral}$$
$$\text{converges}$$

15. For n = positive integer, use integration by parts $(u = x^n,\ dv = e^{-x}dx)$.

$$\int x^n e^{-x}dx = -x^n e^{-x} + n \int x^{n-1} e^{-x}dx$$

Then $\displaystyle\int_0^{+\infty} x^n e^{-x}dx = \lim_{b\to +\infty} (-x^n e^{-x})\Big|_0^b + n\int_0^{+\infty} x^{n-1} e^{-x}$

$$= \lim_{b\to +\infty} \left(\frac{-b^n}{e^b}\right) + n \int_0^{+\infty} x^{n-1} e^{-x}dx$$

$$= n \int_0^{+\infty} x^{n-1} e^{-x}dx$$

since $\displaystyle\lim_{b\to +\infty} \left(\frac{-b^n}{e^b}\right) = 0$ after n applications of

L'Hospital's rule.

19. $\int \dfrac{\sqrt{1 + x^{1/8}}}{x^{3/4}}$ dx (Substitute $u^8 = x$.) $= \int 8u\sqrt{1 + u}$ du

(Substitute $v = 1 + u$.)

$\quad = 8\int (v-1)\sqrt{v}\ dv = 8\int (v^{3/2} - v^{1/2})dv$

$\quad = \dfrac{16}{5} v^{5/2} - \dfrac{16}{3} v^{3/2} + C$

$\quad = \dfrac{16}{5}(1+u)^{5/2} - \dfrac{16}{3}(1+u)^{3/2} + C$

$\quad = \dfrac{16}{5}(\sqrt{1 + x^{1/8}})^5 - \dfrac{16}{3}(\sqrt{1 + x^{1/8}})^3 + C$

$\int_0^{+\infty} \dfrac{\sqrt{1 + x^{1/8}}}{x^{3/4}}$ dx $= \lim\limits_{b \to +\infty} \ [\dfrac{16}{5}(\sqrt{1 + x^{1/8}})^5$

$\qquad\qquad\qquad\qquad\qquad - \dfrac{16}{3}(\sqrt{1 + x^{1/8}})^3]\ \big|_0^b$

$\quad = \lim\limits_{b \to +\infty} \ [\dfrac{16}{5}(\sqrt{1 + b^{1/8}})^5 - \dfrac{16}{3}(\sqrt{1 + b^{1/8}})^3 - \dfrac{16}{5} + \dfrac{16}{3}]$

$\quad = \lim\limits_{b \to +\infty} \ \{16(\sqrt{1 + b^{1/8}})^3 \ [\dfrac{1}{5}(1 + b^{1/8}) - \dfrac{1}{3}] + \dfrac{32}{15}\} = +\infty \ ;$

integral diverges

23. $\lim\limits_{n \to +\infty} n(\sqrt[n]{x} - 1) = \lim\limits_{n \to +\infty} \dfrac{x^{1/n} - 1}{1/n}$ (Now apply
L'Hospital's rule.)

$\overset{*}{=} \lim\limits_{n \to +\infty} \dfrac{x^{1/n}(\ln x)(-1/n^2)}{-1/n^2} = \lim\limits_{n \to +\infty} x^{1/n} \ln x$

$= (1)\ln x = \ln x$ (for $x > 0$)

27. $0 \leq \dfrac{1}{\sqrt{3}} \leq \dfrac{1}{\sqrt{2 + \sin x}}$ and $\int_0^{+\infty} \dfrac{1}{\sqrt{3}}$ dx $= \lim\limits_{b \to +\infty} \int_0^b \dfrac{1}{\sqrt{3}}$ dx

$= \lim\limits_{b \to +\infty} (\dfrac{x}{\sqrt{3}}) \big|_0^b = \lim\limits_{b \to +\infty} (\dfrac{b}{\sqrt{3}}) = +\infty .$ Thus, by Problem

26, $\int_0^{+\infty} \dfrac{1}{\sqrt{2 + \sin x}}$ dx diverges.

31. In order to apply L'Hospital's rule to $\lim\limits_{x \to 0} \dfrac{f(x)}{g(x)}$, we must have $\lim\limits_{x \to 0} \dfrac{f'(x)}{g'(x)} = L$ where L is a real number or $+\infty$ or $-\infty$. In this case $\lim\limits_{x \to 0} \dfrac{f'(x)}{g'(x)}$

$= \lim\limits_{x \to 0} \dfrac{2x \sin(1/x) + x^2 \cos(1/x)(-1/x^2)}{\cos(x)}$

$= \lim\limits_{x \to 0} \dfrac{2x \sin(1/x) - \cos(1/x)}{\cos(x)}$, which fails to exist due to $\cos(1/x)$ oscillating between -1 and $+1$.

35. (a) $\lim\limits_{h \to 0} \dfrac{f(x+2h) - 2f(x+h) + f(x)}{h^2}$

$\overset{*}{=} \lim\limits_{h \to 0} \dfrac{2f'(x+2h) - 2f'(x+h)}{2h}$

$\overset{*}{=} \lim\limits_{h \to 0} \dfrac{4f''(x+2h) - 2f''(x+h)}{2}$

$= f''(x)$ (Use L'Hospital's rule twice and differentiate with respect to h.)

(b) $\lim\limits_{h \to 0} \dfrac{f(x+3h) - 3f(x+2h) + 3f(x+h) - f(x)}{h^3}$

$\overset{*}{=} \lim\limits_{h \to 0} \dfrac{3f'(x+3h) - 6f'(x+2h) + 3f'(x+h)}{3h^2}$

$\overset{*}{=} \lim\limits_{h \to 0} \dfrac{9f''(x+3h) - 12f''(x+2h) + 3f''(x+h)}{6h}$

$\overset{*}{=} \lim\limits_{h \to 0} \dfrac{27f^{(3)}(x+3h) - 24f^{(3)}(x+2h) + 3f^{(3)}(x+h)}{6}$

$= f^{(3)}(x)$ (Use L'Hospital's rule three times and differentiate with respect to h.)

(c) $f^{(n)}(x)$

$$= \lim_{h \to 0} \frac{f(x+nh) - \frac{n}{1!} f(x+(n-1)h) - \cdots + (-1)^n f(x)}{h^n}$$

39. $L\{1\} = \int_0^{+\infty} e^{-sx}(1)dx = \lim_{b \to +\infty} \int_0^b e^{-sx}dx$

$$= \lim_{b \to +\infty} (-\frac{1}{s} e^{-sx}) \Big|_0^b$$

$$= \lim_{b \to +\infty} (-\frac{1}{s} e^{-sb} + \frac{1}{s}) = \frac{1}{s}$$

if $s > 0$ and diverges if $s \le 0$.

43. $L\{e^x\} = \int_0^{+\infty} e^{-sx} e^x dx = \lim_{b \to +\infty} \int_0^b e^{(1-s)x}dx$

$$= \lim_{b \to +\infty} [\frac{1}{1-s} e^{(1-s)x}] \Big|_0^b$$

$$= \lim_{b \to +\infty} [(\frac{1}{1-s}) \frac{1}{e^{(s-1)b}} - \frac{1}{1-s}]$$

$$= -\frac{1}{1-s} = \frac{1}{s-1} , \quad \text{if } s > 1 \text{ and diverges if } s \le 1.$$

47. $\mu = \int_a^b x(\frac{1}{b-a})dx = \frac{1}{2} x^2 (\frac{1}{b-a}) \Big|_a^b = (\frac{1}{2} b^2 - \frac{1}{2} a^2)\frac{1}{b-a}$

$$= \frac{b^2 - a^2}{2(b-a)} = \frac{b+a}{2}$$

Chapter Eleven

Exercise 1, pp. 578-580

3. $s_1 = \ln(1) = 0$; $s_2 = \ln 2$; $s_3 = \ln 3$; $s_4 = \ln 4$

7. $s_1 = \dfrac{1 - (-1)^1}{2} = \dfrac{2}{2} = 1$; $s_2 = \dfrac{1 - (-1)^2}{2} = \dfrac{0}{2} = 0$;

$s_3 = \dfrac{1 - (-1)^3}{2} = 1$; $s_4 = \dfrac{1 - (-1)^4}{2} = 0$

11. $\lim\limits_{n \to +\infty} (2 - \dfrac{4}{n}) = 2 - 0 = 2$; sequence converges

15. $\lim\limits_{n \to +\infty} (0.5)^n = \lim\limits_{n \to +\infty} (\dfrac{1}{2})^n = \lim\limits_{n \to +\infty} \dfrac{1}{2^n} = 0$;

sequence converges

19. $\lim\limits_{n \to +\infty} (\dfrac{n+(-1)^n}{n}) = \lim\limits_{n \to +\infty} \dfrac{n}{n} + \lim\limits_{n \to +\infty} \dfrac{(-1)^n}{n} = 1 + 0 = 1$;

sequence converges

23. $\lim\limits_{n \to +\infty} (\dfrac{n^2}{2n + 1} - \dfrac{n^2}{2n - 1}) = \lim\limits_{n \to +\infty} (\dfrac{n^2(2n-1) - n^2(2n+1)}{(2n+1)(2n-1)})$

$= \lim\limits_{n \to +\infty} (\dfrac{2n^3 - n^2 - 2n^3 - n^2}{4n^2 - 1}) = \lim\limits_{n \to +\infty} \dfrac{-2n^2}{4n^2 - 1}$

$= \lim\limits_{n \to +\infty} \dfrac{n^2(-2)}{n^2(4 - \frac{1}{n^2})} = \lim\limits_{n \to +\infty} \dfrac{-2}{4 - \frac{1}{n^2}} = \dfrac{-2}{4 - 0} = \dfrac{-1}{2}$;

sequence converges

27. $\lim\limits_{n \to +\infty} (2 - \dfrac{1}{2^n}) = 2 - 0 = 2$; sequence converges

31. For $s_n = \cos(n\pi + \pi/2)$, $s_1 = \cos(\pi + \pi/2) = 0$,
$s_2 = \cos(2\pi + \pi/2) = 0$, $s_3 = \cos(3\pi + \pi/2) = 0$,
$\cdots$; $\lim\limits_{n \to +\infty} s_n = 0$; sequence converges

35. $\lim\limits_{n\to+\infty}[\ell n\ n - \ell n\ (n+1)] = \lim\limits_{n\to+\infty}(\ell n\ \frac{n}{n+1})$

$= \ell n(\lim\limits_{n\to+\infty}\frac{n}{n+1})$ (Apply L'Hospital's rule.)

$\overset{*}{=}\ell n(\lim\limits_{n\to+\infty}\frac{1}{1}) = \ell n(1) = 0$; sequence converges

39. $\lim\limits_{n\to+\infty}\dfrac{n+\sin\ n}{n+\cos\ 4n} = \lim\limits_{n\to+\infty}\dfrac{n[1+(\sin\ n)/n]}{n[1+(\cos\ 4n)/n]}$

$= \lim\limits_{n\to+\infty}\dfrac{1+(\sin\ n)/n}{1+(\cos\ 4n)/n} = \dfrac{1+0}{1+0} = 1$; sequence converges

43. Convergent; $\lim\limits_{n\to+\infty}s_n = \lim\dfrac{1+1/3^n}{4^n/3^n} = \lim\dfrac{1+1/3^n}{(4/3)^n} = 0$

(Use $\lim\limits_{n\to+\infty}(4/3)^n = +\infty$ since $4/3 > 1$.)

47. Convergent; $\lim\limits_{n\to+\infty}s_n = \lim\limits_{n\to+\infty}(\dfrac{8}{10})^n = \lim\limits_{n\to+\infty}\dfrac{1}{(\frac{10}{8})^n} = 0$

(Since $\dfrac{10}{8} > 1$, $\lim\limits_{n\to+\infty}(\dfrac{10}{8})^n = +\infty$.)

51. Divergent; sequence oscillates between $+1$ and -1.

55. Divergent; $\lim\limits_{n\to+\infty}s_n = \lim\limits_{n\to+\infty}\dfrac{(-3)^n}{n^{100}} = \lim\limits_{n\to+\infty}\dfrac{(-1)^n 3^n}{n^{100}}$

fails to exist. Apply L'Hospital's rule 100 times to

$\lim\limits_{n\to+\infty}\dfrac{3^n}{n^{100}}$ and we get $+\infty$. The $(-1)^n$ then

causes the original limit to not exist.

59. Convergent; $\lim\limits_{n\to+\infty}s_n = \lim\limits_{n\to+\infty}\sqrt{8-1/n}$

$= \sqrt{8-0} = \sqrt{8} = 2\sqrt{2}$

63. Find n such that $\left|\frac{1}{n} - \frac{1}{50}\right| < \frac{1}{100}$ or

$-\frac{1}{100} < \frac{1}{n} - \frac{1}{50} < \frac{1}{100} \Rightarrow \frac{1}{100} < \frac{1}{n} < \frac{3}{100}$.

$\frac{1}{100} < \frac{1}{n} \Rightarrow 100 > n$ or $99 \geq n$ and $\frac{1}{n} < \frac{3}{100} \Rightarrow$

$n > \frac{100}{3}$ or $n \geq 34$. Thus, $34 \leq n \leq 99$.

67. Let r be a number between 0 and 1, then
$r = \frac{1}{1 + p}$ for $p > 0$. Using the binomial expansion,

$(1+p)^n = 1 + np + n\frac{(n-1)}{2}p^2 + \cdots + p^n > np$, so

$\frac{1}{(1+p)^n} < \frac{1}{np}$. Thus, for $0 < r < 1$, $\lim_{n \to +\infty} r^n =$

$\lim_{n \to +\infty} \frac{1}{(1+p)^n} < \lim_{n \to +\infty} \frac{1}{np} = 0$. But since $r > 0$, we

also have $\lim_{n \to +\infty} r^n \geq 0$. Then by the squeezing

theorem (2.20) we have $\lim_{n \to +\infty} r^n = 0$ (for $0 < r < 1$)

71. If $\lim_{n \to +\infty} s_n = L$, then by Theorem (11.6) $\{s_n\}$ is

bounded and there exists a number K such that
$|s_n| \leq K$ for all n. Using Definition (11.2),
for $\frac{\varepsilon}{K + |L|} > 0$, there exists an N such that

$|s_n - L| < \frac{\varepsilon}{K + |L|}$ for all $n > N$. Now, for every

$\varepsilon > 0$, we have
$|s_n^2 - L^2| = |s_n - L||s_n + L| \leq |s_n - L| (|s_n| + |L|)$

(by the triangle inequality) and from the earlier

inequalities,

$$\left|s_n^2 - L^2\right| \le \left|s_n - L\right| (K + |L|) < \frac{\varepsilon}{K + |L|}(K + |L|) = \varepsilon$$

which is true for all $n > N$. Then by Definition (11.2), $\lim\limits_{n \to +\infty} s_n^2 = L^2$.

Also note that if $\{s_n\}$ is convergent and

$$\lim\limits_{n \to +\infty} s_n = L_1 \qquad \text{then by Theorem (11.4c), p. 571}$$

$$\lim\limits_{n \to +\infty} s_n^2 = (\lim\limits_{n \to +\infty} s_n)(\lim\limits_{n \to +\infty} s_n) = (L)(L) = L^2$$

Thus, $\{s_n^2\}$ is convergent.

75. $u_{n+2} = u_{n+1} + u_n$, $u_1 = 1$, $u_2 = 1$

(a)

i	1	2	3	4	5	6	7	8
u_i	1	1	2	3	5	8	13	21

(b) First note that $(1+\sqrt{5})^2 = 1 + 2\sqrt{5} + 5 = 6 + 2\sqrt{5}$

$$= 2(3+\sqrt{5})$$

and $(1-\sqrt{5})^2 = 1 - 2\sqrt{5} + 5 = 6 - 2\sqrt{5} = 2(3-\sqrt{5})$

Then,

$$u_{n+2} = \frac{(1+\sqrt{5})^{n+2} - (1-\sqrt{5})^{n+2}}{2^{n+2}\sqrt{5}}$$

$$= \frac{(1+\sqrt{5})^2(1+\sqrt{5})^n - (1-\sqrt{5})^2(1-\sqrt{5})^n}{2^{n+2}\sqrt{5}}$$

$$= \frac{2(3+\sqrt{5})(1+\sqrt{5})^n - 2(3-\sqrt{5})(1-\sqrt{5})^n}{2^{n+2}\sqrt{5}}$$

$$= \frac{[2+(1+\sqrt{5})](1+\sqrt{5})^n - [2+(1-\sqrt{5})](1-\sqrt{5})^n}{2^{n+1}\sqrt{5}}$$

$$= \frac{2(1+\sqrt{5})^n + (1+\sqrt{5})^{n+1} - 2(1-\sqrt{5})^n - (1-\sqrt{5})^{n+1}}{2^{n+1}\sqrt{5}}$$

$$= \frac{2[(1+\sqrt{5})^n - (1-\sqrt{5})^n]}{2^{n+1}\sqrt{5}} + \frac{(1+\sqrt{5})^{n+1} - (1-\sqrt{5})^{n+1}}{2^{n+1}\sqrt{5}}$$

$$= u_n + u_{n+1}$$

Also, note that

$$u_1 = \frac{(1+\sqrt{5})^1 - (1-\sqrt{5})^1}{2\sqrt{5}} = \frac{2\sqrt{5}}{2\sqrt{5}} = 1$$

$$U_2 = \frac{(1+\sqrt{5})^2 - (1-\sqrt{5})^2}{2^2\sqrt{5}} = \frac{6 + 2\sqrt{5} - (6-2\sqrt{5})}{4\sqrt{5}} = \frac{4\sqrt{5}}{4\sqrt{5}} = 1$$

<u>Exercise 2</u>, pp. 592-593

3. $S_4 = 1 + 2 + 3 + 4 = 10$

7. Divergent; $a = 1$, but $r = 2 > 1$

11. Divergent; $a = 1$, but $r = \sqrt{2} > 1$

15. Convergent geometric series with $a = \frac{1}{100}$, $r = \frac{1}{100}$;

$$\text{Sum} = \frac{a}{1 - r} = \frac{\frac{1}{100}}{1 - \frac{1}{100}} = \frac{\frac{1}{100}}{\frac{99}{100}} = \frac{1}{99}$$

19. $\lim_{n \to +\infty} a_n = \lim_{n \to +\infty} \frac{n^2 + 1}{4n + 1} \overset{*}{=} \lim_{n \to +\infty} \frac{2n}{4} = +\infty$ (Use

L'Hospital's rule.) Then by (11.18) $\sum_{k=1}^{+\infty} \frac{k^2 + 1}{4k + 1}$

diverges.

23. $\lim_{n \to +\infty} a_n = \lim_{n \to +\infty} (n + \frac{1}{n}) = +\infty$ Then by (11.18),

$\sum_{k=1}^{+\infty} (k + \frac{1}{k})$ diverges.

27. Divergent geometric series with $a = \sqrt[3]{2}$ and $r = \sqrt[3]{2} > 1$.

31. $\lim\limits_{n \to +\infty} a_n = \lim\limits_{n \to +\infty} \frac{n+1}{n-2} \overset{*}{=} \lim\limits_{n \to +\infty} \frac{1}{1} = 1$ (Use L'Hospital's rule.) Then by (11.18), $\sum\limits_{k=3}^{+\infty} \frac{k+1}{k-2}$ diverges.

35. $0.555 \cdots = 0.5 + 0.5(0.1) + 0.5(0.1)^2 + \cdots$

    $= \dfrac{0.5}{1 - 0.1}$, since this is a convergent geometric series with $a = 0.5$ and $r = 0.1$. Thus,

    $0.555 \cdots = \dfrac{1/2}{9/10} = 5/9$.

39. $\sum\limits_{k=1}^{+\infty} \dfrac{1}{x^{k-1}} = 1 + \dfrac{1}{x} + \dfrac{1}{x^2} + \dfrac{1}{x^3} + \cdots = 1 + (\dfrac{1}{x})^1 + (\dfrac{1}{x})^2$

    $+ \cdots$ which is a geometric series with $a = 1$,

    $r = \dfrac{1}{x}$, and it converges to $\dfrac{1}{1 - \dfrac{1}{x}} = \dfrac{1}{\dfrac{x-1}{x}} = \dfrac{x}{x-1}$

    if $\left|\dfrac{1}{x}\right| < 1$. Thus, $\sum\limits_{k=1}^{+\infty} \dfrac{1}{x^{k-1}} = \dfrac{x}{x-1}$ if $\left|\dfrac{1}{x}\right| < 1$

    or equivalently $1 < |x|$.

43. $\dfrac{1}{2} + \dfrac{1}{8} + \dfrac{1}{32} + \cdots + \dfrac{1}{2^{2n-1}} + \cdots$

    $= \dfrac{1}{2} + \dfrac{1}{2}(\dfrac{1}{4}) + \dfrac{1}{2}(\dfrac{1}{4})^2 + \cdots + \dfrac{1}{2}(\dfrac{1}{4})^{n-1} + \cdots$

    is a geometric series with $a = \dfrac{1}{2}$, $r = \dfrac{1}{4}$.

    Thus, the sum $= \dfrac{a}{1-r} = \dfrac{\dfrac{1}{2}}{1 - \dfrac{1}{4}} = \dfrac{\dfrac{1}{2}}{\dfrac{3}{4}} = \dfrac{2}{3}$.

47. Let $\sum\limits_{k=1}^{+\infty} a_k$, $\sum\limits_{k=1}^{+\infty} b_k$ be convergent series such that $\sum\limits_{k=1}^{+\infty} a_k = $ 1

and $\sum\limits_{k=1}^{+\infty} b_k = M$. Then by (11.13), the sequences of

partial sums $S_n = a_1 + a_2 + \cdots + a_n$ and $T_n = b_1 + b_2 + \cdots + b_n$

satisfy $\lim\limits_{n \to +\infty} S_n = L$ and $\lim\limits_{n \to +\infty} T_n = M$. Then, if we define

$Q_n = (a_1 + b_1) + (a_2 + b_2) + \cdots (a_n + b_n) = (a_1 + \cdots + a_n) + (b_1 + \cdots b_n)$

$= S_n + T_n$, we have by limit theorems that $\lim\limits_{n \to +\infty} Q_n$

$= \lim\limits_{n \to +\infty} S_n + \lim\limits_{n \to +\infty} T_n = L + M$. Thus, $\sum\limits_{k=1}^{\infty} (a_k + b_k) = \lim\limits_{n \to +\infty} Q_n$

$= L + M = \sum\limits_{k=1}^{+\infty} a_k + \sum\limits_{k=1}^{+\infty} b_k$

Exercise 3, pp. 606-607

3. For $k \geq 2$, $\sqrt{k} - \dfrac{1}{\sqrt{k}} < \sqrt{k} \Rightarrow \dfrac{1}{\sqrt{k} - \dfrac{1}{\sqrt{k}}} > \dfrac{1}{\sqrt{k}}$.

But $\dfrac{1}{\sqrt{k} - \dfrac{1}{\sqrt{k}}} = \dfrac{1}{\dfrac{k-1}{\sqrt{k}}} = \dfrac{\sqrt{k}}{k-1}$, thus $\dfrac{\sqrt{k}}{k-1} > \dfrac{1}{\sqrt{k}}$.

Since $\sum\limits_{k=2}^{+\infty} \dfrac{1}{\sqrt{k}}$ is a divergent p-series ($p = \dfrac{1}{2}$),

by (11.24) $\sum\limits_{k=2}^{+\infty} \dfrac{\sqrt{k}}{k-1}$ diverges.

7. Compare $\sum\limits_{k=1}^{+\infty} \dfrac{1}{(k+1)(k+2)}$ with $\sum\limits_{k=1}^{+\infty} \dfrac{1}{k^2}$ (convergent

p-series). $\lim\limits_{n \to \infty} \left[\dfrac{1/(n+1)(n+2)}{1/n^2} \right] = \lim\limits_{n \to \infty} \dfrac{n^2}{n^2 + 3n + 2} = 1$

(Use L'Hospital's rule twice.)

Then by (11.26), $\displaystyle\sum_{k=1}^{+\infty} \frac{1}{(k+1)(k+2)}$ also converges.

11. Compare $\displaystyle\sum_{k=1}^{+\infty} \frac{3\sqrt{k} + 2}{2k^2 + 5}$ with $\displaystyle\sum_{k=1}^{+\infty} \frac{1}{k^{3/2}}$ (convergent

p-series). $\displaystyle\lim_{n \to +\infty} \{[(3\sqrt{n} + 2)/(2n^2 + 5)]/[1/(\sqrt{n})^3]\}$

$= \displaystyle\lim_{n \to +\infty} \frac{3n^2 + 2(\sqrt{n})^3}{2n^2 + 5}$ (Use L'Hospital's rule.)

$\overset{*}{=} \displaystyle\lim_{n \to +\infty} \frac{6n + 3\sqrt{n}}{4n} = \displaystyle\lim_{n \to +\infty} (\frac{6}{4} + \frac{3}{4\sqrt{n}}) = \frac{6}{4} = \frac{3}{2}$

Then by (11.26), $\displaystyle\sum_{k=1}^{+\infty} \frac{3\sqrt{k} + 2}{2k^2 + 5}$ also converges.

15. Compare $\displaystyle\sum_{k=1}^{+\infty} \frac{3k + 4}{k2^k}$ with $\displaystyle\sum_{k=1}^{+\infty} \frac{1}{2^k}$ (convergent

geometric series, $a = 1/2$, $r = 1/2$).

$\displaystyle\lim_{n \to +\infty} \{[(3n+4)/n2^n]/(1/2^n)]\} = \displaystyle\lim_{n \to +\infty} \frac{3n + 4}{n}$

$= \displaystyle\lim_{n \to +\infty} (3 + \frac{4}{n}) = 3$

Then by (11.26), $\displaystyle\sum_{k=1}^{+\infty} \frac{3k + 4}{k2^k}$ also converges.

19. Compare $\displaystyle\sum_{k=1}^{+\infty} \frac{k + 5}{k^{k+1}}$ with $\displaystyle\sum_{k=1}^{+\infty} \frac{1}{k^k}$ (convergent by

Example 1, p. 574). $\displaystyle\lim_{n \to +\infty} \{[(n+5)/n^{n+1}]/(1/n^n)\}$

$= \displaystyle\lim_{n \to +\infty} (\frac{n+5}{n}) = \displaystyle\lim_{n \to +\infty} (1 + \frac{5}{n}) = 1$

Then by (11.26), $\displaystyle\sum_{k=1}^{+\infty} \frac{k + 5}{k^{k+1}}$ also converges.

Note that for Problems 23-36, all integrands must be checked to be sure they are continuous, positive, and nonincreasing in order to apply the integral test.

23. $\int_1^{+\infty} \frac{1}{x^{1.01}} dx = \lim_{b \to +\infty} \int_1^b \frac{1}{x^{1.01}} dx$

$$= \lim_{b \to +\infty} \left(\frac{-1}{0.01x^{0.01}} \right) \Big|_1^b$$

$$= \lim_{b \to +\infty} \left(\frac{-1}{0.01b^{0.01}} + \frac{1}{0.01} \right) = 100$$

Thus, $\sum_{k=1}^{+\infty} \frac{1}{k^{1.01}}$ converges.

27. $\int_1^{+\infty} xe^{-x} dx = \lim_{b \to +\infty} \int_1^b xe^{-x} dx$ (Use integration by

parts with $u = x$, $dv = e^{-x} dx$.) $= \lim_{b \to +\infty} (-xe^{-x} - e^{-x}) \Big|_1^b$

$= \lim_{b \to +\infty} \left(\frac{-b}{e^b} - \frac{1}{e^b} + \frac{1}{e} + \frac{1}{e} \right) = 0 + 0 + \frac{1}{e} + \frac{1}{e} = \frac{2}{e}$ (Use

L'Hospital's rule for b/e^b.)

Thus, $\sum_{k=1}^{+\infty} ke^{-k}$ converges.

31. $\int_2^{+\infty} \frac{1}{x(\ln x)} dx = \lim_{b \to +\infty} \int_2^b \frac{1}{x(\ln x)} dx$ (Use

$u = \ln x$.) $= \lim_{b \to +\infty} [\ln(\ln x)] \Big|_2^b$

$$= \lim_{b \to +\infty} [\ln(\ln b) - \ln(\ln 2)] = +\infty$$

Thus, $\sum_{k=2}^{+\infty} \frac{1}{k(\ln k)}$ diverges.

35. $\int_1^{+\infty} \frac{\ln x}{x^2}\, dx = \lim_{b \to +\infty} \int_1^b \frac{\ln x}{x^2}\, dx$ (Use integration

by parts with $u = \ln x$, $dv = \frac{1}{x^2}\, dx$.)

$= \lim_{b \to +\infty} \left(\frac{-\ln x - 1}{x} \right) \Big|_1^b = \lim_{b \to +\infty} \left(\frac{-\ln b - 1}{b} + 1 \right) = 1$

(Use L'Hospital's rule to show that

$\lim_{b \to +\infty} \frac{-\ln b - 1}{b} = 0$.)

Thus, $\sum_{k=1}^{+\infty} \frac{\ln k}{k^2}$ converges.

39. $\lim_{n \to +\infty} \frac{a_{n+1}}{a_n} \quad \lim_{n \to +\infty} \left[\frac{(n+1)(2/3)^{n+1}}{n(2/3)^n} \right]$

$= \lim_{n \to +\infty} \left(1 + \frac{1}{n} \right) \frac{2}{3} = \frac{2}{3}$

By the ratio test, $\sum_{k=1}^{+\infty} k \left(\frac{2}{3} \right)^k$ converges.

43. $\lim_{n \to +\infty} \frac{a_{n+1}}{a_n} = \lim_{n \to +\infty} \left[\frac{n+1/(2n)!}{n/(2n-2)!} \right] = \lim_{n \to +\infty} \left[\frac{(n+1)}{n(2n)(2n-1)} \right]$

$= \lim_{n \to +\infty} \frac{n+1}{4n^3 - 2n^2} = 0$ (Use L'Hospital's rule.)

By the ratio test, $\sum_{k=1}^{+\infty} \frac{k}{(2k-2)!}$ converges.

47. $\lim_{n \to +\infty} \frac{a_{n+1}}{a_n} = \lim_{n \to +\infty} \left[\frac{(n+1)^3/(n+1)!}{n^3/n!} \right] = \lim_{n \to +\infty} \left[\frac{(n+1)^3}{n^3(n+1)} \right]$

$= \lim_{n \to +\infty} \left[\frac{(n+1)^2}{n^3} \right] = \lim_{n \to +\infty} \left[\frac{1}{n} + \frac{2}{n^2} + \frac{1}{n^3} \right] = 0$

By the ratio test, $\sum_{k=1}^{+\infty} \frac{k^3}{k!}$ converges.

51. $\lim\limits_{n \to +\infty} \dfrac{a_{n+1}}{a_n} = \lim\limits_{n \to +\infty} \left[\dfrac{(n+1)/e^{n+1}}{n/e^n}\right] = \lim\limits_{n \to +\infty} \dfrac{n+1}{ne}$

$= \lim\limits_{n \to +\infty} \left[\dfrac{1}{e} + \dfrac{1}{ne}\right] = \dfrac{1}{e}$

By the ratio test, $\sum\limits_{k=1}^{+\infty} \dfrac{k}{e^k}$ converges.

55. $\int \dfrac{1}{x(\ln x)^p}\, dx = \begin{cases} \dfrac{1}{(1-p)(\ln x)^{p-1}} + C & \text{for } p \neq 1 \\[2mm] \ln(\ln x) + C & \text{for } p = 1 \end{cases}$

(In both cases, use $u = \ln x$, $du = \dfrac{1}{x}\, dx$.)

Then for $p \neq 1$, $\int_2^{+\infty} \dfrac{1}{x(\ln x)^p}\, dx$

$= \lim\limits_{b \to +\infty} \left[-\dfrac{1}{(1-p)(\ln x)^{p-1}}\right]\Big|_2^b$

$= \lim\limits_{b \to +\infty} \left[\dfrac{1}{(1-p)(\ln b)^{p-1}} - \dfrac{1}{(1-p)(\ln 2)^{p-1}}\right]$

For $p > 1$, $\int_2^{+\infty} \dfrac{1}{x(\ln x)^p}\, dx = 0 - \dfrac{1}{(1-p)(\ln 2)^{p-1}}$

$= \dfrac{1}{(p-1)(\ln 2)^{p-1}}$ and therefore $\sum\limits_{k=2}^{+\infty} \dfrac{1}{k(\ln k)^p}$

converges by the integral test. For $p < 1$,

$\int_2^{+\infty} \dfrac{1}{x(\ln x)^p}\, dx = +\infty$ and therefore $\sum\limits_{k=2}^{+\infty} \dfrac{1}{k(\ln k)^p}$

diverges by the integral test. This series also diverges for $p = 1$, since then,

$\int_2^{+\infty} \dfrac{1}{x(\ln x)^p}\, dx = \lim\limits_{b \to +\infty} \left[\ln(\ln x)\right]\Big|_2^b$

$= \lim\limits_{n \to +\infty} \left[\ln(\ln b) - \ln(\ln 2)\right] = +\infty.$

59. For the divergent harmonic series $\sum\limits_{k=1}^{+\infty} \dfrac{1}{k}$,

$$\lim_{n \to +\infty} \frac{a_{n+1}}{a_n} = \lim_{n \to +\infty} \frac{1/(n+1)}{1/n} = \lim_{n \to +\infty} \frac{n}{n+1} = 1$$

(Use L'Hospital's rule.)

63. $\lim\limits_{n \to +\infty} \dfrac{a_n}{d_n} = p > 0 \Rightarrow$ for some $\varepsilon > 0$, $\dfrac{a_n}{d_n} \geq \varepsilon$ for all

$n > $ some $N \Rightarrow a_n \geq \varepsilon\, d_n$ for all $n > N$. Then by

comparison test II (11.24), the divergence of

$\sum\limits_{k=1}^{+\infty} d_k \Rightarrow$ divergence of $\sum\limits_{k=1}^{+\infty} a_k$. Similarly, if

$\lim\limits_{n \to +\infty} \dfrac{a_n}{d_n} = +\infty \Rightarrow$ for all $n > $ some N, $\dfrac{a_n}{d_n} > 1 \Rightarrow$

$a_n > d_n$ for all $n > N$. Then again by comparison

test II, $\sum\limits_{k=1}^{+\infty} a_k$ diverges.

67. $\lim\limits_{n \to +\infty} \sqrt[n]{a_n} = \lim\limits_{n \to +\infty} \sqrt[n]{1/(2n)^n} = \lim\limits_{n \to +\infty} \dfrac{1}{2n} = 0$

By the root test, $\sum\limits_{k=1}^{+\infty} \left[\dfrac{1}{(2k)^k}\right]$ converges.

Exercise 4, pp. 615-616

3. $\lim\limits_{n \to +\infty} a_n = \lim\limits_{n \to +\infty} \dfrac{1}{n\, \ell n(n)} = 0$ and $(n+1)\ell n(n+1)$

$> n\, \ell n(n) \Rightarrow a_{n+1} = \dfrac{1}{(n+1)\ell n(n+1)} < \dfrac{1}{n\, \ell n(n)} = a_n$

By the alternating series test, $\sum\limits_{k=2}^{+\infty} (-1)^k \dfrac{1}{k\, \ell n\, k}$

converges.

7. $\lim\limits_{n \to +\infty} a_n = \lim\limits_{n \to +\infty} \dfrac{1}{(n+1)2^n} = 0$ and $(n+2)2^{n+1} > (n+1)2^n$

$\Rightarrow a_{n+1} = \dfrac{1}{(n+2)2^{n+1}} < \dfrac{1}{(n+1)2^n} = a_n$

Thus, by the alternating series test, $\displaystyle\sum_{k=1}^{+\infty} \dfrac{(-1)^{k+1}}{(k+1)2^k}$

converges.

Note: For Problem 11, by (11.34), the error from using terms through a_n is E_n, and $E_n < a_{n+1}$ (the (n+1)st term), so solve for n in $a_{n+1} \le 0.001$.

11. $a_{n+1} = \dfrac{1}{(n+1)^4} \le 0.001 \Rightarrow (n+1)^4 \ge 1000$

$\Rightarrow n + 1 \ge \sqrt[4]{1000} \Rightarrow n \ge 4.62 \Rightarrow$ choose $n = 5$

$\text{Sum} \approx \dfrac{(-1)^2}{1^4} + \dfrac{(-1)^3}{2^4} + \dfrac{(-1)^4}{3^4} + \dfrac{(-1)^5}{4^4} + \dfrac{(-1)^6}{5^4}$

$\approx 1 - 0.0625 + 0.0123 - 0.0039 + 0.0016 \approx 0.9475$

Note: For Problem 15, by (11.34), the error from using terms through a_n is E_n, and $E_n < a_{n+1}$. We want to solve for n so that $a_{n+1} \le 0.0001$.

15. $a_{n+1} = \dfrac{1}{(n+1)!} \left(\dfrac{1}{3}\right)^{n+1} \le 0.0001 \Rightarrow (n+1)!\, 3^{n+1}$

$\ge 10{,}000 \quad 4!(3)^4 = 1944, \; 5!(3)^5 = 29{,}160$

$\Rightarrow n + 1 = 5 \Rightarrow n = 4$

19. $\lim\limits_{n \to +\infty} a_n = \lim\limits_{n \to +\infty} \dfrac{1}{2n} = 0$ and $a_{n+1} = \dfrac{1}{2(n+1)} < \dfrac{1}{2n} = a_n$

Thus, by the alternating series test, $\displaystyle\sum_{k=1}^{+\infty} \dfrac{(-1)^{k+1}}{2k}$

converges. But, $\displaystyle\sum_{k=1}^{+\infty} \dfrac{1}{2k} = \dfrac{1}{2} \sum_{k=1}^{+\infty} \dfrac{1}{k}$ diverges,

since the harmonic series diverges. Therefore, $\displaystyle\sum_{k=1}^{+\infty} \dfrac{(-1)^{k+1}}{2k}$ is conditionally convergent.

23. $\lim\limits_{n \to +\infty} [\dfrac{1/n^{3/2}}{1/n\sqrt{n+3}}] = \lim\limits_{n \to +\infty} (\dfrac{n\sqrt{n+3}}{n^{3/2}}) = \lim\limits_{n \to +\infty} \sqrt{\dfrac{n+3}{n}}$

$= \lim\limits_{n \to +\infty} \sqrt{1 + 3/n} = 1 \Rightarrow \sum\limits_{k=1}^{+\infty} \dfrac{1}{k\sqrt{k+3}}$ converges by

the limit comparison test with $\sum\limits_{k=1}^{+\infty} \dfrac{1}{k^{3/2}}$, a

convergent p-series. Thus, $\sum\limits_{k=1}^{+\infty} \dfrac{(-1)^{k+1}}{k\sqrt{k+3}}$ is

absolutely convergent.

27. $\sum\limits_{k=1}^{+\infty} (\dfrac{1}{5})^k$ is a convergent geometric series with

$a = \dfrac{1}{5}$, $r = \dfrac{1}{5}$. Therefore, $\sum\limits_{k=1}^{+\infty} (-1)^{k+1}(\dfrac{1}{5})^k$ is

absolutely convergent.

31. $\sum\limits_{k=1}^{+\infty} \left| (-1)^{k+1} \dfrac{\sin k}{k^2 + 1} \right| \leq \sum\limits_{k=1}^{+\infty} \dfrac{1}{k^2 + 1} \leq \sum\limits_{k=1}^{+\infty} \dfrac{1}{k^2}$,

which is a convergent p-series. Then using

comparison test I (11.23), we have

$\sum\limits_{k=1}^{+\infty} \left| (-1)^{k+1} \dfrac{\sin k}{k^2 + 1} \right|$ is convergent and hence

$\sum\limits_{k=1}^{+\infty} (-1)^{k+1} \dfrac{\sin k}{k^2 + 1}$ is absolutely convergent.

35. Compare $a_{n+1} = \dfrac{(n+1)!}{(n+1)^2 2^{n+1}}$ to $a_n = \dfrac{n!}{n^2 2^n}$

Rewrite $a_{n+1} = \dfrac{n!}{2(n+1)2^n}$. Now, since $2(n+1) < n^2$

for all $n \geq 3$, we have $\dfrac{1}{2(n+1)} > \dfrac{1}{n^2}$ and

$$\frac{n!}{2(n+1)2^n} > \frac{n!}{n^2 2^n} \,. \quad \text{That is,} \quad a_{n+1} > a_n \quad \text{for all}$$

$n \geq 3$. The terms are increasing, so $\displaystyle\sum_{k=1}^{+\infty} \frac{(-1)^{k+1} k!}{k^2 2^k}$

diverges.

39. $\displaystyle\lim_{n \to +\infty} a_n = \lim_{n \to +\infty} \frac{1}{\sqrt[3]{n}} = 0$ and $a_{n+1} = \frac{1}{\sqrt[3]{n+1}} < \frac{1}{\sqrt[3]{n}} = a_n$

$\Rightarrow$ by the alternating series test $\displaystyle\sum_{k=1}^{+\infty} \frac{(-1)^{k+1}}{k^{1/3}}$

converges, but $\displaystyle\sum_{k=1}^{+\infty} \frac{1}{k^{1/3}}$ is a divergent p-series

$(p = \frac{1}{3})$, so $\displaystyle\sum_{k=1}^{+\infty} \frac{(-1)^{k+1}}{k^{1/3}}$ converges conditionally.

43. The series is $\displaystyle\sum_{k=1}^{+\infty} \frac{(-1)^{k+1}}{k} = 1 - \frac{1}{2} + \frac{1}{3} - \frac{1}{4}$

$+ \frac{1}{5} - \frac{1}{6} \cdots$. Similarly to Problem 42, rearrange the

series into alternating groups of positive and

negative terms, so that the sums are near 2.

47. If we rearrange a series of all positive terms (or
all negative), the rearranged series sums to the
same value as the original series. This is not true
for series of positive *and* negative terms, as we can
see from Problems 42, 43, and 44.

Exercise 5, pp. 618-619

3. $6 + 2 + \frac{2}{3} + \frac{2}{9} + \cdots = 6 + 6(\frac{1}{3}) + 6(\frac{1}{3})^2 + \cdots$

(absolutely) convergent geometric series (with
$a = 6$, $r = \frac{1}{3}$)

7. $1 + \frac{2^2 + 1}{2^3 + 1} + \frac{3^2 + 1}{3^3 + 1} + \frac{4^2 + 1}{4^3 + 1} + \cdots = \displaystyle\sum_{k=1}^{+\infty} \frac{k^2 + 1}{k^3 + 1}$

$$\underset{k\to+\infty}{\text{limit}} \frac{\frac{k^2+1}{k^3+1}}{\frac{1}{k}} = \underset{k\to+\infty}{\text{limit}} \frac{k^3+k}{k^3+1} = \underset{k\to+\infty}{\lim} \frac{1+\frac{1}{k^2}}{1+\frac{1}{k^3}} = 1 > 0$$

Then, since $\Sigma\frac{1}{k}$ diverges (harmonic series), the Limit Comparison Test III says that $\overset{+\infty}{\underset{k=1}{\Sigma}} \frac{k^2+1}{k^3+1}$ diverges.

11. $\underset{k\to+\infty}{\lim} \frac{a_{k+1}}{a_k} = \underset{k\to+\infty}{\lim} \left[\frac{\frac{3^{2(k+1)-1}}{(k+1)^2+2(k+1)}}{\frac{3^{2k-1}}{k^2+2k}} \right]$

$$= \underset{k\to+\infty}{\lim} \left[\frac{3^{2k+1}}{3^{2k-1}} \cdot \frac{k^2+2k}{k^2+4k+3} \right]$$

$$= 3^2 \underset{k\to+\infty}{\lim} \left[\frac{1+\frac{2}{k}}{1+\frac{4}{k}+\frac{3}{k^2}} \right] = 9 > 1$$

Then, by the Ratio Test

$\overset{+\infty}{\underset{k=1}{\Sigma}} \frac{3^{2k-1}}{k^2+2k}$ diverges.

15. $\frac{2}{3} - \frac{3}{4} \cdot \frac{1}{2} + \frac{4}{5} \cdot \frac{1}{3} - \frac{5}{6} \cdot \frac{1}{4} + \cdots$

$$= \overset{+\infty}{\underset{k=1}{\Sigma}} (-1)^{k+1} (\frac{k+1}{k+2}) \cdot \frac{1}{k} = \overset{+\infty}{\underset{k=1}{\Sigma}} (-1)^{k+1} a_k$$

$\underset{k\to+\infty}{\lim} a_k = 0$ and $a_{k+1} = (\frac{k+2}{k+3}) \cdot (\frac{1}{k+1}) < \frac{k+1}{k+2} \cdot \frac{1}{k} = a_k$

Then the alternating series converges.

However, $\overset{+\infty}{\underset{k=1}{\Sigma}} (\frac{k+1}{k+2})\frac{1}{k}$ diverges

(Use the Limit Comparison Test II with the divergent

harmonic series $\Sigma \frac{1}{k}$.)

Therefore, the original series is conditionally convergent

19. $1 + \frac{1 \cdot 3}{2!} + \frac{1 \cdot 3 \cdot 5}{3!} + \frac{1 \cdot 3 \cdot 5 \cdot 7}{4!} + \cdots = \Sigma \frac{(2k)!}{2(k)!(k)!}$

$$\lim_{k \to +\infty} \left(\frac{a_{k+1}}{a_k} \right) = \lim_{k \to +\infty} \left[\frac{\frac{(2k+2)!}{2(k+1)!(k+1)!}}{\frac{(2k)!}{2(k)!(k)!}} \right]$$

$$= \lim_{k \to +\infty} \left[\frac{2(k)!(k)!}{2(k+1)!(k+1)!} \cdot \frac{(2k+2)!}{(2k)!} \right]$$

$$= \lim_{k \to +\infty} \left[\frac{(2k+2)(2k+1)}{(k+1)(k+1)} \right]$$

$$= \lim_{k \to +\infty} \left[\frac{4k^2 + 6k + 2}{k^2 + 2k + 1} \right] = \lim_{k \to +\infty} \left[\frac{4 + \frac{6}{k} + \frac{2}{k^2}}{1 + \frac{2}{k} + \frac{1}{k^2}} \right]$$

$$= 4 > 1. \quad \text{Therefore, by the Ratio Test,}$$
the series diverges.

23. $\lim_{k \to +\infty} \frac{k^2 + 5k}{3 + 5k^2} = \lim_{k \to +\infty} \frac{1 + \frac{5}{k}}{\frac{3}{k^2} + 5} = \frac{1}{5} > 0$

Thus $\displaystyle\sum_{k=1}^{+\infty} \frac{k^2 + 5k}{3 + 5k^2}$ diverges

27. $1 - \frac{2!}{1 \cdot 3} + \frac{3!}{1 \cdot 3 \cdot 5} - \frac{4!}{1 \cdot 3 \cdot 5 \cdot 7} + \cdots$

$$= \sum_{k=1}^{+\infty} (-1)^{k+1} \frac{2(k)!(k)!}{(2k)!} = \sum_{k=1}^{+\infty} (-1)^{k+1} a_k$$

$$\lim_{k \to +\infty} \left(\frac{a_{k+1}}{a_k} \right) = \lim_{k \to +\infty} \left[\frac{\frac{2(k+1)!(k+1)!}{(2k+2)!}}{\frac{2(k)!(k)!}{(2k)!}} \right]$$

$$= \lim_{k \to +\infty} \left[\frac{2(k+1)!(k+1)!}{2(k)!(k)!} \cdot \frac{(2k)!}{(2k+2)!} \right]$$

$$= \lim_{k \to +\infty} \left[\frac{(k+1)(k+1)}{(2k+2)(2k+1)} \right] = \lim_{k \to +\infty} \left[\frac{k^2 + 2k + 1}{4k^2 + 6k + 2} \right]$$

$$= \lim_{k \to +\infty} \left[\frac{1 + \frac{2}{k} + \frac{1}{k^2}}{4 + \frac{6}{k} + \frac{2}{k^2}} \right] = \frac{1}{4} < 1$$

Then, by the Ratio Test,

$$\sum_{k=1}^{+\infty} \frac{2(k)!(k)!}{(2k)!}$$ converges, and therefore the original

alternating series converges absolutely.

31. $\displaystyle \sum_{k=1}^{+\infty} (-\frac{1}{k})^k = \sum_{k=1}^{+\infty} (-1)^k \frac{1}{k^k}$

$$\lim_{k \to +\infty} \frac{\frac{1}{k^k}}{\frac{1}{k^2}} = \lim_{k \to +\infty} \frac{1}{k^{k-2}} = 0$$

Then by problem 62, section 3, since $\sum \frac{1}{k^2}$ is a

convergent p-series, (p=2), we have that $\sum \frac{1}{k^k}$

converges. Hence, the original alternating series is absolutely convergent.

35. $\displaystyle \lim_{k \to +\infty} \frac{\sin^3(\frac{1}{k})}{\frac{1}{k^2}}$ (Apply L'Hopital's Rule)

$$\overset{*}{=} \lim_{k \to +\infty} \frac{3 \sin^2(\frac{1}{k})\cos(\frac{1}{k})\frac{-1}{k^2}}{\frac{-2}{k^3}}$$

$$= \frac{3}{2} \lim_{k \to +\infty} \frac{\sin^2(\frac{1}{k})\cos(\frac{1}{k})}{\frac{1}{k}}$$ (Apply L'Hopital's Rule)

$$\stackrel{*}{=} \frac{3}{3} \lim_{k \to +\infty} \frac{[2\sin(\frac{1}{k})\cos(\frac{1}{k})\cos(\frac{1}{k})(\frac{-1}{k^2}) - \sin^2(\frac{1}{k})\sin(\frac{1}{k})(\frac{-1}{k^2})]}{\frac{-1}{k^2}}$$

$$= \frac{3}{2} \lim_{k \to +\infty} [2\sin(\frac{1}{k})\cos^2(\frac{1}{k}) - \sin^3(\frac{1}{k})] = \frac{3}{2}[0] = 0$$

Since $\Sigma \frac{1}{k^2}$ is a convergent p-series (p=2), we have by problem 62, section 3, that $\Sigma \sin^3(\frac{1}{k})$ converges (absolutely).

Exercise 6, pp. 624-625

3. $\lim\limits_{n \to +\infty} \dfrac{(n+1)^2 |x|^{n+1}/3^{n+1}}{n^2 |x|^n/3^n}$

$= \lim\limits_{n \to +\infty} (\dfrac{n^2+2n+1}{n^2})(\dfrac{1}{3})|x| = \lim\limits_{n \to +\infty} (1 + \dfrac{2}{n}+\dfrac{1}{n^2}) (\dfrac{1}{3})|x|$

$= (1)(\dfrac{1}{3})|x| = \dfrac{|x|}{3}$

Therefore, by the ratio test, $\sum\limits_{k=0}^{+\infty} \dfrac{k^2 x^k}{3^k}$ converges

for $\dfrac{|x|}{3} < 1 \Longrightarrow |x| < 3$. Test of endpoints: $x = 3$

$\Longrightarrow \sum\limits_{k=0}^{+\infty} \dfrac{k^2 3^k}{3^k} = \sum\limits_{k=0}^{+\infty} k^2$, which diverges since

$\lim\limits_{n \to +\infty} a_n = \lim\limits_{n \to +\infty} n^2 \neq 0$; $x = -3 \Longrightarrow \sum\limits_{k=0}^{+\infty} \dfrac{k^2(-3)^k}{3^k}$

$$= \sum_{k=0}^{+\infty} (-1)^k k^2,$$ which diverges for the same reason.

Therefore, the interval of convergence is $-3 < x < 3$.

7. $$\lim_{n \to +\infty} \left[\frac{(n+1)|2x|^{n+1}/3^{n+1}}{n|2x|^n/3^n} \right]$$

$$= \lim_{n \to +\infty} (1 + \frac{1}{n})(\frac{2}{3})|x| = \frac{2}{3}|x|$$

Thus, by the ratio test, $\sum\limits_{k=0}^{+\infty} \dfrac{k(2x)^k}{3^k}$ converges for

$\frac{2}{3}|x| < 1 \Rightarrow |x| < \frac{3}{2}$. Test of endpoints: $x = \frac{3}{2}$

$$\Rightarrow \sum_{k=0}^{+\infty} \frac{k[(2)(\frac{3}{2})]^k}{3^k} = \sum_{k=0}^{+\infty} k,$$ which diverges since

$$\lim_{n \to +\infty} a_n = \lim_{n \to +\infty} n \neq 0; \quad x = -\frac{3}{2} \Rightarrow \sum_{k=0}^{+\infty} \frac{k[(2)(-\frac{3}{2})]^k}{3^k}$$

$$= \sum_{k=0}^{+\infty} (-1)^k k,$$ which diverges for the same reason.

So, the interval of convergence is $-\frac{3}{2} < x < \frac{3}{2}$.

11. $$\lim_{n \to +\infty} \left[\frac{(n+1)|x|^{n+1}/\ln(n+2)}{n|x|^n/\ln(n+1)} \right]$$

$$= \lim_{n \to +\infty} (\frac{n+1}{n})[\frac{\ln(n+1)}{\ln(n+2)}]|x| = |x|$$

(Use L'Hospital's rule to show that $\lim\limits_{n \to +\infty} \dfrac{n+1}{n} = 1$

and $\lim\limits_{n \to +\infty} \dfrac{\ln(n+1)}{\ln(n+2)} = 1$.) Then by the ratio test,

$$\sum_{k=1}^{+\infty} \frac{k \, x^k}{\ln(k+1)}$$ converges for $|x| < 1$. Test of

endpoints: $x = 1 \Rightarrow \sum\limits_{k=1}^{+\infty} \dfrac{k \, 1^k}{\ell n(k+1)} = \sum\limits_{k=1}^{+\infty} \dfrac{k}{\ell n(k+1)}$,

which diverges since, by L'Hospital's rule, we can

show that $\lim\limits_{n \to +\infty} a_n = \lim\limits_{n \to +\infty} \dfrac{n}{\ell n(n+1)} \neq 0$; $x = -1$

$\Rightarrow \sum\limits_{k=1}^{+\infty} \dfrac{k(-1)^k}{\ell n(k+1)}$, which diverges for the same reason.

Hence, the interval of convergence is $-1 < x < 1$.

15. $\lim\limits_{n \to +\infty} \left[\dfrac{|x|^{n+1}/\ell n(n+2)}{|x|^n/\ell n(n+1)} \right] = \lim\limits_{n \to +\infty} \dfrac{\ell n(n+1)}{\ell n(n+2)} |x| = |x|$

(Use L'Hospital's rule to show that $\lim\limits_{n \to +\infty} \dfrac{\ell n(n+1)}{\ell n(n+2)} = 1$

So, by the ratio test, $\sum\limits_{k=1}^{+\infty} \dfrac{x^k}{\ell n(k+1)}$ converges for

$|x| < 1$. Test of endpoints: $x = 1 \Rightarrow \sum\limits_{k=1}^{+\infty} \dfrac{1^k}{\ell n(k+1)}$

$= \sum\limits_{k=1}^{+\infty} \dfrac{1}{\ell n(k+1)}$, which diverges by the limit

comparison test with the divergent harmonic series

$\sum\limits_{k=1}^{+\infty} \dfrac{1}{k}$; $x = -1 \Rightarrow \sum\limits_{k=1}^{+\infty} \dfrac{(-1)^k}{\ell n(k+1)}$, which converges by

the alternating series test. Thus, the interval of

convergence is $-1 \leq x < 1$.

19. $\lim\limits_{n \to +\infty} \left[\dfrac{|-1|^{n+1}|x-1|^{4n+4}/(n+1)!}{|-1|^n|x-1|^{4n}/n!} \right]$

$= \lim\limits_{n \to +\infty} \dfrac{n!}{(n+1)!} |x-1|^4 = \lim\limits_{n \to +\infty} \dfrac{|x-1|^4}{n+1}$

= 0 for all x. Thus, by the ratio test,

$$\sum_{k=0}^{+\infty} \frac{(-1)^k (x-1)^{4k}}{k!}$$ converges for all x, so the

interval of convergence is $-\infty < x < +\infty$.

23. $$\lim_{n \to +\infty} \left[\frac{1/(n+1)|x|^{n+1}}{1/n|x|^n}\right] = \lim_{n \to +\infty} \left(\frac{n}{n+1}\right) \frac{1}{|x|}$$

$$= \lim_{n \to +\infty} \left[\frac{1}{1 + (\frac{1}{n})}\right] \frac{1}{|x|} = \frac{1}{|x|}.$$ Thus, by the ratio

test, $$\sum_{k=1}^{+\infty} \frac{1}{k \, x^k}$$ converges for $\frac{1}{|x|} < 1 \Rightarrow |x| > 1$

Test of endpoints: $x = 1 \Rightarrow \sum_{k=1}^{+\infty} \frac{1}{k \, 1^k} = \sum_{k=1}^{+\infty} \frac{1}{k}$,

which is the divergent harmonic series; $x = -1$

$\Rightarrow \sum_{k=1}^{+\infty} \frac{1}{k(-1)^k}$, which is the convergent alternating

harmonic series. Thus, the original series converges

for $x > 1$ or $x \leq -1$.

27. $$\lim_{n \to +\infty} \left[\frac{|-1|^{n+1} 2^{n+1} |\sin x|^{n+1}}{|-1|^n 2^n |\sin x|^n}\right]$$

$$= \lim_{n \to +\infty} 2|\sin x| = 2|\sin x|.$$ So by the ratio test,

$$\sum_{k=0}^{+\infty} (-1)^k 2^k (\sin x)^k$$ converges for $2|\sin x| < 1$

$\Rightarrow |\sin x| < \frac{1}{2}$. For both endpoints, $\sin x = \frac{1}{2}$

and $\sin x = -\frac{1}{2}$, we get divergent series whose

nth terms fail to go to zero. Thus, our convergence is for $-\frac{1}{2} < \sin x < +\frac{1}{2}$; that is, $-\frac{\pi}{6} < x < \frac{\pi}{6}$ (original restriction had x such that $-\frac{\pi}{2} \le x \le \frac{\pi}{2}$)

31. If $\sum\limits_{k=1}^{+\infty} a_k x^k$ converges for $x = 6$ and diverges

for $x = -8$, then by Theorem (11.36), it converges absolutely for $|x| < 6$ and diverges for $|x| > 8$. Thus statements a, e, and f are true and we can make no conclusion about b, c, and d.

35. Let $\sum\limits_{k=1}^{+\infty} a_k x^k$ converge for $|x| < x_0$. If one of

the endpoint series $\sum\limits_{k=1}^{+\infty} a_k x_0^k$ or $\sum\limits_{k=1}^{+\infty} a_k (-x_0)^k$

converges absolutely, then by Theorem (11.35), both

series converge, so $\sum\limits_{k=1}^{+\infty} a_k x^k$ is absolutely

convergent at both of the endpoints.

Exercise 7, pp. 632-633

3. $f(x) = e^x$ $\qquad\qquad$ $f(1) = e$

$f'(x) = e^x$ $\qquad\qquad$ $f'(1) = e$

$f''(x) = e^x$ $\qquad\qquad$ $f''(1) = e$

$\vdots$ $\qquad\qquad\qquad$ $\vdots$

$e^x = e + \dfrac{e}{1!}(x-1) + \dfrac{e}{2!}(x-1)^2 + \dfrac{e}{3!}(x-1)^3 + \cdots$

$= \sum\limits_{k=0}^{+\infty} \dfrac{e}{k!}(x-1)^k$

7. $f(x) = \dfrac{1}{x}$ $f(1) = 1$

$f'(x) = -\dfrac{1}{x^2}$ $f'(1) = -1$

$f''(x) = \dfrac{2}{x^3}$ $f''(1) = 2$

$\vdots$ $\vdots$

$f^{(n)}(x) = (-1)^n \dfrac{n!}{x^n}$ / $f^{(n)}(1) = (-1)^n n!$

$\vdots$ $\vdots$

$\dfrac{1}{x} = 1 + \dfrac{-1}{1!}(x-1) + \dfrac{2}{2!}(x-1)^2 + \cdots + \dfrac{(-1)^n n!}{n!}(x-1)^n + \cdots$

$= \displaystyle\sum_{k=0}^{+\infty} (-1)^k (x-1)^k$

11. $f(x) = \dfrac{1}{1 - 3x}$ $f(0) = 1$

$f'(x) = \dfrac{3}{(1-3x)^2} = \dfrac{(3)1!}{(1-3x)^2}$ $f'(0) = (3)1!$

$f''(x) = \dfrac{18}{(1-3x)^3} = \dfrac{(3)^2 2!}{(1-3x)^3}$ $f''(0) = (3)^2 2!$

$f^{(3)}(x) = \dfrac{(3)^3 3!}{(1-3x)^4}$ $f^{(3)}(0) = (3)^3 3!$

$\vdots$ $\vdots$

$\dfrac{1}{1 - 3x} = 1 + \dfrac{(3)1!}{1!}(x-0) + \dfrac{(3)^2 2!}{2!}(x-0)^2$

$+ \dfrac{(3)^3 3!}{3!}(x-0)^3 + \cdots$

$= 1 + 3x + (3)^2 x^2 + (3)^3 x^3 + \cdots$

$= \displaystyle\sum_{k=0}^{+\infty} 3^k x^k$

15. $f(x) = 3x^3 + 2x^2 + 5x - 6$ $f(0) = -6$

   $f'(x) = 9x^2 + 4x + 5$ $f'(0) = 5$

   $f''(x) = 18x + 4$ $f''(0) = 4$

   $f^{(3)}(x) = 18$ $f^{(3)}(0) = 18$

   $f^{(4)}(x) = 0$ $f^{(4)}(0) = 0$

   $\vdots$ $\vdots$

   $f(x) = -6 + \dfrac{5}{1!}(x-0) + \dfrac{4}{2!}(x-0)^2 + \dfrac{18}{3!}(x-0)^3$

   $\qquad + \dfrac{0}{4!}(x-0)^4 + \cdots$

   $= -6 + 5x + 2x^2 + 3x^3$

19. $\dfrac{1}{1-x} = 1 + x + x^2 + \cdots + x^n + \cdots$ for $-1 < x < 1$

   Replacing x with x^2 yields $\dfrac{1}{1-x^2} = 1 + x^2 + x^4$

   $+ \cdots = \displaystyle\sum_{k=0}^{+\infty} x^{2k}$, which converges for $-1 < x^2 < 1$

   $\Rightarrow -1 < x < 1$. Checking the endpoints, $x = 1$ and $x = -1$, both yield a divergent series whose nth term fails to go to zero. So we have convergence only for $-1 < x < 1$.

23.   Using Gregory's series,

   $\tan^{-1}(\tfrac{1}{2}) = \dfrac{1}{2} - \dfrac{(1/2)^3}{3} + \dfrac{(1/2)^5}{5} - \dfrac{(1/2)^7}{7} + \cdots$

   $\approx 0.5 - 0.04167 + 0.00625 - 0.00112$

   $\qquad + 0.00022 - 0.00004 + \cdots$

   and

   $\tan^{-1}(\tfrac{1}{3}) = \dfrac{1}{3} - \dfrac{(1/3)^3}{3} + \dfrac{(1/3)^5}{5} - \dfrac{(1/3)^7}{7} + \cdots$

   $\approx 0.33333 - 0.01235 + 0.00082 - 0.00007 + \cdots$

For accuracy within 0.0001, $\tan^{-1}(\frac{1}{2}) \approx 0.46364$
(using 6 terms) and $\tan^{-1}(\frac{1}{3}) \approx 0.32173$ (using 4
terms. Using the identity $\tan^{-1}1 = \tan^{-1}(\frac{1}{2})$
$+ \tan^{-1}(\frac{1}{3})$, we have $\frac{\pi}{4} = \tan^{-1}1 \approx 0.46364 + 0.32173$
$= 0.78537$ and hence $\pi = 4(\frac{\pi}{4}) \approx 4(0.78537) = 3.14148$
or $\pi \approx 3.1415$, accurate within 0.0001.

Exercise 8, pp. 637-638

3. From Problem 13, Exercise 7, $\ell n(1+x)$
$= \sum\limits_{k=1}^{+\infty} (-1)^{k+1} \frac{x^k}{k}$. Then substituting x^2 for x
yields $\ell n(1+x^2) = \sum\limits_{k=1}^{+\infty} (-1)^{k+1} \frac{x^{2k}}{k}$

7. Since from (11.51), $e^x = 1 + x + \frac{x^2}{2!} + \frac{x^3}{3!} + \cdots$
and from (11.56), $\sin x = x - \frac{x^3}{3!} + \frac{x^5}{5!} - \frac{x^7}{7!} + \cdots$
we can multiply the two series, term by term to yield
$$e^x \sin x = 1(x - \frac{x^3}{3!} + \frac{x^5}{5!} - \frac{x^7}{7!} + \cdots)$$
$$+ x(x - \frac{x^3}{3!} + \frac{x^5}{5!} - \frac{x^7}{7!} + \cdots)$$
$$+ \frac{x^2}{2!}(x - \frac{x^3}{3!} + \frac{x^5}{5!} - \frac{x^7}{7!} + \cdots)$$
$$+ \frac{x^3}{3!}(x - \frac{x^3}{3!} + \frac{x^5}{5!} - \cdots) + \cdots$$
$$e^x \sin x = x + x^2 + x^3(-\frac{1}{3!} + \frac{1}{2!}) + x^4(-\frac{1}{3!} + \frac{1}{3!})$$
$$+ x^5(\frac{1}{5!} - \frac{1}{2!3!} + \frac{1}{4!}) + \cdots$$

$$= x + x^2 + \left(-\frac{1}{6} + \frac{1}{2}\right)x^3 + \left(-\frac{1}{6} + \frac{1}{6}\right)x^4$$

$$+ \left(\frac{1}{120} - \frac{1}{12} + \frac{1}{24}\right)x^5 + \cdots$$

$$= x + x^2 + \frac{1}{3}x^3 - \frac{1}{30}x^5 + \cdots$$

11. We found in Problem 6 that $\sin x^2 = \sum\limits_{k=0}^{+\infty} \frac{(-1)^k x^{4k+2}}{(2k+1)!}$, which by the result for $\sin x$, will converge for all x. Now, integrating both sides using (11.37),

we get $\int_0^x \sin t^2 dt = \sum\limits_{k=0}^{+\infty} \frac{(-1)^k x^{4k+3}}{(4k+3)(2k+1)!}$, which also

converges for all x. Thus, $\int_0^1 \sin t^2 dt$

$= \sum\limits_{k=0}^{+\infty} \frac{(-1)^k}{(4k+3)(2k+1)!}$. Since this is an alternating

series, by (11.34), the error when summing the first n terms is less than a_{n+1}. Therefore, an estimate accurate within 0.001 would be found using only a_0 and a_1, since $\left|a_2\right| = \left|\frac{1}{1320}\right| < 0.001$.

$$\int_0^1 \sin t^2 dt \approx \frac{1}{3} - \frac{1}{42} \approx 0.3095 \approx 0.310$$

15. From (11.56), $\sin x = \sum\limits_{k=0}^{+\infty} \int_0^1 \frac{(-1)^k x^{2k+1}}{(2k+1)!}$ so $\frac{\sin x}{x}$

$= \sum\limits_{k=0}^{+\infty} \frac{(-1)^k x^{2k}}{(2k+1)!}$. Then, by integrating both sides,

we have $\int_0^1 \frac{\sin x}{x} dx = \sum\limits_{k=0}^{+\infty} \int_0^1 \frac{(-1)^k x^{2k}}{(2k+1)!} dx$

$= \sum\limits_{k=0}^{+\infty} \frac{(-1)^k x^{2k+1}}{(2k+1)(2k+1)!}\Big|_0^1 = 1 - \frac{1}{18} + \frac{1}{600} - \frac{1}{35,280} + \cdots$

Since $|a_3| = \left|\frac{1}{35,280}\right| < 0.001$, we have 0.001

accuracy with $\int_0^1 \frac{\sin x}{x}\,dx \approx 1 - \frac{1}{18} + \frac{1}{600} \approx 0.946$

19. $\sin x = x - \frac{x^3}{3!} + \frac{x^5}{5!} - \frac{x^7}{7!} + \cdots$,   convergent
            for all x

    $x \sin x = x^2 - \frac{x^4}{3!} + \frac{x^6}{5!} - \frac{x^8}{7!} + \cdots$

      $\cos x = 1 - \frac{x^2}{2!} + \frac{x^4}{4!} - \frac{x^6}{6!} + \cdots$,   convergent
            for all x

$1 - \cos x = \frac{x^2}{2!} - \frac{x^4}{4!} + \frac{x^6}{6!} - \cdots$

Thus,

$$\lim_{x \to 0} \frac{x \sin x}{1 - \cos x} = \lim_{x \to 0} \left[\frac{x^2 - \frac{x^4}{3!} + \frac{x^6}{5!} - \frac{x^8}{7!} + \cdots}{\frac{x^2}{2!} - \frac{x^4}{4!} + \frac{x^6}{6!} - \frac{x^8}{8!} + \cdots}\right]$$

Then dividing both numerator and denominator by x^2,

$$\lim_{x \to 0} \frac{x \sin x}{1 - \cos x} = \lim_{x \to 0} \left[\frac{1 - \frac{x^2}{3!} + \frac{x^4}{5!} - \frac{x^6}{7!} + \cdots}{\frac{1}{2!} - \frac{x^2}{4!} + \frac{x^4}{6!} - \frac{x^6}{8!} + \cdots}\right]$$

$$= \frac{1}{\frac{1}{2}} = 2$$

23. $f(x) = \cos x$ $f(0) = 1$

      $f'(x) = -\sin x$ $f'(0) = 0$

      $f''(x) = -\cos x$ $f''(0) = -1$

    $f^{(3)}(x) = \sin x$ $f^{(3)}(0) = 0$

    $f^{(4)}(x) = \cos x$ $f^{(4)}(0) = 1$

        $\vdots$ $\vdots$

By Taylor's formula with remainder,

$$\cos x = 1 + \frac{0}{1!} x + \frac{-1}{2!} x^2 + \frac{0}{3!} x^3 + \frac{1}{4!} x^4 + \cdots$$

$$+ (-1)^n \frac{x^{2n}}{(2n)!} + R_{n+1}(x)$$

where $R_{n+1}(x) = (-1)^{n+1} \dfrac{f^{(2n+2)}(u)x^{2n+2}}{(2n+2)!}$,

where u is between 0 and x. From the pattern above, $f^{(2n+2)}(u)$ is either $\cos u$ or $-\cos u$ (since $f^{(2n+2)}$ is an even drivative). Either way

$$\left| R_{n+1}(x) \right| = \frac{|x|^{2n+2}}{(2n+2)!} \left| \cos u \right| \leq \frac{|x|^{2n+2}}{(2n+2)!} \quad \text{(since}$$

$|\cos u| \leq 1)$. Using the ratio test, the series

$\displaystyle\sum_{k=0}^{+\infty} \frac{|x|^{2k+2}}{(2k+2)!}$ easily can be shown to converge for

all x, hence the limit of the nth term must go

to 0. Thus, $\displaystyle\lim_{n \to +\infty} \left| R_{n+1}(x) \right| \leq \lim_{n \to +\infty} \frac{|x|^{2n+2}}{(2n+2)!} = 0$.

Therefore, $\lim R_{n+1}(x) = 0$ by the squeezing theorem (2.20). Then by (11.49), the series converges to $\cos x$ for all x, that is,

$$\cos x = 1 - \frac{x^2}{2!} + \frac{x^4}{4!} - \frac{x^6}{6!} + \cdots \text{ for all } x$$

Exercise 9, pp. 640-641

3. $(1+x)^{1/5} = \displaystyle\sum_{k=0}^{+\infty} \binom{\frac{1}{5}}{k} x^k$, the binomial series for

$m = \frac{1}{5}$. By (11.69), part (d), we have convergence

for $-1 \leq x \leq 1$.

$$(1+x)^{1/5} = 1 + \frac{1}{5} x + \frac{1/5(1/5 - 1)}{2!} x^2$$

$$+ \frac{1/5(1/5-1)(1/5-2)}{3!} x^3 + \cdots$$

$$= 1 + \frac{1}{5} x - \frac{4}{(5)^2 2!} x^2 + \frac{(4)(9)}{(5)^3 3!} x^3$$

$$- \frac{(4)(9)(14)}{(5)^4 4!} x^4 + \cdots$$

$$= 1 + \frac{1}{5} x + \sum_{k=2}^{+\infty} \frac{(-1)^{k+1}(4)(9)\cdots(5k-6)}{(5)^k k!} x^k$$

7. First, using the binomial series with $m = \frac{1}{3}$ and x^4 substituted for x, we have

$$\sqrt[3]{1 + x^4} = 1 + \frac{1}{3} x^4 + \frac{\frac{1}{3}(\frac{1}{3} - 1)}{2!} x^8$$

$$+ \frac{\frac{1}{3}(\frac{1}{3} - 1)(\frac{1}{3} - 2)}{3!} x^{12} + \cdots$$

$$= 1 + \frac{1}{3} x^4 - \frac{2}{(3)^2 2!} x^8 + \frac{(2)(5)}{(3)^3 3!} x^{12}$$

$$- \frac{(2)(5)(8)}{(3)^4 4!} x^{16} + \cdots$$

which, by (11.69), converges for $-1 \le x^4 \le 1$
$\Rightarrow -1 \le x \le 1$. Now, integrating both sides term by term,

$$\int_0^{0.2} \sqrt[3]{1 + x^4}\, dx = (x + \frac{1}{(3)(5)} x^5 - \frac{2}{(3)^2 2! 9} x^9$$

$$+ \frac{(2)(5)}{(3)^3 3! 13} x^{13} - \cdots)\Big|_0^{0.2}$$

$$= (0.2) + \frac{1}{15}(0.2)^5 - \frac{1}{81}(0.2)^9 + \frac{5}{1053}(0.2)^{13} - \cdots$$

To obtain 0.001 accuracy, since the series alternates (after the first term), we need use only terms through a_n where $|a_{n+1}| < 0.001$ [By (11.34)]. Since $a_1 = \frac{1}{15}(0.2)^5 \approx 0.00002 < 0.001$, we use only a_0. Therefore, $\int_0^{0.2} \sqrt[3]{1 + x^4}\, dx \approx 0.2$.

11. The binomial series with $x = \frac{1}{n}$ and $n > 0$ is

$$(1 + \frac{1}{n})^n = 1 + 1 + \frac{n(n-1)}{2!}(\frac{1}{n})^2 + \frac{n(n-1)(n-2)}{3!}(\frac{1}{n})^3 + \cdots$$

$$= 1 + 1 + \frac{1}{2!}(\frac{n-1}{n}) + \frac{1}{3!}(\frac{(n-1)(n-2)}{n^2}) + \cdots$$

Comparing this term by term with

$$\sum_{k=0}^{+\infty} \frac{1}{k!} = 1 + 1 + \frac{1}{2!} + \frac{1}{3!} + \cdots \quad \text{we see that}$$

$$(1 + \frac{1}{n})^n < \sum_{k=0}^{+\infty} \frac{1}{k!} \quad \text{(equality if } n = 1 \text{ or } 2).$$

Since $e^x = \sum_{k=0}^{+\infty} \frac{x^k}{k!}$ (by 11.51) for all x, then

if $x = 1$, $e^1 = \sum_{k=0}^{+\infty} \frac{1}{k!}$. Thus, $(1 + \frac{1}{n})^n < e$ for

all $n > 0$.

Miscellaneous Exercises, pp. 641-645

3. In general, no conclusion can be made about the convergence or divergence of a series formed by taking the term-by-term difference (or sum) of two divergent series. For example,

$$\sum_{k=1}^{+\infty} (-1)^k = -1 + 1 - 1 + 1 - 1 + 1 - \cdots ,$$

$$\sum_{k=1}^{+\infty} (-1)^{k+1} = 1 - 1 + 1 - 1 + 1 - 1 + \cdots , \quad \text{and}$$

$$\sum_{k=1}^{+\infty} 1^k = 1 + 1 + 1 + 1 + \cdots \quad \text{are all divergent and}$$

yet, adding term by term produces

$$\sum_{k=1}^{+\infty} (-1)^k + \sum_{k=1}^{+\infty} (-1)^{k+1} = 0 + 0 + 0 + \cdots ,$$

a convergent series, but

$$\sum_{k=1}^{+\infty} (-1)^k + \sum_{k=1}^{+\infty} 1^k = 0 + 2 + 0 + 2 + 0 + 2 + \cdots,$$

is a divergent series.

7. (a) From the figure, the sum of the areas in the n - 1 circumscribed rectangles is greater (or equal, for infinitely large n) than the area under the curve. For $f(x) = \dfrac{1}{x}$, the sum of the rectangular areas is $f(1) + f(2) + \cdots + f(n-1) = 1 + \dfrac{1}{2} + \cdots + \dfrac{1}{n-1}$ and the area under the curve is $\int_1^n f(x)\,dx = \int_1^n \dfrac{1}{x}\,dx$

$$= \ell n\, x \Big|_1^n = \ell n\, n - 0.$$ Thus, $\int_1^n f(x)\,dx \le f(1)$ $+ f(2) + \cdots + f(n-1) \Rightarrow \ell n\, n \le 1 + \dfrac{1}{2} + \cdots$

$+ \dfrac{1}{n-1} \Rightarrow 0 \le 1 + \dfrac{1}{2} + \cdots + \dfrac{1}{n-1} - \ell n\, n$

$\Rightarrow \dfrac{1}{n} \le 1 + \dfrac{1}{2} + \cdots + \dfrac{1}{n-1} + \dfrac{1}{n} - \ell n\, n \ \Big(= C_n\Big),$

$\Rightarrow \dfrac{1}{n} \le C_n.$ Clearly $\dfrac{1}{n} > 0,$ so $0 < \dfrac{1}{n} \le C_n.$

(b) From the figure, the area in the inscribed rectangle from n to n + 1 [area = (1) f(n+1)] is less than the area under the curve from n to n + 1, namely $\int_n^{n+1} f(x)\,dx.$ Thus,

$\int_n^{n+1} f(x)\,dx > f(n+1).$ Therefore, $\int_n^{n+1} f(x)\,dx$

$= \int_n^{n+1} \dfrac{1}{x}\,dx = \ell n\, x \Big|_n^{n+1} = \ell n(n+1) - \ell n\, n > f(n+1)$

or $\ell n(n+1) - \ell n\, n > \dfrac{1}{n+1}.$ Now, if we use this result in calculating the difference $C_n - C_{n+1},$ we have

$$C_n - C_{n+1} = (1 + \frac{1}{2} + \cdots + \frac{1}{n} - \ln n)$$

$$- [1 + \frac{1}{2} + \cdots + \frac{1}{n} + \frac{1}{n+1} - \ln(n+1)]$$

$$= \ln(n+1) - \ln n - \frac{1}{n+1} > \frac{1}{n+1} - \frac{1}{n+1}$$

$$= 0$$

Thus, $C_n > C_{n+1}$, which makes $\{C_n\}$ a decreasing sequence, bounded below by 0 (part(a) gave us $C_n > 0$). By (11.9), this makes $\{C_n\}$ convergent. We define

$$\lim_{n \to +\infty} C_n = \lim_{n \to +\infty} \{1 + \frac{1}{2} + \cdots + \frac{1}{n} - \ln n\} = \gamma ,$$

Euler's number.

(c) Solve $M = \gamma + \ln n$ for n (an integer).

Using $\gamma \approx 0.5772$, we have $\ln n \approx M - 0.5772$
or $n \approx e^{M-0.5772}$ (round n to an integer)

For $M = 3$, $n \approx 11.3 \approx 11$

$M = 4$, $n \approx 30.7 \approx 31$

$M = 10$, $n \approx 12{,}367.2 \approx 12{,}367$

$M = 20$, $n \approx 2.724 \times 10^8$

11. Using the partial fractions technique,

$\frac{1}{k(k+2)} = \frac{1/2}{k} - \frac{1/2}{k+2}$. Then the nth partial sum

S_n of $\sum\limits_{k=1}^{+\infty} \frac{1}{k(k+2)}$ is $S_n = \frac{1}{2}(\frac{1}{1} - \frac{1}{3}) + \frac{1}{2}(\frac{1}{2} - \frac{1}{4})$

$+ \frac{1}{2}(\frac{1}{3} - \frac{1}{5}) + \cdots + \frac{1}{2}(\frac{1}{n-1} - \frac{1}{n+1}) + \frac{1}{2}(\frac{1}{n} - \frac{1}{n+2})$

$= \frac{1}{2}(1 + \frac{1}{2} - \frac{1}{n+1} - \frac{1}{n+2})$

Now, letting $n \to +\infty$, we have

$$\sum_{k=1}^{+\infty} \frac{1}{k(k+2)} = \lim_{n \to +\infty} S_n = \lim_{n \to +\infty} \frac{1}{2}(1 + \frac{1}{2} - \frac{1}{n+1} - \frac{1}{n+2})$$

$$= \frac{1}{2}(1 + \frac{1}{2} - 0 - 0] = \frac{3}{4}$$

15. Since $e^x = 1 + \frac{x}{1!} + \frac{x^2}{2!} + \cdots$ for all x, then

$$\frac{e^x - 1}{x} = 1 + \frac{x}{2!} + \frac{x^2}{3!} + \frac{x^3}{4!} + \cdots , \quad \text{for all x.}$$

Differentiating both sides with respect to x yields

$$\frac{xe^x - (e^x-1)(1)}{x^2} = \frac{1}{2!} + \frac{2x}{3!} + \frac{3x^2}{4!} + \frac{4x^3}{5!} + \cdots , \quad \text{for}$$

all x, or just

$$\frac{xe^x - e^x + 1}{x^2} = \sum_{k=1}^{+\infty} \frac{kx^{k-1}}{(k+1)!} \quad \text{for all x. Thus, for}$$

$$x = 1, \quad \frac{1e^1 - e^1 + 1}{(1)^2} = 1 = \sum_{k=1}^{+\infty} \frac{k}{(k+1)!}$$

19. $1 + \frac{2}{2^2} + \frac{3}{3^3} + \frac{1}{4^4} + \frac{2}{5^5} + \frac{3}{6^6} + \cdots$ is, term by term,

less than or equal to the series $\sum_{k=1}^{+\infty} \frac{k!}{k^k}$

$$= 1 + \frac{2}{2^2} + \frac{6}{3^3} + \frac{24}{4^4} + \frac{120}{5^5} + \cdots , \quad \text{which was shown to}$$

be convergent in Example 9, Section 3. Thus the given series converges.

23. The partial sum S_n for $\sum_{k=2}^{+\infty} \ell n(\frac{k}{k+1})$

$$= \sum_{k=2}^{+\infty} [\ell n\, k - \ell n(k+1)] \quad \text{is} \quad S_n = (\ell n\, 2 - \ell n\, 3)$$

$+ (\ell n\, 3 - \ell n\, 4) + (\ell n\, 4 - \ell n\, 5) + \cdots + [\ell n\, n - \ell n(n+1)]$

$= \ell n\, 2 - \ell n(n+1)$

Since $\lim_{n \to +\infty} S_n = \lim_{n \to +\infty} [\ell n\, 2 - \ell n(n+1)] = -\infty$, the series diverges.

27. Applying the limit comparison test (11.26) with the convergent p-series $\displaystyle\sum_{k=1}^{+\infty} \frac{1}{k^2}$, we have

$$\lim_{n \to +\infty} \left[\frac{\tan^{-1}n/n^2}{1/n^2}\right] = \lim_{n \to +\infty} \tan^{-1}n = \frac{\pi}{2} > 0; \quad \text{thus,}$$

$\displaystyle\sum_{k=1}^{+\infty} \frac{\tan^{-1}k}{k^2}$ also converges.

31. (a) Following Figure 11, section 3 (integral test), we see that for $f(x) = \frac{1}{1 + x^2}$ ($f(1)$ is the area of a rectangle not shown in Figure 11.)

$$f(1) + f(2) + \cdots + f(n) \le \int_0^n \frac{1}{1 + x^2}\, dx$$

$$= \tan^{-1}x \Big|_0^n = \tan^{-1}n - \tan^{-1}0$$

Thus, $\displaystyle S_n = \sum_{k=1}^{+\infty} \frac{1}{1 + k^2} \le \tan^{-1}n$ for $n \ge 1$.

(b) Since the terms $\frac{1}{1 + k^2}$ are all positive, $S_{n+1} > S_n$. Thus, S_n is bounded above by $\frac{\pi}{2}$

($\displaystyle\lim_{n \to +\infty} \tan^{-1}n = \frac{\pi}{2}$) and monotonic (increasing) and must have a limit by (11.9). Therefore,

$\displaystyle\sum_{k=1}^{+\infty} \frac{1}{1 + k^2}$ converges.

(c) Using Figure 12, section 3, with circumscribed rectangles,

$$S_n \ge \int_1^n \frac{1}{1 + x^2}\, dx = \tan^{-1}x \Big|_1^n = \tan^{-1}n - \tan^{-1}1$$

or $S_n \ge \tan^{-1}n - \frac{\pi}{4}$

From part (b), we also have $S_n \le \tan^{-1}n$.

Thus, taking the limit as $n \to +\infty$ yields

$$\lim_{n \to +\infty} (\tan^{-1} n - \frac{\pi}{4}) \leq \lim_{n \to +\infty} S_n \leq \lim_{n \to +\infty} (\tan^{-1} n)$$

$$\frac{\pi}{4} \leq \sum_{k=1}^{+\infty} \frac{1}{1 + k^2} \leq \frac{\pi}{2}$$

35. From (11.39),

$$\ln(\frac{1}{1-x}) = x + \frac{x^2}{2} + \frac{x^3}{3} + \cdots + \frac{x^n}{n} + \cdots , \quad \text{for}$$

$-1 < x < 1$. Then, integrating both sides term by term, we have

$$\int_0^x \ln(\frac{1}{1-t}) dt = (\frac{t^2}{2} + \frac{t^3}{2(3)} + \frac{t^4}{3(4)} + \cdots + \frac{t^{n+1}}{n(n+1)} + \cdots)\Big|_0^x$$

The left integral can be done by parts when it is rewritten:

$$\int_0^x \ln(\frac{1}{1-t}) dt = -\int_0^x \ln(1-t) dt \quad (\text{Use} \quad u = \ln(1-t),$$
$$dv = dt.)$$

$$= -t \ln(1-t)\Big|_0^x - \int_0^x \frac{t}{1-t} dt = -x\ln(1-x)$$

$$- \int_0^x (-1 + \frac{1}{1-t}) dt$$

$$= -x\ln(1-x) + [t + \ln(1-t)]\Big|_0^x = -x\ln(1-x) + x + \ln(1-x)$$
$$= (1-x)\ln(1-x) + x$$

Thus,

$$(1-x)\ln(1-x) + x = \frac{x^2}{1(2)} + \frac{x^3}{2(3)} + \frac{x^4}{3(4)} + \cdots$$

$$= \sum_{k=1}^{+\infty} \frac{x^{k+1}}{k(k+1)} , \quad \text{for} \quad -1 < x < 1$$

39. (a) $f(x) = 3 \sin(\frac{x}{2})$. Using (11.56),

$$\sin x = x - \frac{x^3}{3!} + \frac{x^5}{5!} - \cdots = \sum_{k=0}^{+\infty} (-1)^k \frac{x^{2k+1}}{(2k+1)!} ,$$

for all x. Now, substituting $\frac{x}{2}$ for x and multiplying both sides by 3, yields

$$f(x) = 3 \sin \frac{x}{2} = 3(\frac{x}{2}) - \frac{3}{3!}(\frac{x}{2})^3 + \frac{3}{5!}(\frac{x}{2})^5 - \cdots$$

$$= \sum_{k=0}^{+\infty} \frac{(-1)^k 3(x/2)^{2k+1}}{(2k+1)!}$$

$$= \frac{3}{2} x - \frac{1}{16} x^3 + \frac{1}{1280} x^5 - \cdots$$

$$= \frac{3}{2} \sum_{k=0}^{+\infty} \frac{(-1)^k x^{2k+1}}{4^k (2k+1)!}$$

(b) Since the series for $\sin x$ converges for all x, the series for $3 \sin \frac{x}{2}$ will similarly have an interval of convergence of $-\infty < x < +\infty$. This can also be easily verified using the ratio test.

(c) To obtain 0.1 accuracy on an interval of $-2 < x < 2$ from this alternating series, we need to find n such that $a_{n+1} < 0.1$ for the "worst" case of convergence; in this situation, that is $x = 2$ or $x = -2$. Since

$$|a_1| = \frac{2^3}{16} = 0.5 \quad \text{and} \quad |a_2| = \frac{2^5}{1280} = 0.025 < 0.1,$$

we need use only the first two terms,

$$\frac{3}{2} x - \frac{1}{16} x^3.$$

Chapter Twelve

Exercise 2, pp. 653-655

3. $a = 2$, focus on left $\implies y^2 = -8x$

7. $a = 3$, focus above origin $\implies x^2 = 12y$

11. $a = 3$, focus on left $\implies y^2 = -12x$

15. Parabola opens to left, thus $y^2 = -4ax$.
    As $a = 3$, $y^2 = -12x$.

19. $x^2 = -12y = -4ay \implies a = 3$
    Vertex $(0,0)$
    Focus $(0,-3)$
    Directrix $y = 3$

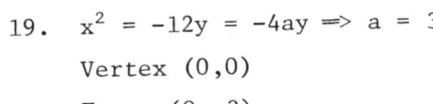

23. $x^2 = 8y = 4ay \implies a = 2$
    Vertex $(0,0)$
    Focus $(0,2)$
    Directrix $y = -2$

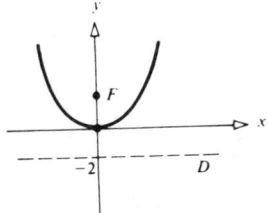

27. Area $= \int_0^4 [2 - \frac{x^2}{8}]dx = (2x - \frac{x^3}{24})\Big|_0^4 = 8 - \frac{64}{24} = \frac{16}{3}$

31. Here $y = Cx^2 \implies 75 = C(150)^2$

    $\implies C = \frac{1}{300}$ and $y = \frac{1}{300} x^2$.

    80 meters from center,

    height $y = \frac{80^2}{300}$ m $= \frac{64}{3}$ m

35. Without loss of generality
    we may orient the parabola
    along the positive y-axis
    with vertex at the origin.
    The equation is then

    $y = cx^2$.

    Area of circumscribed
    rectangle = $(2x)(cx^2) = 2cx^3$.

    Area of parabolic segment
    = (Area of circumscribed
    rectangle) — (Area under
    the parabola)

    $= 2cx^3 - 2 \int_0^x ct^2 dt = 2cx^3 - \dfrac{2}{3} cx^3 = \dfrac{4}{3} cx^3$. This
    gives the desired result.

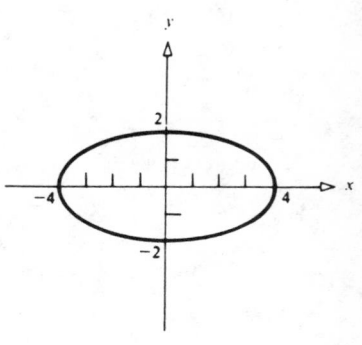

Exercise 3, pp. 661-662

3. $c = 2$, $a = 3$, $b^2 = a^2 - c^2 = 5$
   Equation is $\dfrac{x^2}{5} + \dfrac{y^2}{9} = 1$

7. $c = 4$, $b = 1$, $a^2 = b^2 + c^2 = 17$
   Equation is $\dfrac{x^2}{17} + \dfrac{y^2}{1} = 1$

11. $\dfrac{x^2}{16} + \dfrac{y^2}{4} = 1$

    Since the denominator for x^2
    is larger, the major axis is
    along the x-axis.

    $a = 4$, $b = 2$,

    $c = \sqrt{a^2 - b^2} = 2\sqrt{3}$

    Vertices at $(\pm 4, 0)$

    Covertices at $(0, \pm 2)$

    Foci at $(\pm 2\sqrt{3}, 0)$

15. $x^2 + y^2 = 16$ (or $\dfrac{x^2}{16} + \dfrac{y^2}{16} = 1$)

Equation represents a circle
of radius 4 centered at (0,0).
Often, this is called a
degenerate ellipse,
with a = b = 4 and
focal distance c = 0.

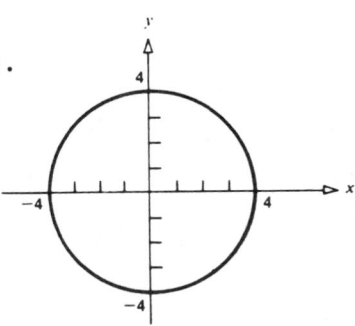

19. (a) Nearly circular (e $\approx$ 0 => c $\approx$ 0)

(b) Intermediate in shape, with ratio of minor
axis to major axis equalling $\sqrt{3}/2 \approx 0.866$

(c) Thin and elongated (c $\approx$ a => b $\approx$ 0)

23. Equation is $\dfrac{x^2}{400} + \dfrac{y^2}{225} = 1$

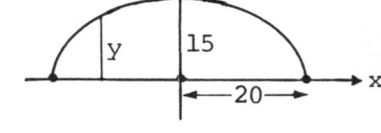

or $y = 15\sqrt{1 - x^2/400}$

Hence from left base to right base at 10 ft intervals,

height is 0 ft, $\dfrac{15\sqrt{3}}{2}$ ft, 15 ft, $\dfrac{15\sqrt{3}}{2}$ ft, and 0 ft.

27. $\dfrac{x^2}{a^2} + \dfrac{y^2}{b^2} = 1$. Implicit differentation with respect

to x yields $\dfrac{2x}{a^2} + \dfrac{2yy'}{b^2} = 0$ or $y' = -\dfrac{b^2 x}{a^2 y}$

Hence, at (x_0, y_0), the tangent line has slope

$-\dfrac{b^2 x_0}{a^2 y_0}$. Its equation is therefore

$$(y - y_0) = -\dfrac{b^2 x_0}{a^2 y_0}(x - x_0)$$

Multiplying by $\dfrac{y_0}{b^2}$, $\dfrac{yy_0}{b^2} - \dfrac{y_0^2}{b^2} = -\dfrac{xx_0}{a^2} + \dfrac{x_0^2}{a^2}$

or $\dfrac{xx_0}{a^2} + \dfrac{yy_0}{b^2} = \dfrac{x_0^2}{a^2} + \dfrac{y_0^2}{b^2} = 1$

3. $a = 2$, $c = 3$, $b^2 = c^2 - a^2 = 5$

   Equation is $y^2/4 - x^2/5 = 1$

7. $c = 4$ and $\dfrac{b}{a} = 2$. Since $b^2 + a^2 = c^2 = 16$,

   $4a^2 + a^2 = 16$, $a^2 = 16/5$. Then $b^2 = c^2 - a^2$

   $= 64/5$. Equation is $\dfrac{5x^2}{16} - \dfrac{5y^2}{64} = 1$ or $\dfrac{x^2}{(16/5)} - \dfrac{y^2}{(64/5)} =$

11. $\dfrac{x^2}{16} - \dfrac{y^2}{4} = 1$ Hyperbola with

   $a = 4$, $b = 2$, so $c = \sqrt{a^2 + b^2}$

   $= 2\sqrt{5}$. Transverse axis is the
   x-axis, with center at origin;
   vertices at $(\pm 4, 0)$; foci at
   $(\pm 2\sqrt{5}, 0)$; asymptotes $y = \pm \dfrac{1}{2} x$.

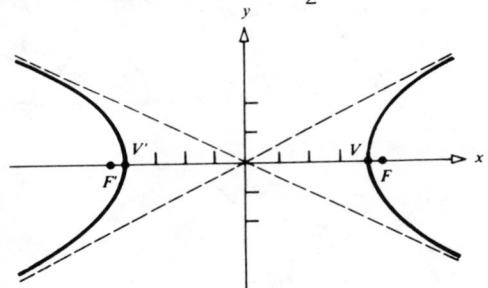

15. $\dfrac{x^2}{16} - \dfrac{y^2}{16} = 1$ Hyperbola with

   $a = 4$, $b = 4$, so $c = \sqrt{a^2 + b^2}$

   $= 4\sqrt{2}$. Transverse axis is the
   x-axis, with center at origin;

   vertices at $(\pm 4, 0)$; foci at

   $(\pm 4\sqrt{2}, 0)$; asymptotes $y = \pm x$.

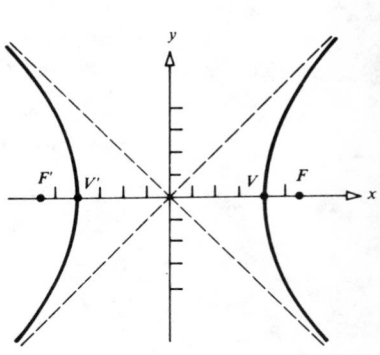

19. $c^2 = a^2 + b^2 = 2a^2$ here. Hence, eccentricity

$= c/a = \sqrt{2}$.

23. (a) For $\dfrac{x^2}{25} - \dfrac{y^2}{16} = 1$, $a = 5$ and $b = 4$, so

asymptotes are $y = \pm \dfrac{4}{5} x$.

(b) For $\dfrac{y^2}{2} - \dfrac{x^2}{18} = 1$, $a = \sqrt{2}$ and $b = 3\sqrt{2}$, so

asymptotes are $y = \pm \dfrac{1}{3} x$.

27. With the given center and vertex, we know the

equation has the form $\dfrac{x^2}{(147)^2} - \dfrac{y^2}{b^2} = 1$

Since the hyperbola goes through (155,123), we know

$\dfrac{(155)^2}{(147)^2} - \dfrac{(123)^2}{b^2} = 1$

Solving this yields $b^2 = \dfrac{(123)^2 (147)^2}{(155)^2 - (147)^2}$

$= \dfrac{(123)^2 (147)^2}{2416}$

Thus, the equation of the hyperbola is

$\dfrac{x^2}{21,609} - \dfrac{y^2}{135,316} = 1$

Exercise 5, p. 677

3. $4x^2 - 4(y^2 - 2y + 1) = 1$

$x^2/(1/4) - (y-1)^2/(1/4) = 1$

Hyperbola centered at $(0,1)$;
line $y = 1$ is transverse
axis; vertices at $(\pm 1/2, 1)$;
foci at $(\pm\sqrt{2}/2, 1)$; asymptotes
are $y = 1 \pm x$.

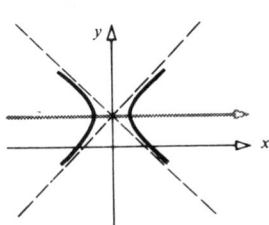

7. $9(x^2+6x+9) + 16(y^2-2y+1) = 47 + 97$

$9(x+3)^2 + 16(y-1)^2 = 144$

$$\frac{(x+3)^2}{16} + \frac{(y-1)^2}{9} = 1$$

Ellipse centered at $(-3,1)$;
vertices at $(1,1)$, $(-7,1)$,
$(-3,4)$, $(-3,-2)$; foci at
$(-3 \pm \sqrt{7},1)$.

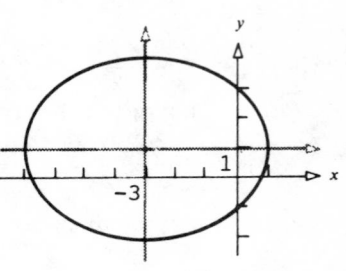

11. $4(x^2+2x+1) - 3(y^2-4y+4) = 20 - 8$

$4(x+1)^2 - 3(y-2)^2 = 12$

$$\frac{(x+1)^2}{3} - \frac{(y-2)^2}{4} = 1$$

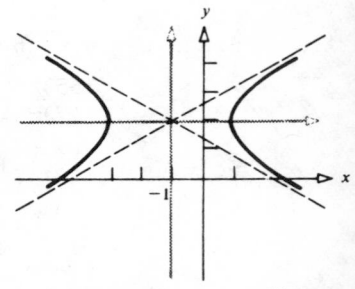

Hyperbola centered at $(-1,2)$;
line $y = 2$ is transverse axis
vertices at $(-1 \pm \sqrt{3},2)$; foci
at $(-1 \pm \sqrt{7},2)$; asymptotes are
$y - 2 = \pm \frac{2\sqrt{3}}{3}(x+1)$.

15. $3x^2 + 5\sqrt{3}xy + 8y^2 - 10 = 0$ $A = 3$, $C = 8$, $B = 5\sqrt{3}$,
so the angle θ of rotation satisfies
$\cot 2\theta = \frac{A - C}{B} = -\frac{\sqrt{3}}{3} \Rightarrow 2\theta = \frac{2\pi}{3} \Rightarrow \theta = \frac{\pi}{3}$. Using
(s,t) in the rotated system, $x = \frac{1}{2} s - \frac{\sqrt{3}}{2} t$,
$y = \frac{\sqrt{3}}{2} s + \frac{1}{2} t$. New equation is

$$3(\frac{1}{2} s - \frac{\sqrt{3}}{2} t)^2 + 5\sqrt{3}(\frac{1}{2} s - \frac{\sqrt{3}}{2} t)(\frac{\sqrt{3}}{2} s + \frac{1}{2} t)$$

$$+ 8(\frac{\sqrt{3}}{2} s + \frac{1}{2} t)^2 = 10$$

$$3[\frac{1}{4} s^2 + \frac{3}{4} t^2 - \frac{\sqrt{3}}{2} st] + 5\sqrt{3}[\frac{\sqrt{3}}{4} s^2 - \frac{\sqrt{3}}{4} t^2 - \frac{1}{2} st]$$

$$+ 8[\frac{3}{4} s^2 + \frac{1}{4} t^2 + \frac{\sqrt{3}}{2} st] = 10$$

$$\frac{21}{2} s^2 + \frac{1}{2} t^2 = 10 \quad \text{or} \quad \frac{21}{20} s^2 + \frac{1}{20} t^2 = 1$$

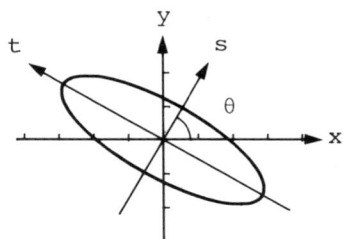

The conic is thus an ellipse with minor axis along
$y = \sqrt{3}x$ (or $\theta = \frac{\pi}{3}$), centered at $(0,0)$. In the
(s,t) coordinate system, vertices are at $(0, \pm 2\sqrt{5})$
and $(\pm 2\sqrt{5/21}, 0)$; foci at $(0, \pm 20/\sqrt{21})$.

19. $9x^2 - 6xy + y^2 - 3x + y - 2 = 0$

   $A = 9$, $B = -6$, $C = 1$

   $B^2 - 4AC = 36 - 36 = 0 \Rightarrow$ graph is a parabola

23. We may use the formulas derived just before (12.15)
   for "A" and "C". If θ is the angle of rotation,
   angle of rotation,

   $A'' + C'' = A \cos^2\theta + B \sin \theta \cos \theta + C \sin^2\theta$

   $\qquad\qquad + A \sin^2\theta - B \sin \theta \cos \theta + C \cos^2\theta$

   $\qquad = (A+C)(\cos^2\theta + \sin^2\theta) = A + C$

   Hence, $A'' + C'' = A + C$

Miscellaneous Exercises, pp. 678-679

3. $P(x,y)$ must satisfy

   $\sqrt{(x-5)^2 + y^2} = \frac{5}{3}|x - \frac{9}{5}|$

   $(x-5)^2 + y^2 = \frac{25}{9}(x^2 - \frac{18}{5}x + \frac{81}{25})$

   $\frac{16}{9}x^2 - y^2 = 16$ or $\frac{x^2}{9} - \frac{y^2}{16} = 1$

   The graph is a hyperbola centered at the origin
   with vertices at $(\pm 3, 0)$; x-axis as transverse axis;
   foci at $(\pm 5, 0)$; and asymptotes $y = \pm 4/3 \ x$.

7. $\frac{d}{dx}(x^2+4y^2) = \frac{d}{dx}(4)$ or $2x + 8yy' = 0 \implies y' = -\frac{x}{4y}$.

Hence, at the points of tangency (x_0, y_0), slope of tangent line is also the slope from (x_0, y_0) to $(4,1)$:

$$-\frac{x_0}{4y_0} = \frac{y_0 - 1}{x_0 - 4} \implies 4y_0^2 - 4y_0 = 4x_0 - x_0^2$$

Also, we have that $4y_0^2 = 4 - x_0^2$ since (x_0, y_0) lies on the ellipse. Substituting this in the equation above yields:
$1 - y_0 = x_0$.

Thus, $(1-y_0^2) + 4y_0^2 = 4 \implies 5y_0^2 - 2y_0 - 3 = 0$

or $(5y_0+3)(y_0-1) = 0 \implies y_0 = 1, -3/5$
The points of tangency are $(0,1)$ and $(\frac{8}{5}, -\frac{3}{5})$,
with slopes 0 and $\frac{2}{3}$, respectively.
Equations are $y = 1$ and $y + \frac{3}{5} = \frac{2}{3}(x - \frac{8}{5})$
$\implies y = \frac{2}{3}x - \frac{5}{3} \implies 2x - 3y = 5$

11. If $AC > 0$, then x^2 and y^2 have the same sign. This means that (except for degenerate cases or equations with no solution) one may complete the squares to obtain an equation of the form

$\frac{(x-h)^2}{a^2} + \frac{(y-k)^2}{b^2} = 1$, which represents an ellipse

when $A \neq C$ (or $a \neq b$) and a circle when $A = C$ (or $a = b$).

If $AC = 0$, then one of A or C is 0, so the equation is quadratic in one variable and linear in the other (discounting degenerate cases). The graph is thus a parabola.

If $AC < 0$, x^2 and y^2 have opposite signs, so we may complete the squares to find

$\frac{(x-h)^2}{a^2} - \frac{(y-k)^2}{b^2} = \pm 1$, either of which represents

a hyperbola.

15. Equation is $\dfrac{x^2}{9} + \dfrac{y^2}{4} = 1$. Thus, the base of a cross section at x is a segment of length $2y = 4\sqrt{1 - x^2/9}$.

(a) Cross-sectional area $= \dfrac{1}{2}\pi r^2 = \dfrac{1}{2}\pi[2\sqrt{1 - x^2/9}]^2$

$$= 2\pi(1 - \dfrac{x^2}{9})$$

Volume $= \int_{-3}^{3} 2\pi(1 - \dfrac{x^2}{9})\,dx = 4\pi \int_{0}^{3}(1 - \dfrac{x^2}{9})\,dx$

$$= 4\pi(x - \dfrac{x^3}{27})\Big|_{0}^{3} = 8\pi$$

(b) Cross-sectional area $= [4\sqrt{1 - x^2/9}]^2$

$$= 16(1 - \dfrac{x^2}{9})$$

Volume $= \int_{-3}^{3} 16(1 - \dfrac{x^2}{9})\,dx = 32 \int_{0}^{3}(1 - \dfrac{x^2}{9})\,dx$

$$= 32(x - \dfrac{x^3}{27})\Big|_{0}^{3} = 64$$

19. If $e < 1$, Problem 17 showed the condition characterized an ellipse, while if $e > 1$ Problem 18 showed it characterized a hyperbola. When $e = 1$, we have $d(F,P) = d(D,P)$, which is the geometric definition of a parabola, for the point P.

23. Without loss of generality, suppose the ellipse is oriented along the x-axis, i.e., $a > b$. Consider the vertical chord through the focus $F(c,0)$. Its endpoints have abscissa c, so from the equation $\dfrac{x^2}{a^2} + \dfrac{y^2}{b^2} = 1$, the ordinates are $\pm b\sqrt{1 - c^2/a^2}$. The length of the chord is thus

$$2b\sqrt{\dfrac{a^2 - c^2}{a^2}} = 2b\sqrt{\dfrac{b^2}{a^2}} = 2\,\dfrac{b^2}{a}\ .$$

27. Since the hyperbola $\frac{x^2}{a^2} - \frac{y^2}{b^2} = 1$ is symmetric with respect to both coordinate axes, it suffices to consider the perpendicular from $F(c,0)$ to the upper asymptote $y = \frac{b}{a} x$. This perpendicular has slope $-\frac{a}{b}$, so its equation is $y = -\frac{a}{b}(x-c)$. The asymptote and the perpendicular thus intersect where

$$\frac{b}{a} x = -\frac{a}{b}(x-c) \quad \text{or} \quad x(\frac{b}{a} + \frac{a}{b}) = \frac{ac}{b} \implies x(\frac{b^2 + a^2}{a}) = ac$$

This becomes $x \frac{c^2}{a} = ac$ or $x = \frac{a^2}{c} = \frac{a}{e}$. The intersection therefore occurs on the directrix $x = \frac{a}{e}$.

Chapter Thirteen

Exercise 1, pp. 686-687

3. $(-4, -\pi/3)$; $(4, 2\pi/3)$; $(-4, 5\pi/3)$

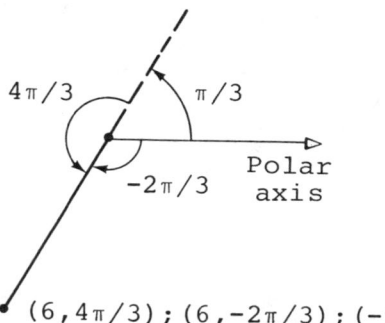

7. $(6, 4\pi/3)$; $(6, -2\pi/3)$; $(-6, \pi/3)$

11. $x = -6 \cos(\frac{-\pi}{6}) = -6(\frac{\sqrt{3}}{2}) = -3\sqrt{3};$

   $y = -6 \sin(\frac{-\pi}{6}) = -6(-\frac{1}{2}) = 3;$ $(-3\sqrt{3}, 3)$

15. $x = 2\sqrt{2} \cos(\frac{-\pi}{4}) = 2\sqrt{2}(\frac{\sqrt{2}}{2}) = 2;$

   $y = 2\sqrt{2} \sin(\frac{-\pi}{4}) = 2\sqrt{2}(\frac{-\sqrt{2}}{2}) = -2;$ $(2, -2)$

For Problems 17-22, we want $r > 0$, so choose
$r = +\sqrt{x^2 + y^2}$; then find θ such that $0 \le \theta < 2\pi$,
with $\tan \theta = \frac{y}{x}$ if $x \ne 0$, and $\theta = \frac{\pi}{2}$ or $\frac{3\pi}{2}$ if $x = 0$.

19. $r = \sqrt{(-2)^2 + (-2\sqrt{3})^2} = \sqrt{16} = 4$;

$\tan \theta = \dfrac{-2\sqrt{3}}{-2} = \sqrt{3} \Longrightarrow \theta = \dfrac{\pi}{3}$ or $\dfrac{4\pi}{3}$; choose

$\theta = \dfrac{4\pi}{3}$, since $r > 0$ and $(-2, -2\sqrt{3})$ is in the third quadrant.

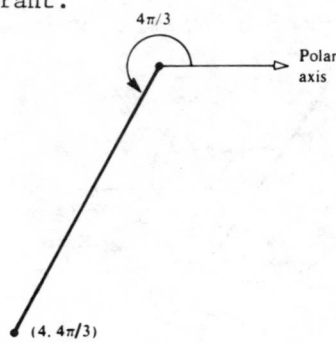

4π/3

Polar
axis

(4, 4π/3)

23. $\dfrac{x^2}{4} + \dfrac{y^2}{9} = 1 \Longrightarrow \dfrac{(r \cos \theta)^2}{4} + \dfrac{(r \sin \theta)^2}{9} = 1$

$\Longrightarrow r^2 \left(\dfrac{\cos^2 \theta}{4} + \dfrac{\sin^2 \theta}{9} \right) = 1$

27. $x^2 = 1 - 4y \Longrightarrow (r \cos \theta)^2 = 1 - 4r \sin \theta$

$\Longrightarrow r^2 \cos^2 \theta + 4r \sin \theta = 1$

31. $r = \cos \theta \Longrightarrow r^2 = r \cos \theta \Longrightarrow x^2 + y^2 = x$

35. $r = \dfrac{4}{1 - \cos \theta} \Longrightarrow r - r \cos \theta = 4 \Longrightarrow r - x = 4$

$\Longrightarrow r = 4 + x \Longrightarrow r^2 = (4+x)^2 \Longrightarrow x^2 + y^2 = 16 + 8x + x^2$

$\Longrightarrow y^2 = 8(x+2)$

39. $r = 2 \Longrightarrow r^2 = 4$ (which gives the same set of points in polar coordinates as just $r = 2$) $\Longrightarrow x^2 + y^2 = 4$

Exercise 2, p. 695

3. $\theta = \dfrac{11\pi}{6}$ (straight line through the pole, angle $= \dfrac{11\pi}{6}$)

Intercept: at the pole

Symmetry: origin (pole), since replacing r with
          $-r$ leaves the equation unchanged

Tangent at pole: $r = 0 \Rightarrow \theta = \dfrac{11\pi}{6}$

The whole line is included, since r can be both positive and negative.

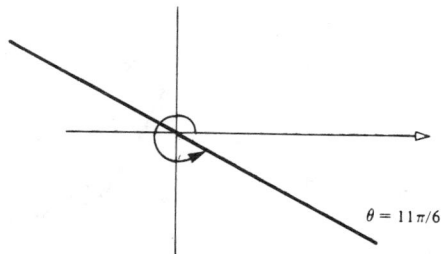

$\theta = 11\pi/6$

7. $r = 1 - \cos\theta$ (cardioid)

   Intercepts: $\theta = 0 \Rightarrow r = 0,\quad \theta = \pi \Rightarrow r = 2;$

   $\qquad\qquad \theta = \dfrac{\pi}{2} \Rightarrow r = 1,\quad \theta = \dfrac{3\pi}{2} \Rightarrow r = 1$

   Symmetry: polar axis $[1 - \cos(-\theta) = 1 - \cos\theta]$

   Tangent at pole: $1 - \cos\theta = 0 \Rightarrow \cos\theta = 1 \Rightarrow \theta = 0$

θ	$\cos\theta$	r
$0 \to \dfrac{\pi}{2}$	$1 \to 0$	$0 \to 1$
$\dfrac{\pi}{2} \to \pi$	$0 \to -1$	$1 \to 2$
$\pi \to \dfrac{3\pi}{2}$	(Use symmetry)	
$\dfrac{3\pi}{2} \to 2\pi$		

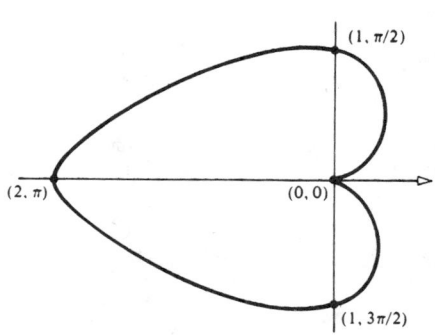

$(1, \pi/2)$

$(2, \pi)$ $(0, 0)$

$(1, 3\pi/2)$

1. $r = 4 - 3\sin\theta$ (limaçon)

   Intercepts: $\theta = 0 \Rightarrow r = 4,\quad \theta = \pi \Rightarrow r = 4;$

   $\theta = \dfrac{\pi}{2} \Rightarrow r = 1,\quad \theta = \dfrac{3\pi}{2} \Rightarrow r = 7;$ unlike Problem 10, there is no point at the pole, since $4 - 3\sin\theta = 0$ $\Rightarrow \sin\theta = \dfrac{4}{3}$, which is impossible.

   Symmetry: $\dfrac{\pi}{2}$ axis, since $\sin(\pi-\theta) = \sin\theta$ (Use

   $\qquad\qquad \sin(\pi-\theta) = \sin\pi\cos\theta - \cos\pi\sin\theta)$

Tangent at pole: none

θ	$\sin \theta$	r
$0 \rightarrow \dfrac{\pi}{2}$	$0 \rightarrow 1$	$4 \rightarrow 1$
$\dfrac{\pi}{2} \rightarrow \pi$	(Use symmetry)	
$\pi \rightarrow \dfrac{3\pi}{2}$	$0 \rightarrow -1$	$4 \rightarrow 7$
$\dfrac{3\pi}{2} \rightarrow 2\pi$	(Use symmetry)	

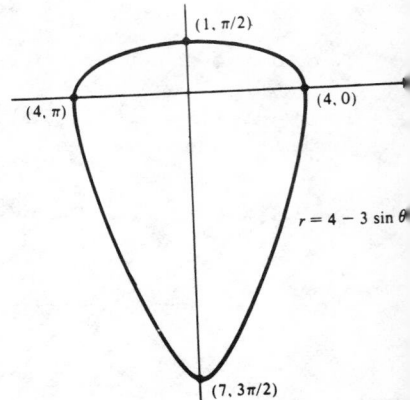

$(1, \pi/2)$

$(4, \pi)$ $(4, 0)$

$r = 4 - 3 \sin \theta$

$(7, 3\pi/2)$

15. $r = 3 \cos 3\theta$ (three-leaved rose)

Intercepts: $\theta = 0 \Rightarrow r = 3$ ($\theta=\pi \Rightarrow r=-3$ represents the same point); $\theta = \dfrac{\pi}{2} \Rightarrow r = 0$ (this point at the pole is repeated for $\theta = \dfrac{3\pi}{2}$, as well as for $\theta = \dfrac{\pi}{6}$, $\dfrac{5\pi}{6}$, $\dfrac{7\pi}{6}$)

Symmetry: polar axis, since $\cos(-3\theta) = \cos 3\theta$

Tangents at pole: $3 \cos 3\theta = 0 \Rightarrow \cos 3\theta = 0$
$\Rightarrow 3\theta = \dfrac{\pi}{2}$, $\dfrac{3\pi}{2}$, etc. $\Rightarrow \theta = \dfrac{\pi}{6}$, $\dfrac{3\pi}{6}$ (or $\dfrac{\pi}{2}$), $\dfrac{5\pi}{6}$,
$\dfrac{7\pi}{6}$, $\dfrac{9\pi}{6}$ (or $\dfrac{3\pi}{2}$), and $\dfrac{11\pi}{6}$

Note: The numbers on the leaves identify the order in which the leaves are produced.

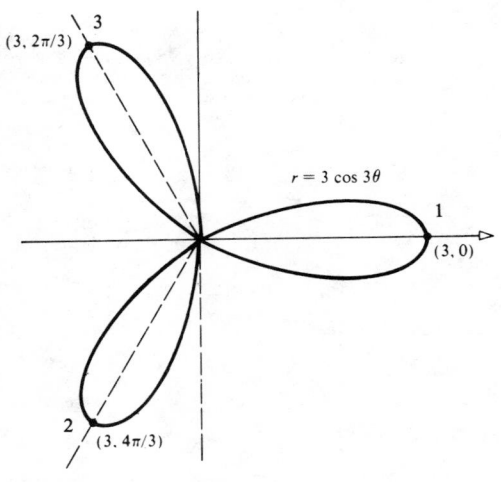

$(3, 2\pi/3)$ 3

$r = 3 \cos 3\theta$

1

$(3, 0)$

2

$(3, 4\pi/3)$

θ	3θ	$\cos 3\theta$	r	Leaf
$0 \to \dfrac{\pi}{6}$	$0 \to \dfrac{\pi}{2}$	$1 \to 0$	$3 \to 0$	1 (top)
$\dfrac{\pi}{6} \to \dfrac{\pi}{3}$	$\dfrac{\pi}{2} \to \pi$	$0 \to -1$	$0 \to -3$	2
$\dfrac{\pi}{3} \to \dfrac{\pi}{2}$	$\pi \to \dfrac{3\pi}{2}$	$-1 \to 0$	$-3 \to 0$	
$\dfrac{\pi}{2} \to \dfrac{2\pi}{3}$	$\dfrac{3\pi}{2} \to 2\pi$	$0 \to 1$	$0 \to 3$	3
$\dfrac{2\pi}{3} \to \dfrac{5\pi}{6}$	$2\pi \to \dfrac{5\pi}{2}$	$1 \to 0$	$3 \to 0$	
$\dfrac{5\pi}{6} \to \pi$	$\dfrac{5\pi}{2} \to 3\pi$	$0 \to -1$	$0 \to -3$	1 (bottom)

The graph repeats itself from $\theta = \pi$ to $\theta = 2\pi$.

19. $r^2 = 16 \cos 2\theta$ (lemniscate)

Intercepts: $\theta = 0 \Rightarrow r = \pm 4$ ($\theta = \pi$ produces the same points); the graph is not defined for $\theta = \dfrac{\pi}{2}$ or $\dfrac{3\pi}{2}$; the pole, $r = 0$, is produced for $\theta = \dfrac{\pi}{4}$ and $\dfrac{3\pi}{4}$

Symmetry: polar axis, since $\cos(-2\theta) = \cos 2\theta$; $\dfrac{\pi}{2}$ axis, since $\cos 2(\pi - \theta) = \cos 2\pi \cos 2\theta$ $+ \sin 2\pi \sin 2\theta = \cos 2\theta$; and the pole, since $(-r)^2 = r^2$

Tangents at pole: $16 \cos 2\theta = 0 \Rightarrow \cos 2\theta = 0$ $\Rightarrow 2\theta = \dfrac{\pi}{2}, \dfrac{3\pi}{2}, \dfrac{5\pi}{2}, \dfrac{7\pi}{2} \Rightarrow \theta = \dfrac{\pi}{4}, \dfrac{3\pi}{4}, \dfrac{5\pi}{4}, \dfrac{7\pi}{4}$

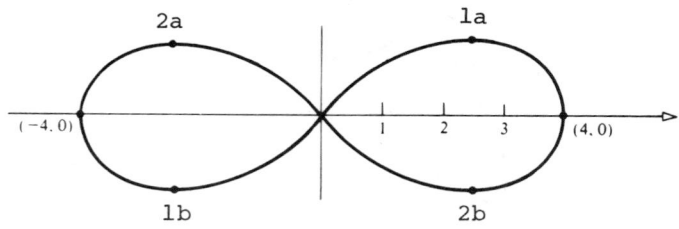

2a   1a

$(-4, 0)$ 1   2   3 $(4, 0)$

1b   2b

θ	2θ	$\cos 2\theta$	r	Graph section
$0 \to \dfrac{\pi}{4}$	$0 \to \dfrac{\pi}{2}$	$1 \to 0$	$\begin{array}{c} +4 \to 0 \\ \text{and} \\ -4 \to 0 \end{array}$	$\begin{array}{c} 1a \\ \\ 1b \end{array}$
$\dfrac{\pi}{4} \to \dfrac{\pi}{2}$	$\dfrac{\pi}{2} \to \pi$	$0 \to -1$	(Undefined)	
$\dfrac{\pi}{2} \to \dfrac{3\pi}{4}$	$\pi \to \dfrac{3\pi}{2}$	$-1 \to 0$	(Undefined)	
$\dfrac{3\pi}{4} \to \pi$	$\dfrac{3\pi}{2} \to 2\pi$	$0 \to 1$	$\begin{array}{c} 0 \to +4 \\ \text{and} \\ 0 \to -4 \end{array}$	$\begin{array}{c} 2a \\ \\ 2b \end{array}$

23. $r^2 = 4 \sin \theta$ (lemniscate)

Intercepts: $\theta = 0 \Rightarrow r = 0$ ($\theta = \pi$ produces the same point); $\theta = \dfrac{\pi}{2} \Rightarrow r = \pm 2$; the graph is not defined for $\theta = \dfrac{3\pi}{2}$.

Symmetry: $\dfrac{\pi}{2}$ axis, since $\sin(\pi - \theta) = \sin \theta$; pole, since $(-r)^2 = r^2$; and polar axis, since it is symmetric with respect to the other two.

Tangents at pole: $0 = 4 \sin \theta \Rightarrow \sin \theta = 0$ $\Rightarrow \theta = 0, \pi$

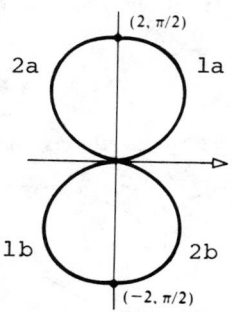

θ	$\sin\theta$	r	Graph section
$0 \to \dfrac{\pi}{2}$	$0 \to 1$	$0 \to +2$ and $0 \to -2$	1a 1b
$\dfrac{\pi}{2} \to \pi$	$1 \to 0$	$+2 \to 0$ and $-2 \to 0$	2a 2b
$\pi \to \dfrac{3\pi}{2}$ $\dfrac{3\pi}{2} \to 2\pi$	(Graph is undefined)		

27.

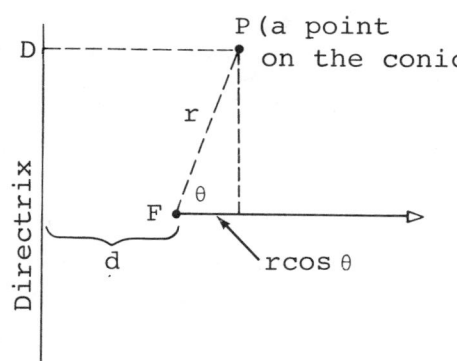

D

Directrix

d

r

θ

F

$r\cos\theta$

P (a point on the conic)

By Problems 17-19 of the Miscellaneous Exercises for Chapter 12 a conic can be defined by

$$\frac{|FP|}{|DP|} = e$$

where $|FP|$ is the distance from a point P on the conic to the focus F (that is, $|FP| = r$) and $|DP|$ is the distance from a point P on the conic to the directrix D. Clearly, $|DP|$ must be a distance perpendicular to the directrix, so $|DP| = d + r\cos\theta$. Thus,

$$\frac{|FP|}{|DP|} = e \Rightarrow \frac{r}{d + r\cos\theta} = e \Rightarrow r = ed + er\cos\theta$$

$$\Rightarrow r(1 - e\cos\theta) = ed \Rightarrow r = \frac{ed}{1 - e\cos\theta}$$

Now converting this to rectangular coordinates and squaring,

$$r = \frac{ed}{1 - e\cos\theta} \implies r - er\cos\theta = ed$$

$$\implies r = ed + er\cos\theta \implies \sqrt{x^2 + y^2} = ed + ex$$

$$\implies (\sqrt{x^2 + y^2})^2 = (ed + ex)^2$$

$$\implies x^2 + y^2 = e^2d^2 + 2e^2dx + e^2x^2$$

For part(b), if $e = 1$, we get $y^2 = d^2 + 2dx$, a parabola that opens around the x-axis.

For parts(a) and (c), we get
$$x^2(1-e^2) - 2e^2dx + y^2 = e^2d^2$$

When the square is completed, we will have an ellipse if $1 - e^2 > 0$ (that is, $e < 1$) and we will have a hyperbola if $1 - e^2 < 0$ (that is, $e > 1$). Remember that by definition, $e > 0$, so no negative values are considered for these inequalities.

<u>Exercise 3</u>, p. 698

3.

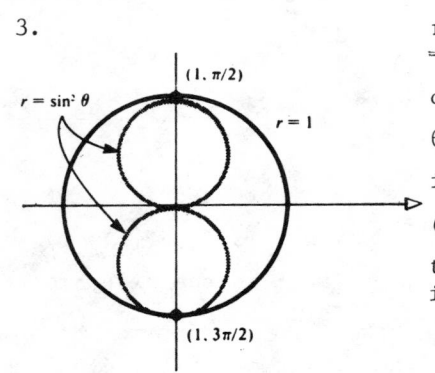

$r = 1$, $r = \sin^2\theta$
$\implies 1 = \sin^2\theta \implies \sin\theta = +1$
or $\sin\theta = -1 \implies \theta = \dfrac{\pi}{2}$ or
$\theta = \dfrac{3\pi}{2} \implies$ points of
intersection are $(1, \dfrac{\pi}{2})$,
$(1, \dfrac{3\pi}{2})$. Since $r = 1$,
the pole is not a point of intersection.

7. $r = -4\cos\theta$, $r = 2(1-\cos\theta) \implies -4\cos\theta = 2(1-\cos\theta)$
$\implies 0 = 2 + 2\cos\theta \implies -2 = 2\cos\theta \implies -1 = \cos\theta$
$\implies \theta = \pi \implies r = -4\cos\pi = -4(-1) = 4 \implies$ a point of
intersection is $(4, \pi)$. Check the pole (set $r = 0$):

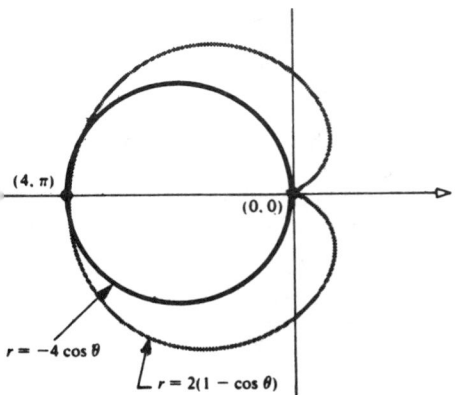

$0 = -4 \cos \theta \Rightarrow \theta = \dfrac{\pi}{2}$

and $0 = 2(1 - \cos \theta)$

$\Rightarrow 0 = 1 - \cos \theta \Rightarrow \cos \theta = 1$

$\Rightarrow \theta = 0$. Thus, the pole

is on each graph, so $(0,0)$

is another point of

intersection.

11.

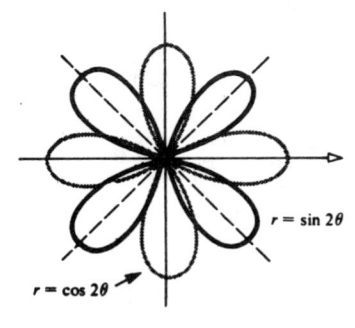

$r = \cos 2\theta$, $r = \sin 2\theta$

$\Rightarrow \cos 2\theta = \sin 2\theta$

$\Rightarrow 1 = \dfrac{\sin 2\theta}{\cos 2\theta} \Rightarrow 1 = \tan 2\theta$

$\Rightarrow 2\theta = \dfrac{\pi}{4} + n\pi$

$\Rightarrow \theta = \dfrac{\pi}{8} + \dfrac{n\pi}{2} \Rightarrow \theta = \dfrac{\pi}{8}$,

$\dfrac{5\pi}{8}$, $\dfrac{9\pi}{8}$, $\dfrac{13\pi}{8}$ and (using

negative n values)

$\theta = -\dfrac{3\pi}{8}$, $-\dfrac{7\pi}{8}$, $-\dfrac{11\pi}{8}$,

$-\dfrac{15\pi}{8}$. All of these

θ values give us $|r| = \dfrac{\sqrt{2}}{2}$. For convenience of

plotting, we will use all positive r values and,

therefore, some θ values altered by $180°$. Eight

points of intersection are $(\dfrac{\sqrt{2}}{2}, \dfrac{\pi}{8})$, $(\dfrac{\sqrt{2}}{2}, \dfrac{3\pi}{8})$,

$(\dfrac{\sqrt{2}}{2}, \dfrac{5\pi}{8})$, $(\dfrac{\sqrt{2}}{2}, \dfrac{7\pi}{8})$, $(\dfrac{\sqrt{2}}{2}, \dfrac{9\pi}{8})$, $(\dfrac{\sqrt{2}}{2}, \dfrac{11\pi}{8})$,

$(\dfrac{\sqrt{2}}{2}, \dfrac{13\pi}{8})$, $(\dfrac{\sqrt{2}}{2}, \dfrac{15\pi}{8})$. Check the pole (set r = 0):

$0 = \cos 2\theta \Rightarrow 2\theta = \dfrac{\pi}{2} \Rightarrow \theta = \dfrac{\pi}{4}$ and $0 = \sin 2\theta$

$\Rightarrow 2\theta = 0 \Rightarrow \theta = 0$. Thus, the pole is on each graph, so (0,0) is another point of intersection.

15. By inspection, r is a maximum when $\cos\theta$ is a maximum (that is, $\theta = 0 \Rightarrow \cos\theta = 1$), and r is a minimum when $\cos\theta$ is a minimum (that is, $\theta = \pi \Rightarrow \cos\theta = -1$). [These θ values can also be found by setting $\dfrac{dr}{d\theta} = 0$ and solving for θ.] At aphelion, $r \approx \dfrac{3.442 \times 10^7}{1 - 0.206(1)} = \dfrac{3.442 \times 10^7}{0.794}$

$\approx 4.335 \times 10^7$ miles.

At perihelion, $r \approx \dfrac{3.442 \times 10^7}{1 - 0.206(-1)} = \dfrac{3.442 \times 10^7}{1.206}$

$\approx 2.8541 \times 10^7$ miles.

Exercise 4, pp. 702-703

3.

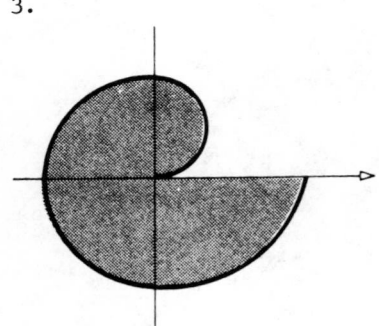

$A = \int_0^{2\pi} \dfrac{1}{2}(a\theta)^2 d\theta$

$= \dfrac{a^2}{2} \int_0^{2\pi} \theta^2 d\theta$

$= \dfrac{a^2}{6} \theta^3 \Big|_0^{2\pi}$

$= \dfrac{4a^2\pi^3}{3}$

7.

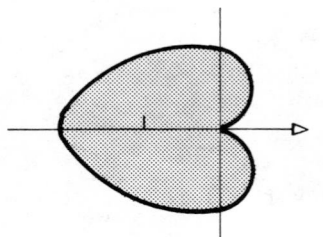

$r = 2 \sin^2\left(\dfrac{\theta}{2}\right) = 2\left(\dfrac{1 - \cos\theta}{2}\right)$

$= 1 - \cos\theta$. By symmetry, double the area from $\theta = 0$ to $\theta = \pi$.

$A = 2 \int_0^{\pi} \dfrac{1}{2}(1 - \cos\theta)^2 d\theta$

$= \int_0^{\pi} [1 - 2\cos\theta + \cos^2\theta] d\theta$

$$A = \int_0^{\pi} (1 - 2 \cos \theta + \frac{1 + \cos 2\theta}{2}) d\theta$$

$$= \int_0^{\pi} (\frac{3}{2} - 2 \cos \theta + \frac{1}{2} \cos 2\theta) d\theta$$

$$= (\frac{3}{2} \theta - 2 \sin \theta + \frac{1}{4} \sin 2\theta) \Big|_0^{\pi} = \frac{3\pi}{2}$$

The result should be the same as in Problem 5, since the figures are just mirror reflections of each other.

11.

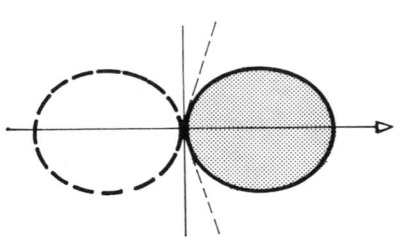

Technically, both halves of the figure are created from $\theta = -\frac{\pi}{4}$ to $\theta = \frac{\pi}{4}$. To look at only the right loop we look only at $r = +\sqrt{4 \cos 2\theta}$. Symmetry will allow us to just double the area from $\theta = 0$ to $\theta = \frac{\pi}{4}$.

$$A = 2 \int_0^{\pi/4} \frac{1}{2} [\sqrt{4 \cos 2\theta}]^2 d\theta$$

$$= \int_0^{\pi/4} 4 \cos 2\theta \, d\theta$$

$$= 2 \sin 2\theta \Big|_0^{\pi/4} = 2$$

15. Find points of intersection: $r = 1 - \cos \theta$, $r = \sin \theta \implies 1 - \cos \theta = \sin \theta \implies 1 = \cos \theta + \sin \theta$
$\implies \theta = 0$ and $\theta = \frac{\pi}{2}$.

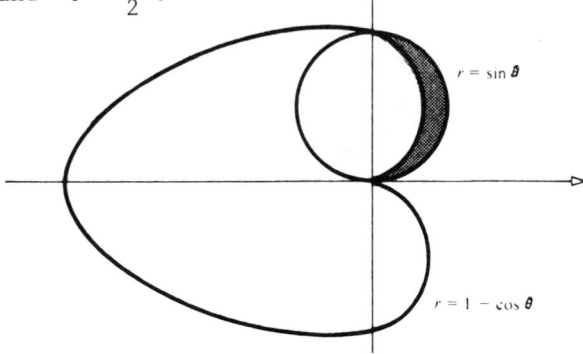

$r = \sin \theta$

$r = 1 - \cos \theta$

$$A = \int_0^{\pi/2} \frac{1}{2}[(\sin \theta)^2 - (1 - \cos \theta)^2]d\theta$$

$$= \frac{1}{2} \int_0^{\pi/2} (\sin^2\theta - 1 + 2 \cos \theta - \cos^2\theta)d\theta$$

$$= \frac{1}{2} \int_0^{\pi/2} [-1 + 2 \cos \theta - (\cos^2\theta - \sin^2\theta)]d\theta$$

$$= \frac{1}{2} \int_0^{\pi/2} (-1 + 2 \cos \theta - \cos 2\theta)d\theta$$

$$= \frac{1}{2}(-\theta + 2 \sin \theta - \frac{1}{2} \sin 2\theta)\Big|_0^{\pi/2}$$

$$= \frac{1}{2}[-\frac{\pi}{2} + 2(1)] = -\frac{\pi}{4} + 1$$

19.

$r = 8 \cos \theta$

$r = 2 \sec \theta$

Find points of intersection: $r = 2 \sec \theta$, $r = 8 \cos \theta$
$\Rightarrow 2 \sec \theta = 8 \cos \theta$
$\Rightarrow \frac{1}{4} = \cos^2\theta \Rightarrow \cos \theta = \pm \frac{1}{2}$
$\Rightarrow \theta = \frac{\pi}{3}, -\frac{\pi}{3}$. (By the picture, these are the only values needed.) Use symmetry and double the area from $\theta = 0$ to $\theta = \frac{\pi}{3}$.

$$A = 2 \int_0^{\pi/3} \frac{1}{2}[(8 \cos \theta)^2 - (2 \sec \theta)^2]d\theta$$

$$= \int_0^{\pi/3} (64 \cos^2\theta - 4 \sec^2\theta)d\theta$$

$$= \int_0^{\pi/3} [64(\frac{1 + \cos 2\theta}{2}) - 4 \sec^2\theta]d\theta$$

$$= \int_0^{\pi/3} (32 + 32 \cos 2\theta - 4 \sec^2\theta)d\theta$$

$$= (32\theta + 16 \sin 2\theta - 4 \tan \theta)\Big|_0^{\pi/3}$$

$$= \frac{32\pi}{3} + 16(\frac{\sqrt{3}}{2}) - 4(\sqrt{3}) = \frac{32\pi}{3} + 4\sqrt{3}$$

23.

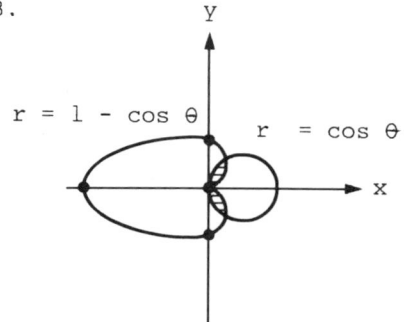

Find points of intersection:
$r = 1 - \cos \theta$, $r = \cos \theta$
$\Rightarrow \cos \theta = 1 - \cos \theta$

$\Rightarrow \cos \theta = \dfrac{1}{2}$

$\Rightarrow \theta = \dfrac{\pi}{3}, \dfrac{-\pi}{3}$

The pole is also on both graphs.

Enlarged picture of desired area

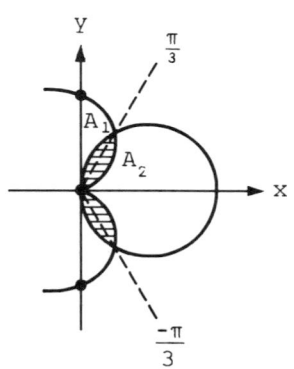

Notes:
(1) We will find the area above the polar axis and double it.
(2) The area above the polar axis has two parts:
   A_1 = area inside the circle ($r = \cos \theta$) for $\dfrac{\pi}{3} \le \theta \le \dfrac{\pi}{2}$
   A_2 = area inside the cardiod ($r = 1 - \cos \theta$) for $0 \le \theta \le \dfrac{\pi}{3}$.

$A_1 = \int_{\pi/3}^{\pi/2} \dfrac{1}{2} [\cos \theta]^2 \, d\theta = \int_{\pi/3}^{\pi/2} \dfrac{1}{2} [\dfrac{1 + \cos 2\theta}{2}] \, d\theta$

$= \dfrac{1}{4} \int_{\pi/3}^{\pi/2} [1 + \cos 2\theta] \, d\theta = [\dfrac{1}{4} \theta + \dfrac{1}{8} \sin 2\theta] \Big|_{\pi/3}^{\pi/2}$

$= (\dfrac{\pi}{8} + 0) - (\dfrac{\pi}{12} + \dfrac{\sqrt{3}}{16}) = \dfrac{\pi}{24} - \dfrac{\sqrt{3}}{16}$

$A_2 = \int_0^{\pi/3} \dfrac{1}{2} [1 - \cos \theta]^2 \, d\theta = \int_0^{\pi/3} [\dfrac{1}{2} - \cos \theta + \dfrac{1}{2} \cos^2 \theta] \, d\theta$

$$= \int_0^{\pi/3} [\frac{1}{2} - \cos \theta + \frac{1}{2}(\frac{1 + \cos 2\theta}{2})]\, d\theta$$

$$= \int_0^{\pi/3} [\frac{3}{4} - \cos \theta + \frac{1}{4} \cos 2\theta]\, d\theta$$

$$= [\frac{3}{4} \theta - \sin \theta + \frac{1}{8} \sin 2\theta]\Big|_0^{\pi/3}$$

$$= (\frac{\pi}{4} - \frac{\sqrt{3}}{2} + \frac{\sqrt{3}}{16}) - (0-0+0)$$

$$= \frac{\pi}{4} - \frac{7\sqrt{3}}{16}$$

Total area = $2(A_1 + A_2)$

$$= 2(\frac{\pi}{24} - \frac{\sqrt{3}}{16} + \frac{\pi}{4} - \frac{7\sqrt{3}}{16})$$

$$= \frac{7\pi}{12} - \sqrt{3}$$

Exercise 5, p. 706

3. $\tan \psi = \dfrac{r}{dr/d\theta} = \dfrac{4 \cos \theta}{-4 \sin \theta}\Big|_{\theta = \pi/4} = \dfrac{\sqrt{2}/2}{-\sqrt{2}/2} = -1$

7. $\tan \psi = \dfrac{r}{dr/d\theta} = \dfrac{a\theta}{a}\Big|_{\theta = -\pi/2} = -\dfrac{\pi}{2}$

11. For $r = 2$, formula (13.10) fails to apply, since $dr/d\theta = 0$. $r = 2 \Rightarrow$ circle, center at pole, radius = 2 $\Rightarrow$ horizontal tangents: $(2, \pi/2)$, $(2, 3\pi/2)$; vertical tangents: $(2,0)$, $(2,\pi)$

15. $\tan \psi = \dfrac{r}{dr/d\theta} = \dfrac{3(1 + \cos\theta)}{-3 \sin\theta} = \dfrac{-1 -\cos\theta}{\sin\theta}$ As in (13.7),

    with $\emptyset = \theta + \psi$

    $$\tan \emptyset = \frac{\tan \theta + \tan \psi}{1 - \tan \theta \tan \psi} = \frac{\dfrac{\sin\theta}{\cos\theta} + \dfrac{-1 - \cos\theta}{\sin\theta}}{1 - \dfrac{\sin\theta}{\cos\theta}(\dfrac{-1-\cos\theta}{\sin\theta})}$$

$$= \frac{\dfrac{\sin^2\theta - \cos\theta - \cos^2\theta}{\sin\theta \, \cos\theta}}{\dfrac{\cos\theta \, \sin\theta + \sin\theta + \sin\theta \, \cos\theta}{\sin\theta \, \cos\theta}}$$

$$= \frac{(1-\cos^2\theta) - \cos\theta - \cos^2}{2\sin\theta\cos\theta + \sin\theta} = \frac{-2\cos^2\theta - \cos\theta + 1}{\sin\theta \, (2\cos\theta + 1)}$$

Horizontal tangents: $\emptyset = k\pi \Rightarrow \tan\emptyset = 0 \Rightarrow -2\cos^2\theta - \cos\theta + 1 = 0 \Rightarrow (-2\cos\theta + 1)(\cos\theta + 1) = 0 \Rightarrow \cos\theta = 1/2$ or $\cos\theta = -1 \Rightarrow \theta = \pi/3, 5\pi/3, \pi* \Rightarrow$ distinct points: $(9/2, \pi/3), (9/2, 5\pi/3)$ (or $(9/2, -\pi/3)$) and $(0, \pi)$; vertical tangents: $\emptyset = \pi/2 + k\pi \Rightarrow \tan\emptyset$ undefined $\Rightarrow \sin\theta(2\cos\theta + 1) = 0 \Rightarrow \sin\theta = 0$ or $\cos\theta = -1/2 \Rightarrow \theta = 0, \theta = 2\pi/3, 4\pi/3 \Rightarrow$ distinct points: $(6, 0), (3/2, 2\pi/3), (3/2, 4\pi/3)$

*Note: $\theta = \pi$ actually gives us $\tan\emptyset = 0/0$. However, when we let $\theta \to \pi$ and use L'Hopital's rule, we get $\tan\emptyset = 0$.

It is also possible to manipulate $\tan\psi$ as was done in Example 1 to arrive at $\tan\psi = -\cot\theta/2$; then using identities, $\tan\psi = -\cot\theta/2 = -\tan(\pi/2 - \theta/2) = \tan(\theta/2 - \pi/2) \Rightarrow \psi = \theta/2 - \pi/2$ and $\emptyset = \psi + \theta \Rightarrow \emptyset = (\theta/2 - \pi/2) + \theta \Rightarrow \theta = (2\emptyset + \pi)/3$.

$$\emptyset = k\pi \Rightarrow \theta = \frac{(2k+1)\pi}{3} \Rightarrow \theta = \pi/3, \pi, 5\pi/3$$

$$\emptyset = \pi/2 + k\pi \Rightarrow \theta = \frac{(2k+2)\pi}{3} \Rightarrow \theta = 0, 2\pi/3, 4\pi/3$$

$$\lim_{\theta \to \pi} \tan\emptyset = \lim_{\theta \to \pi} \frac{-2\cos^2\theta - \cos\theta + 1}{2\sin\theta \, \cos\theta + \sin\theta}$$

$$= \lim_{\theta \to \pi} \frac{4\cos\theta \, \sin\theta + \sin\theta}{2\cos^2\theta - 2\sin^2\theta + \cos\theta} = \frac{0}{1} = 0$$

19. Find the points of intersection: $r = 3\cos\theta$, $r = 1 + \cos\theta \Rightarrow 3\cos\theta = 1 + \cos\theta = 2\cos\theta = 1 \Rightarrow \cos\theta = \frac{1}{2} \Rightarrow \theta = \frac{\pi}{3}, \frac{-\pi}{3}$
Also, the pole is on both graphs.

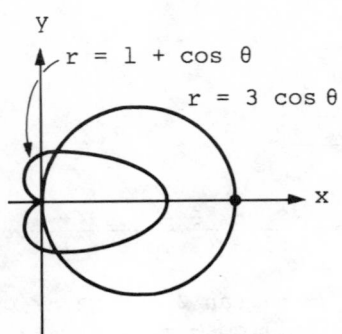

r = 1 + cos θ

r = 3 cos θ

Points of intersection:
$(\frac{3}{2}, \frac{\pi}{3})$, $(\frac{3}{2}, \frac{-\pi}{3})$

and the pole

$(0, \frac{\pi}{2})$ on r = 3 cos θ

$(0, \pi)$ on r = 1 + cos θ

For r = 3 cosθ, $\tan\psi_1 = \dfrac{r}{\dfrac{dr}{d\theta}} = \dfrac{3\cos\theta}{-3\sin\theta} = -\cot\theta$

For r = 1 + cosθ, $\tan\psi_2 = \dfrac{1 + \cos\theta}{-\sin\theta} = -\csc\theta - \cot\theta$

For the pole:

$\tan\psi_1 = -\cot\theta\Big|_{\theta=\frac{\pi}{2}} = 0 \Rightarrow \psi_1 = \frac{\pi}{2}$

$\tan\psi_2 = -\csc\theta - \cot\theta\Big|_{\theta=\pi} = $ undefined $\Rightarrow \psi_2 = 0$

=> the graphs intersect with a right angle at the pole.

For the point $(\frac{3}{2}, \frac{\pi}{3})$:

$\tan\psi_1 = -\cot\theta\Big|_{\theta=\frac{\pi}{3}} = \dfrac{-1}{\sqrt{3}} \Rightarrow \psi_1 = \dfrac{-\pi}{6}$

$\tan\psi_2 = -\csc\theta - \cot\theta\Big|_{\theta=\frac{\pi}{3}} = \dfrac{-2}{\sqrt{3}} - \dfrac{1}{\sqrt{3}} = \dfrac{-3}{\sqrt{3}} \Rightarrow \psi_2 = \dfrac{\pi}{3}$

=> the graphs intersect at a $\frac{\pi}{6}$ angle at $(\frac{3}{2}, \frac{\pi}{3})$.

For the point $(\frac{3}{2}, \frac{-\pi}{3})$:

$\tan\psi_1 = -\cot\theta\Big|_{\theta=\frac{-\pi}{3}} = \dfrac{1}{\sqrt{3}} \Rightarrow \psi_1 = \dfrac{\pi}{6}$

$\tan\psi_2 = -\csc\theta - \cot\theta\Big|_{\theta=\frac{-\pi}{3}} = \dfrac{2}{\sqrt{3}} + \dfrac{1}{\sqrt{3}} = \dfrac{3}{\sqrt{3}} \Rightarrow \psi_2 = \dfrac{\pi}{3}$

=> the graphs intersect at a $\frac{\pi}{6}$ angle at $(\frac{3}{2}, \frac{-\pi}{3})$.

23. $\tan \psi = \dfrac{r}{dr/d\theta} = \dfrac{ke^{\alpha\theta}}{k\alpha e^{\alpha\theta}} = \dfrac{1}{\alpha}$, and α is given as a positive

constant. Therefore ψ has a constant value $\tan^{-1}(\dfrac{1}{\alpha})$.

Exercise 6, pp. 715-717

3. $x = 2$, $y = t^2 + 4$, $0 \le t < +\infty \Rightarrow x = 2$, for
   $y \ge 4$. Similar graph to Problem 1, except it starts
   at $y = 4$ and rises faster than in Problem 1.

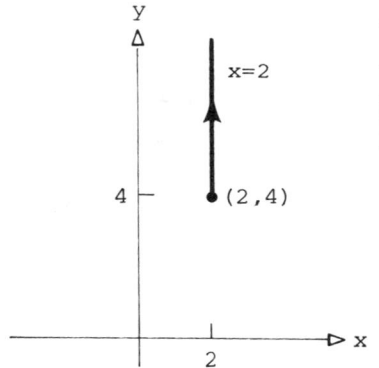

t	0	1	2
x	2	2	2
y	4	5	8

7. $x = t^{1/2} + 1$, $y = t^{3/2}$, $t \ge 1 \Rightarrow x - 1 = t^{1/2}$
   $\Rightarrow (x-1)^3 = t^{3/2} = y \Rightarrow y = (x-1)^3$, $x \ge 2$

t	1	4	9
x	2	3	4
y	1	8	27

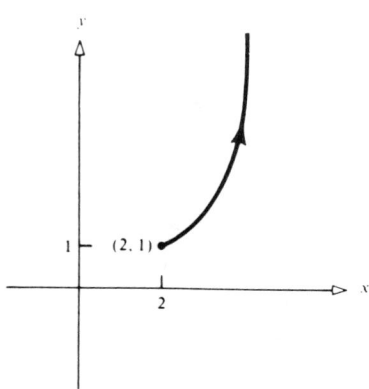

11. $x = \sec t,\ y = \tan t,\ -\dfrac{\pi}{2} < t < \dfrac{\pi}{2} \Rightarrow x^2 = \sec^2 t,$

$y^2 = \tan^2 t$ and $1 + \tan^2 t = \sec^2 t \Rightarrow 1 + y^2 = x^2$

$\Rightarrow x^2 - y^2 = 1$ and since $x = \sec t,\ y = \tan t,$

$-\dfrac{\pi}{2} < t < \dfrac{\pi}{2} \Rightarrow x \geq 1,\ -\infty < y < +\infty$

t	$-\pi/4$	0	$\pi/4$
x	$\sqrt{2}$	1	$\sqrt{2}$
y	-1	0	1

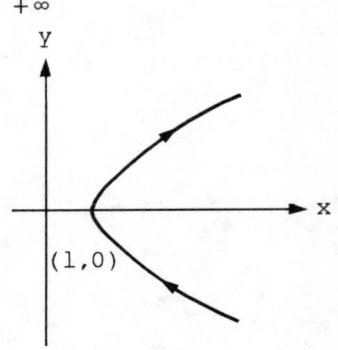

15. $y = 4x^3$

    (1) $x = t$ (2) $x = \sqrt[3]{t}$

          $y = 4t^3$ $y = 4t$

19. Following the work in Example 5, for $\dfrac{x^2}{9} + \dfrac{y^2}{4} = 1$,

$x = 3 \cos t,\ y = 2 \sin t$ will give counterclockwise
motion starting at $(3,0)$, and to get 3 seconds for
a complete revolution, define $x = 3 \cos(\dfrac{2\pi}{3} t),\ y = 2 \sin(\dfrac{2}{3}$
$0 \leq t \leq 3.$

For Problems 21–32, use $\dfrac{dy}{dx} = \dfrac{dy/dt}{dx/dt}$.

23. $\dfrac{dy}{dx} = \dfrac{2e^t}{2t}\Big|_{t=2} = \dfrac{e^2}{2}$

27. $x = 3t^2,\ y = t^3 - 4t \Rightarrow \dfrac{dy}{dx} = \dfrac{3t^2 - 4}{6t}$

    (a) $\dfrac{dy}{dx} = 0 \Rightarrow 3t^2 - 4 = 0 \Rightarrow t = \pm\sqrt{\dfrac{4}{3}} = \dfrac{\pm 2\sqrt{3}}{3}$

           $\Rightarrow$ horizontal tangents at $(4,\ \pm\dfrac{16\sqrt{3}}{9})$

(b) $(\frac{dy}{dx} \Rightarrow +\infty) \Rightarrow 6t = 0 \Rightarrow t = 0 \Rightarrow$ vertical

tangent at $(0,0)$

31. $x = 2 \cos t$, $y = 1 + \cos 2t \Rightarrow \dfrac{dy}{dx} = \dfrac{-2 \sin 2t}{-2 \sin t}$

$= \dfrac{-2(2 \cos t \sin t)}{-2 \sin t} = 2 \cos t$

(a) $\dfrac{dy}{dx} = 0 \Rightarrow 2 \cos t = 0 \Rightarrow t = \dfrac{\pi}{2} + k\pi$,

k = integer; horizontal tangent: $t = \dfrac{\pi}{2}$

$\Rightarrow (x,y) = (0,0)$, only distinct point.

(b) $\dfrac{dy}{dx}$ defined for all t suggests no vertical

tangents. Eliminate the parameter $\Rightarrow (\dfrac{x}{2})^2$

$= \cos^2 t = \dfrac{1 + \cos 2t}{2} = \dfrac{y}{2} \Rightarrow y = \dfrac{x^2}{2}$ for

$-1 \le x \le 1 \Rightarrow$ just a portion of a parabola.

35. $x = \cos 2t$, $y = \sin t \Rightarrow \dfrac{dy}{dx} = \dfrac{\cos t}{-2 \sin 2t}\Big|_{t=\pi/4}$

$= \dfrac{\sqrt{2}/2}{-2(1)} = \dfrac{-\sqrt{2}}{4} = $ slope; $t = \dfrac{\pi}{4} \Rightarrow x = 0$, $y = \dfrac{\sqrt{2}}{2}$

Tangent line $\Rightarrow y - \dfrac{\sqrt{2}}{2} = \dfrac{-\sqrt{2}}{4}(x-0) \Rightarrow y = \dfrac{-\sqrt{2}}{4}x + \dfrac{\sqrt{2}}{2}$

39. From (13.13), $y' = dy/dx = \dfrac{dy/dt}{dx/dt}$, thus,

$\dfrac{d^2y}{dx^2} = \dfrac{d}{dx}(\dfrac{dy}{dx}) = \dfrac{dy'}{dx} = \dfrac{dy'/dt}{dx/dt}$

$x = t + \dfrac{1}{t} = t + t^{-1}$, $y = 4 + t$

$\dfrac{dy}{dx} = \dfrac{1}{1 - t^{-2}} = \dfrac{t^2}{t^2 - 1}$

$\dfrac{d^2y}{dx^2} = \dfrac{\dfrac{d}{dt}(\dfrac{t^2}{t^2-1})}{\dfrac{d}{dt}(t+t^{-1})} = \dfrac{\dfrac{(t^2-1)(2t) - t^2(2t)}{(t^2-1)^2}}{1 - t^{-2}}$

$$= \frac{\frac{2t^3 - 2t - 2t^3}{(t^2-1)^2}}{\frac{t^2 - 1}{t^2}} = \frac{-2t^3}{(t^2-1)^3}$$

43. (a) $t = \frac{x}{v_0 \cos \theta} \Rightarrow (v_0 \sin\theta)(\frac{x}{v_0 \cos \theta}) - 16(\frac{x}{v_0 \cos\theta})^2$

$= y = (\tan\theta)x - \frac{16}{v_0^2 \cos^2\theta} x^2$; since θ and v_0 are

constants, we have an equation of the form
$y = Ax + Bx^2$, which is a parabola. Since $B < 0$,
we have a parabolic arch above the x-axis,
opening downward with x-intercepts of $x = 0$ and
$x = -\frac{A}{B}$.

(b) $v(t) = [(\frac{dx}{dt})^2 + (\frac{dy}{dt})^2]^{1/2}$

$= [(v_0 \cos\theta)^2 + (v_0 \sin\theta - 32t)^2]^{1/2}$

$v(t) = (v_0^2\cos^2\theta + v_0^2\sin^2\theta - 64v_0 \sin\theta\ t + 1024t^2)^{1/2}$

$v(t) = (v_0^2 - 64v_0 \sin\theta\ t + 1024t^2)^{1/2}$

$\Rightarrow v(1) = (v_0^2 - 64v_0 \sin\theta + 1024)^{1/2}$

$v(2) = (v_0^2 - 128v_0 \sin\theta + 4096)^{1/2}$

(c) Projectile hits the ground when $y = 0$: $y = v_0 \sin\theta\ t$

$- 16t^2 = 0 \Rightarrow t(v_0 \sin\theta - 16t) = 0 \Rightarrow t = 0$ or

$t = \frac{v_0 \sin\theta}{16}$ ($t = 0$ is our starting point).

Horizontal distance $= x = v_0 \cos\theta\ t$

$\Rightarrow x = v_0 \cos\theta(\frac{v_0 \sin\theta}{16}) = \frac{v_0^2 \cos\theta \sin\theta}{16}$

$= \frac{v_0^2(2 \sin\theta \cos\theta)}{2(16)} = \frac{v_0^2 \sin 2\theta}{32}$

We also could have used part (a) above: the

projectile travels to $x = -\frac{A}{B} = \frac{\tan\theta}{16/v_0^2 \cos^2\theta}$ or

$$x = \frac{v_0{}^2\cos^2\theta \cdot \frac{\sin\theta}{\cos\theta}}{16} = \frac{v_0{}^2 \cos\theta \sin\theta}{16} = \frac{v_0{}^2 \sin 2\theta}{32}$$

7. Following the reasoning using the chain rule for formula (13.13), if we let $y' = \frac{dy}{dx} = \frac{dy/dt}{dx/dt}$, then

$$\frac{d(y')}{dt} = \frac{d(y')}{dx} \cdot \frac{dx}{dt}, \text{ and since } \frac{d^2y}{dx^2} = \frac{d(y')}{dx}, \text{ we have}$$

$$\frac{d^2y}{dx^2} = \frac{d(y')}{dx} = \frac{d(y')/dt}{dx/dt}.$$

For $x = a \cos g(t)$ and $y = b \sin g(t)$, assuming that g is a differentiable (nonconstant) function of t,

$$\frac{dy}{dx} = \frac{b \cos g(t) \cdot g'(t)}{-a \sin g(t) \cdot g'(t)} = \frac{b \cos g(t)}{-a \sin g(t)} = - \frac{b \cos g(t)}{a \sin g(t)}$$

and

$$\frac{d^2y}{dx^2} = \frac{\dfrac{-a \sin g(t)[-b \sin g(t) \cdot g'(t)] - b \cos g(t)[-a \cos g(t)g'(t)]}{a^2 \sin^2 g(t)}}{-a \sin g(t) \cdot g'(t)}$$

$$= \frac{abg'(t)[\sin^2 g(t) + \cos^2 g(t)]}{-a^3 \sin^3 g(t) \cdot g'(t)} = \frac{-b}{a^2 \sin^3 g(t)}$$

Now checking the equation to be proved:

$$xy^2 \frac{d^2y}{dx^2} = a \cos g(t) \cdot b^2 \sin^2 g(t)\left(\frac{-b}{a^2 \sin^3 g(t)}\right)$$

$$= - \frac{b^3 \cos g(t)}{a \sin g(t)} = b^2\left(- \frac{b \cos g(t)}{a \sin g(t)}\right) = b^2 \frac{dy}{dx},$$

as desired.

Note that using $x = a \cos g(t)$ and $y = b \sin g(t)$, we match the answers in the text:

$$\frac{dy}{dx} = \frac{-b \cos g(t)}{a \sin g(t)} = \frac{-b^2}{a^2}\left(\frac{x}{y}\right) \text{ and } \frac{d^2y}{dx^2} = -b/a^2 \sin^3 g(t)$$

$$= - \frac{b^4}{a^2 y^3}$$

Exercise 7, pp. 723-724

For Problems 1-8, use formula (13.21).

3. $s = \int_0^2 \sqrt{(1)^2 + (t)^2} \, dt$ (substitute $t = \tan\theta$,
$$\sqrt{1 + t^2} = \sec\theta,$$
$$dt = \sec^2\theta \, d\theta)$$

$\quad = \int_0^{\tan^{-1}(2)} \sec^3\theta \, d\theta$ (see formula (9.6))

$\quad = (\frac{1}{2}\sec\theta\tan\theta + \frac{1}{2}\ln|\sec\theta + \tan\theta|)\Big|_0^{\tan^{-1}(2)}$

$\quad = \frac{1}{2}(\sqrt{5})(2) + \frac{1}{2}\ln|\sqrt{5} + 2| - \frac{1}{2}(1)(0) - \frac{1}{2}\ln|1 + 0|$

$\quad = \sqrt{5} + \frac{1}{2}\ln(\sqrt{5} + 2)$ Note: For $\sec\theta$, use
$\quad 1 + \tan^2\theta = \sec^2\theta$ where $\theta = \tan^{-1}(2) \Rightarrow \tan\theta = 2$.

7. $s = \int_0^{2\pi} \sqrt{(2\cos t)^2 + (-2\sin t)^2} \, dt$

$\quad = \int_0^{2\pi} 2\sqrt{\cos^2 t + \sin^2 t} \, dt = \int_0^{2\pi} 2 \, dt$

$\quad = 2t\Big|_0^{2\pi} = 4\pi$

For Problems 9-14, use formula (13.24).

11. $s = \int_0^{\pi} \sqrt{(\cos^2\frac{\theta}{2})^2 + [2 \cdot \frac{1}{2}\cos\frac{\theta}{2}(-\sin\frac{\theta}{2})]^2} \, d\theta$

$\quad = \int_0^{\pi} \sqrt{\cos^4\frac{\theta}{2} + \cos^2\frac{\theta}{2}\sin^2\frac{\theta}{2}} \, d\theta$

$\quad = \int_0^{\pi} \sqrt{\cos^2\frac{\theta}{2}} \sqrt{\cos^2\frac{\theta}{2} + \sin^2\frac{\theta}{2}} \, d\theta$
$\quad\quad$ (Note: $\cos\frac{\theta}{2} \geq 0$ for $0 \leq \theta \leq \pi$)

$\quad = \int_0^{\pi} \cos\frac{\theta}{2} \, d\theta = 2\sin\frac{\theta}{2}\Big|_0^{\pi} = 2$

15. Use (13.21).

$$s = \int_0^{\pi/2} \sqrt{(3b \sin^2 t \cos t)^2 + (-3b \cos^2 t \sin t)^2} \; dt$$

$$= \int_0^{\pi/2} \sqrt{9b^2 \sin^4 t \cos^2 t + 9b^2 \cos^4 t \sin^2 t} \; dt$$

$$= \int_0^{\pi/2} \sqrt{9b^2 \sin^2 t \cos^2 t} \; \sqrt{\sin^2 t + \cos^2 t} \; dt$$

$$(\text{for } 0 \le t \le \frac{\pi}{2} \;, \; \sin t \cos t \ge 0)$$

$$= \int_0^{\pi/2} 3b \sin t \cos t \; dt \quad (\text{substitute } u = \sin t,$$
$$du = \cos t \; dt)$$

$$= \int_0^1 3bu\,du = \frac{3bu^2}{2} \Big|_0^1 = \frac{3b}{2}$$

For Problems 17-22, use formula (13.21).

19. $$s = \int_0^2 \sqrt{(t)^2 + [(2t+3)^{1/2}]^2} \; dt = \int_0^2 \sqrt{t^2 + 2t + 3} \; dt$$

$$= \int_0^2 \sqrt{(t+1)^2 + 2} \; dt \quad (\text{substitute } t + 1 = \sqrt{2} \tan \theta,$$

$$\sqrt{(t+1)^2 + 2} = \sqrt{2} \sec \theta, \; dt = \sqrt{2} \sec^2\theta \; d\theta)$$

$$= \int_{\tan^{-1}(1/\sqrt{2})}^{\tan^{-1}(3/\sqrt{2})} 2 \sec^3\theta \; d\theta \quad (\text{see Problem 3})$$

$$= 2(\frac{1}{2} \sec \theta \tan \theta + \frac{1}{2} \ln|\sec \theta + \tan \theta|) \Big|_{\tan^{-1}(1/\sqrt{2})}^{\tan^{-1}(3/\sqrt{2})}$$

$$= \frac{\sqrt{11}}{\sqrt{2}}(\frac{3}{\sqrt{2}}) + \ln(\frac{\sqrt{11}}{\sqrt{2}} + \frac{3}{\sqrt{2}}) - \frac{\sqrt{3}}{\sqrt{2}}(\frac{1}{\sqrt{2}}) - \ln(\frac{\sqrt{3}}{\sqrt{2}} + \frac{1}{\sqrt{2}})$$

$$= \frac{1}{2}(3\sqrt{11} - \sqrt{3}) + \ln(\frac{\sqrt{11} + 3}{\sqrt{3} + 1})$$

23. $t = 1$ to $t = 1.1 \Rightarrow dt = 0.1$

$$x = t^{1/3} \Rightarrow dx = \frac{1}{3} t^{-2/3} dt \Big|_{t=1} = \frac{1}{3}(1)^{-2/3}(0.1) \approx 0.0333$$

$$y = t^2 \Rightarrow dy = 2t \; dt \Big|_{t=1} = 2(1)(0.1) = 0.2$$

$$ds = \sqrt{(dx)^2 + (dy)^2} \approx \sqrt{(0.0333)^2 + (0.2)^2}$$

$$\approx \sqrt{0.04111} \approx 0.2028$$

27.

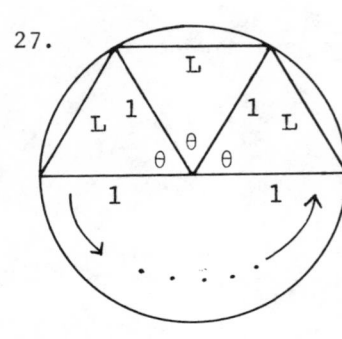

Each $\theta = \dfrac{2\pi}{n}$, n = number of sides of regular inscribed polygon. Use the law of cosines to find L:

$$L^2 = 1^2 + 1^2 - 2(1)(1)\cos\theta$$

$$= 2 - 2\cos\theta$$

$$L = \sqrt{2 - 2\cos\theta} = \sqrt{4\sin^2\frac{\theta}{2}}$$

$$= 2\sin\frac{\theta}{2} = 2\sin\frac{\pi}{n}$$

The circumference C is approximately the perimeter of the polygon, or nL. Then, letting $n \to +\infty$:

$$C = \lim_{n \to +\infty} 2n\sin\frac{\pi}{n}(\text{let } u = \frac{\pi}{n}) = \lim_{u \to 0+} 2\pi \cdot \frac{\sin u}{u}$$

$$= 2\pi \cdot 1 = 2\pi$$

Exercise 8, pp. 729-730

3. $y = x^2 - x^3$, $y' = 2x - 3x^2\big|_{x=1} = -1$,

$y'' = 2 - 6x\big|_{x=1} = -4$

$$K = \frac{|y''|}{[1 + (y')^2]^{3/2}} = \frac{4}{[1 + 1]^{3/2}} = \frac{2}{\sqrt{2}} = \sqrt{2}$$

7. $y = x^{-3/2}$, $y' = -\dfrac{3}{2}x^{-5/2}\big|_{x=1} = -\dfrac{3}{2}$,

$y'' = \dfrac{15}{4}y^{-7/2}\big|_{x=1} = \dfrac{15}{4}$

$$K = \frac{|y''|}{[1 + (y')^2]^{3/2}} = \frac{15/4}{[1 + 9/4]^{3/2}}$$

$$= \frac{15/4}{13\sqrt{13}/8} = \frac{30}{13\sqrt{13}}$$

11. $y = \cos x$, $y' = -\sin x\big|_{x=0} = 0$,

$y'' = -\cos x\big|_{x=0} = -1$

$K = \dfrac{|y''|}{[1 + (y')^2]^{3/2}} = \dfrac{1}{[1 + 0]^{3/2}} = 1$

15. $x = t^2 \implies x' = 2t\big|_{t=t_1} = 2t_1$, $x'' = 2\big|_{t=t_1} = 2$

$y = \dfrac{2}{t} \implies y' = \dfrac{-2}{t^2}\big|_{t=t_1} = \dfrac{-2}{t_1{}^2}$, $y'' = \dfrac{4}{t^3}\big|_{t=t_1} = \dfrac{4}{t_1{}^3}$

$K = \dfrac{|x'y'' - x''y'|}{[(x')^2 + (y')^2]^{3/2}} = \dfrac{|2t_1(4/t_1{}^3) - 2(-2/t_1^2)|}{[(2t_1)^2 + (-2/t_1{}^2)^2]^{3/2}}$

$- \dfrac{12/t_1{}^2}{[4t_1{}^2 + (4/t_1{}^4)]^{3/2}} = \dfrac{12/t_1{}^2}{8(t_1{}^6 + 1)^{3/2}/t_1{}^6} = \dfrac{3t_1{}^4}{2(t_1{}^6+1)^{3/2}}$

19. $y = \sin x \implies y' = \cos x\big|_{x=\pi/2} = 0$,

$y'' = -\sin x\big|_{x=\pi/2} = -1$

$K = \dfrac{|y''|}{[1 + (y')^2]^{3/2}} = \dfrac{1}{[1 + 0]^{3/2}} = 1$; $\rho = \dfrac{1}{K} = 1$

23. $x = 3t^2 \implies x' = 6t\big|_{t=1} = 6$, $x'' = 6\big|_{t=1} = 6$

$y = 3t - t^3 \implies y' = 3 - 3t^2\big|_{t=1} = 0$,

$y'' = -6t\big|_{t=1} = -6$

$K = \dfrac{|x'y'' - x''y'|}{[(x')^2 + (y')^2]^{3/2}} = \dfrac{|6(-6) - 6(0)|}{[36 + 0]^{3/2}}$

$= \dfrac{36}{216} = \dfrac{1}{6}$

$\rho = \dfrac{1}{K} = 6$

27. $y = \ln x \Rightarrow y' = \frac{1}{x}$, $y'' = -\frac{1}{x^2}$

$\kappa = \dfrac{|y''|}{[1 + (y')^2]^{3/2}} = \dfrac{\left|-\frac{1}{x^2}\right|}{[1 + \frac{1}{x^2}]^{3/2}} = \dfrac{\frac{1}{x^2}}{\frac{[x^2 + 1]^{3/2}}{(x^2)^{3/2}}}$

$= \dfrac{|x^3|}{x^2[x^2 + 1]^{3/2}} = \dfrac{|x|}{[x^2 + 1]^{3/2}} = \dfrac{x}{[x^2 + 1]^{3/2}}$,

since for $\ell y = \ln x$, $x > 0$.

$\dfrac{d\kappa}{dx} = \dfrac{[x^2 + 1]^{3/2} \cdot 1 - x \cdot \frac{3}{2}(2x)[x^2 + 1]^{1/2}}{[x^2 + 1]^3}$

$= \dfrac{[x^2 + 1]^{1/2}[x^2 + 1 - 3x^2]}{[x^2 + 1]^3} = \dfrac{1 - 2x^2}{[x^2 + 1]^{5/2}}$

$\dfrac{d\kappa}{dx} = 0 \Rightarrow 1 - 2x^2 = 0 \Rightarrow x = \dfrac{1}{\sqrt{2}}$ (not $-\dfrac{1}{\sqrt{2}}$, since

$x > 0$)

$0 < x < \dfrac{1}{\sqrt{2}} \Rightarrow 1 - 2x^2 > 0 \Rightarrow \dfrac{d\kappa}{dx} > 0 \Rightarrow \kappa$ increasing

$x > \dfrac{1}{\sqrt{2}} \Rightarrow 1 - 2x^2 < 0 \Rightarrow \dfrac{d\kappa}{dx} < 0 \Rightarrow \kappa$ decreasing

Thus, κ is a maximum at the point $(\dfrac{1}{\sqrt{2}}$, $-\ell n\sqrt{2})$

31. $r = 2 \cos 2\theta\Big|_{\theta=\pi/12} = 2(\dfrac{\sqrt{3}}{2}) = \sqrt{3}$. $\dfrac{dr}{d\theta} = -4 \sin 2\theta\Big|_{\theta=\pi/12}$

$= -4(\dfrac{1}{2}) = -2$

$\dfrac{d^2r}{d\theta^2} = -8 \cos 2\theta\Big|_{\theta=\pi/12} = -8(\dfrac{\sqrt{3}}{2}) = -4\sqrt{3}$

From Problem 30,

$\kappa = \dfrac{|r^2 + 2(dr/d\theta)^2 - r(d^2r/d\theta^2)|}{[r^2 + (dr/d\theta)^2]^{3/2}}$

$= \dfrac{|3 + 2(4) - \sqrt{3}(-4\sqrt{3})|}{[3 + 4]^{3/2}} = \dfrac{23}{7\sqrt{7}}$

35. $y = a \cosh(x/a) \Rightarrow y' = \sinh(x/a) \Rightarrow y'' = \dfrac{1}{a} \cosh(x/a)$

$$\kappa = \frac{|y''|}{[1 + (y')^2]^{3/2}} = \frac{\frac{1}{a}\cosh(x/a)}{[1 + \sinh^2(x/a)]^{3/2}}$$

$$= \frac{\frac{1}{a}\cosh(x/a)}{[\cosh^2(x/a)]^{3/2}}$$

(see Chapter 8 for hyperbolic formulas)

$$\kappa = \frac{1}{a\cosh^2(x/a)} = \frac{a}{a^2\cosh^2(x/a)} = \frac{a}{y^2}$$

39. $x = \theta - \sin\theta$, $y = 1 - \cos\theta$
The highest point on an arch of the cycloid occurs
when $\theta = \pi + 2k\pi$ [Let's use $\theta = \pi$]

$x' = 1 - \cos\theta\big|_{\theta=\pi} = 2$

$x'' = \sin\theta\big|_{\theta=\pi} = 0$

$y' = \sin\theta\big|_{\theta=\pi} = 0$

$y'' = \cos\theta\big|_{\theta=\pi} = -1$

Now, using (13.35),

$$\kappa = \frac{|x'y'' - x''y'|}{[(x')^2 + (y')^2]^{3/2}} = \frac{|2(-1) - 0(0)|}{[(2)^2 + (0)^2]^{3/2}}$$

$$\kappa = \frac{2}{8} = \frac{1}{4}$$

Exercise 9, pp. 733-734

3. $x = a\cos^3\theta$, $y = a\sin^3\theta \Rightarrow x' = -3a\cos^2\theta\sin\theta$,
$y' = 3a\sin^2\theta\cos\theta$

$$S = \int_0^{\pi/2} 2\pi y\sqrt{(x')^2 + (y')^2}\ d\theta$$

$$= \int_0^{\pi/2} 2\pi(a\sin^3\theta)\sqrt{9a^2\cos^4\theta\ \sin^2\theta + 9a^2\sin^4\theta\ \cos^2\theta}\ d\theta$$

$$S = 6a^2\pi \int_0^{\pi/2} \sin^3\theta\sqrt{\cos^2\theta \ \sin^2\theta} \ \sqrt{\cos^2\theta + \sin^2\theta} \ d\theta$$

(Note: $\cos\theta \ \sin\theta \geq 0$ for $0 \leq \theta \leq \dfrac{\pi}{2}$)

$$= 6a^2\pi \int_0^{\pi/2} \sin^4\theta \ \cos\theta \ d\theta \quad \text{(use } u = \sin\theta,$$
$$du = \cos\theta \ d\theta)$$

$$= 6a^2\pi \int_0^1 u^4 du = \frac{6a^2\pi}{5} u^5 \Big|_0^1 = \frac{6a^2\pi}{5}$$

7. $y = \dfrac{a}{2}(e^{x/a} + e^{-x/a}) \Rightarrow y' = \dfrac{1}{2} e^{x/a} - \dfrac{1}{2} e^{-x/a}$

$$S = \int_0^a 2\pi y\sqrt{1 + (y')^2} \ dx$$

$$= \int_0^a 2\pi \left[\frac{a}{2}(e^{x/a} + e^{-x/a})\right] \sqrt{1 + \frac{1}{4} e^{2x/a} - \frac{1}{2} + \frac{1}{4} e^{-2x/a}} \ dx$$

$$= a\pi \int_0^a (e^{x/a} + e^{-x/a}) \sqrt{\frac{1}{4} e^{2x/a} + \frac{1}{2} + \frac{1}{4} e^{-2x/a}} \ dx$$

$$= a\pi \int_0^a (e^{x/a} + e^{-x/a}) \sqrt{(\frac{1}{2} e^{x/a} + \frac{1}{2} e^{-x/a})^2} \ dx$$

$$= a\pi \int_0^a (e^{x/a} + e^{-x/a})(\frac{1}{2} e^{x/a} + \frac{1}{2} e^{-x/a}) \ dx$$

$$= \frac{a\pi}{2} \int_0^a (e^{2x/a} + 2 + e^{-2x/a}) dx$$

$$= \frac{a\pi}{2}(\frac{a}{2} e^{2x/a} + 2x - \frac{a}{2} e^{-2x/a}) \Big|_0^a$$

$$= \frac{a\pi}{2}(\frac{a}{2} e^2 + 2a - \frac{a}{2} e^{-2}) = \frac{a^2\pi}{4}(e^2 - e^{-2}) + a^2\pi$$

11. $y = e^x \Rightarrow y' = e^x$

$$S = \int_0^1 2\pi y\sqrt{1 + (y')^2} \ dx = \int_0^1 2\pi e^x\sqrt{1 + e^{2x}} \ dx$$

(use $u = e^x$, $du = e^x dx$)

$$S = 2\pi \int_1^e \sqrt{1 + u^2}\, du \quad [\text{use } u = \tan\theta,\ du = \sec^2\theta\, d\theta$$
$$\text{and Formula (9.6).}]$$

$$= 2\pi\left(\frac{1}{2}\sec\theta\,\tan\theta + \frac{1}{2}\ln\left|\sec\theta + \tan\theta\right|\right)\Big|_{\pi/4}^{\tan^{-1}(e)}$$

$$= \pi\left[(\sqrt{e^2 + 1})(e) + \ln(\sqrt{e^2 + 1} + e)\right.$$
$$\left. - \sqrt{2}(1) - \ln(\sqrt{2} + 1)\right]$$

15. $x = 3t^2,\ y = 2t^3 \implies x' = 6t,\ y' = 6t^2,$ about the y-axis

$$S = \int_0^1 2\pi x\sqrt{((x')^2 + (y')^2)}\, dt$$

$$= \int_0^1 2\pi(3t^2)\sqrt{36t^2 + 36t^4}\, dt$$

$$= 36\pi \int_0^1 t^3\sqrt{1 + t^2}\, dt \quad (\text{use } u = 1 + t^2,\ t^2 = u - 1,$$
$$\frac{1}{2}\,du = t\,dt)$$

$$= 36\pi\left(\frac{1}{2}\right)\int_1^2 (u-1)\sqrt{u}\, du = 18\pi\int_1^2 (u^{3/2} - u^{1/2})\, du$$

$$= 18\pi\left(\frac{2}{5}u^{5/2} - \frac{2}{3}u^{3/2}\right)\Big|_1^2 = 18\pi\left(\frac{8\sqrt{2}}{5} - \frac{4\sqrt{2}}{3} - \frac{2}{5} + \frac{2}{3}\right)$$

$$= \frac{24\pi}{5}(\sqrt{2} + 1)$$

19. $f(\theta) = e^\theta \implies f'(\theta) = e^\theta$

$$S = 2\pi\int_0^\pi f(\theta)\sin\theta\sqrt{[f(\theta)]^2 + [f'(\theta)]^2}\, d\theta$$

$$= 2\pi\int_0^\pi e^\theta\sin\theta\sqrt{e^{2\theta} + e^{2\theta}}\, d\theta$$

$$= 2\sqrt{2}\pi\int_0^\pi e^{2\theta}\sin\theta\, d\theta \quad (\text{use Integration by Parts,}$$

Chapter 9, Section 2)

$$= \frac{2\sqrt{2}\pi e^{2\theta}}{5}(2\sin\theta - \cos\theta)\Big|_0^\pi = \frac{2\sqrt{2}\pi}{5}(e^{2\pi} + 1)$$

23. $y = \cosh x \Rightarrow y' = \sinh x$

$S = \int_a^b 2\pi y\sqrt{1 + (y')^2}\, dx = \int_a^b 2\pi\cosh x\sqrt{1 + \sinh^2 x}\, dx$

$\quad = 2\pi \int_a^b \cosh x\sqrt{\cosh^2 x}\, dx$ (note: $1 + \sinh^2 x$

$\quad = \cosh^2 x$ and $\cosh x \geq 0$ for all x)

$\quad = 2\pi \int_a^b \cosh^2 x\, dx$ (see Chapter 8, Section 6,

to develop a formula analogous to

$\cos^2 x = \dfrac{1 + \cos 2x}{2}$)

$\quad = 2\pi \int_a^b \dfrac{\cosh 2x + 1}{2}\, dx = \pi \int_a^b [\cosh 2x + 1]\, dx$

$\quad = \pi[\tfrac{1}{2}\sinh 2x + x]\Big|_a^b$

$\quad = \dfrac{\pi}{2}[\sinh 2b - \sinh 2a + 2(b-a)]$

Miscellaneous Exercises, pp. 734-736

3. $A = \int_1^\pi \tfrac{1}{2}[f(\theta)]^2 d\theta = \int_1^\pi \tfrac{1}{2}(\tfrac{1}{\theta})^2 d\theta = \tfrac{1}{2}\int_1^\pi \dfrac{1}{\theta^2}\, d\theta$

$\quad = -\dfrac{1}{2} \cdot \dfrac{1}{\theta}\Big|_1^\pi = -\dfrac{1}{2\pi} + \dfrac{1}{2} = \dfrac{1}{2\pi}(-1 + \pi)$

7. $x = \dfrac{1}{t^2}$, $y = \dfrac{2}{t^2 + 1} \Rightarrow t^2 = \dfrac{1}{x} \Rightarrow y = \dfrac{2}{1/x + 1}$

$\quad \Rightarrow y = \dfrac{2x}{1 + x}$ for $x > 0$ only

11. $x = \sin\theta - 2$, $y = 4 - 2\cos\theta \Rightarrow x + 2 = \sin\theta$,

$\dfrac{y - 4}{-2} = \cos\theta \Rightarrow (x+2)^2 = \sin^2\theta$, $\dfrac{(y-4)^2}{4} = \cos^2\theta$

$\quad \Rightarrow (x+2)^2 + \dfrac{(y-4)^2}{4} = \sin^2\theta + \cos^2\theta = 1$

For problems 13-18 use formula (13.13) for

$$y' = \frac{dy}{dx} = \frac{dy/dt}{dx/dt} \quad \text{and} \quad \frac{d^2y}{dx^2} = \frac{d(y')}{dx} = \frac{d(y')/dt}{dx/dt}$$

15. $x = 2 \cos 2\theta$, $y = \sin \theta \implies \dfrac{dy}{dx} = \dfrac{\cos \theta}{-4 \sin 2\theta}$

$$= \frac{\cos \theta}{-8 \sin \theta \cos \theta} \implies \frac{dy}{dx} = \frac{-1}{8 \sin \theta} \implies$$

$$\frac{d^2y}{dx^2} = \frac{8 \cos \theta/(64 \sin^2\theta)}{-8 \sin \theta \cos \theta} = -\frac{8 \cos \theta}{512 \sin^3\theta \cos \theta} = \frac{-1}{64 \sin^3\theta}$$

19.

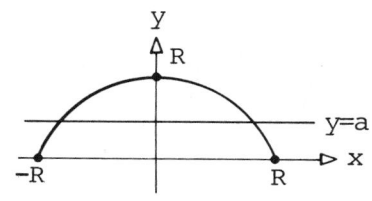

$$y = \sqrt{R^2 - x^2}$$

Points of intersection:

$$a = \sqrt{R^2 - x^2} \implies x^2 = R^2 - a^2$$

$\implies x = \pm\sqrt{R^2 - a^2}$. Rotate the

curve $y = \sqrt{R^2 - x^2}$ for

$$-\sqrt{R^2 - a^2} \le x \le \sqrt{R^2 - a^2}$$

around the x-axis. (Start from $x = 0$ and double
the result.)

$$y = \sqrt{R^2 - x^2} \implies y' = \frac{-x}{\sqrt{R^2 - x^2}} \implies \sqrt{1 + (y')^2}$$

$$= \sqrt{1 + \frac{x^2}{R^2 - x^2}} = \sqrt{\frac{R^2}{R^2 - x^2}} = \frac{R}{\sqrt{R^2 - x^2}}$$

$$S = 2\int_0^{\sqrt{R^2-a^2}} 2\pi y\sqrt{1 + (y')^2} \, dx$$

$$= 4\pi \int_0^{\sqrt{R^2-a^2}} \sqrt{R^2 - x^2} \cdot \frac{R}{\sqrt{R^2 - x^2}} \, dx$$

$$= 4\pi \int_0^{\sqrt{R^2-a^2}} R \, dx = 4\pi Rx \Big|_0^{\sqrt{R^2-a^2}} = 4\pi R\sqrt{R^2 - a^2}$$

23. Use the chain rule, as in Problems 13-18.

$x = f(t)$, $y = g(t)$ $\Rightarrow$ by (13.13), Section 6,

$\dfrac{dy}{dx} = \dfrac{g'(t)}{f'(t)}$ $\Rightarrow$ for $y' = \dfrac{dy}{dx}$, $\dfrac{d(y')}{dt} = \dfrac{d(y')}{dx} \cdot \dfrac{dx}{dt}$

$\Rightarrow \dfrac{d(y')}{dx} = \dfrac{d(y')/dt}{dx/dt}$ and since $\dfrac{d^2y}{dx^2} = \dfrac{d(y')}{dx}$

$\Rightarrow \dfrac{d^2y}{dx^2} = \dfrac{d(y')/dt}{dx/dt} = \dfrac{\dfrac{f'(t)\ g''(t)\ -\ g'(t)\ f''(t)}{[f'(t)]^2}}{f'(t)}$

$= \dfrac{f'(t)\ g''(t)\ -\ g'(t)\ f''(t)}{[f'(t)]^3}$

For Problems 26-30, use the last results from Problem 25.

27. $y = \sin x\big|_{x=\pi/2} = 1$, $y' = \cos x\big|_{x=\pi/2} = 0$,

$y'' = -\sin x\big|_{x=\pi/2} = -1$

$h = \dfrac{\pi}{2} - \dfrac{0(1+0)}{-1} = \dfrac{\pi}{2}$, $k = 1 + \dfrac{(1+0)}{-1} = 0$

$\Rightarrow$ Center $= (\dfrac{\pi}{2}, 0)$

31. $x = t^2 + 2$, $y = t^3 - 4t$

t	-3	-2	-1	0	1	2	3
x	11	6	3	2	3	6	11
y	-15	0	3	0	-3	0	15

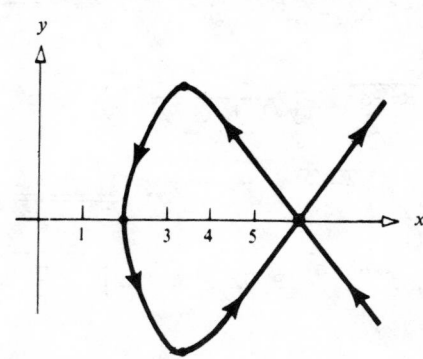

$$\frac{dy}{dx} = \frac{dy/dt}{dx/dt} = \frac{3t^2 - 4}{2t}$$

$$\frac{dy}{dx} = 0 \implies 3t^2 - 4 = 0 \implies t = \pm \frac{2}{\sqrt{3}}$$

Horizontal tangents: $t = \frac{2}{\sqrt{3}} \implies (x,y) = (\frac{10}{3}, \frac{-16}{3\sqrt{3}})$,

$t = \frac{-2}{\sqrt{3}} \implies (x,y) = (\frac{10}{\sqrt{3}}, \frac{16}{3\sqrt{3}})$

$\frac{dy}{dx}$ undefined $\implies 2t = 0 \implies t = 0$

Vertical tangent: $t = 0 \implies (x,y) = (2,0)$

The curve hits $(6,0)$ for $t = -2$ and $t = 2$

For $t = -2$, $\frac{dy}{dx} - -2 \implies$ tangent line: $(y-0) = -2(x-6)$

$\implies y = -2x + 12$

For $t = 2$, $\frac{dy}{dx} = +2 \implies$ tangent line: $(y-0) = 2(x-6)$

$\implies y = 2x - 12$

35.

Slope of line $L = m = \frac{y - 1}{x - 1}$.

Since we want $y = x^2$, try

$$m = \frac{x^2 - 1}{x - 1} \quad \text{or}$$

$$m = \frac{(x-1)(x+1)}{(x-1)} = x + 1.$$

Thus, $x = m - 1$ and

$$y = x^2 = m^2 - 2m + 1.$$

39. $r = a \cos \theta + b \sin \theta \implies r^2 = ar \cos \theta + br \sin \theta$

$\implies x^2 + y^2 = ax + by \implies (x^2 - ax) + (y^2 - by) = 0$

$\implies (x^2 - ax + \frac{a^2}{4}) + (y^2 - by + \frac{b^2}{4}) = \frac{a^2}{4} + \frac{b^2}{4}$

$\implies (x - \frac{a}{2})^2 + (y - \frac{b}{2})^2 = \frac{a^2 + b^2}{4}$. Circle, center

$(a/2, b/2)$, radius $= \sqrt{a^2 + b^2}/2$

43. $x = a \cos \theta, \quad y = b \sin \theta \Rightarrow$ by Problem 23,

$$\frac{dy}{dx} = \frac{b \cos \theta}{-a \sin \theta} \quad \text{and}$$

$$\frac{d^2y}{dx^2} = \frac{-a \sin \theta(-b \sin \theta) - (b \cos \theta)(-a \cos \theta)}{[-a \sin \theta]^3}$$

$$= \frac{ab(\sin^2\theta + \cos^2\theta)}{-a^3\sin^3\theta} = \frac{-b}{a^2\sin^3\theta}$$

Then by Problem 25,

$$h = x - \frac{y'[1 + (y')^2]}{y''}$$

$$= a \cos \theta - \frac{(b \cos \theta/-a \sin \theta)(1 + b^2\cos^2\theta/a^2\sin^2\theta)}{-b/(a^2\sin^3\theta)}$$

$$= a \cos \theta - \frac{\cos \theta}{a}(a^2\sin^2\theta + b^2\cos^2\theta)$$

$$= \frac{a^2\cos \theta - a^2\cos \theta \sin^2\theta - b^2\cos^3\theta}{a}$$

$$= \frac{a^2\cos \theta(1-\sin^2\theta) - b^2\cos^3\theta}{a} = \frac{a^2 - b^2}{a} \cos^3\theta$$

$$k = y + \frac{[1 + (y')^2]}{y''} = b \sin \theta + \frac{(1 + b^2\cos^2\theta/a^2\sin^2\theta)}{-b/(a^2\sin^3\theta)}$$

$$= b \sin \theta - \frac{\sin \theta}{b}(a^2\sin^2\theta + b^2\cos^2\theta)$$

$$= \frac{b^2\sin \theta - a^2\sin^3\theta - b^2\cos^2\theta \sin \theta}{b}$$

$$= \frac{b^2\sin \theta(1-\cos^2\theta) - a^2\sin^3\theta}{b} = \frac{b^2 - a^2}{b} \sin^3\theta$$